TRACTORS AND THEIR POWER UNITS

Fourth Edition

TRACTORS AND THEIR POWER UNITS

John B. Liljedahl
Professor Emeritus
Agricultural Engineering Department
Purdue University

Paul K. Turnquist
Professor and Head
Agricultural Engineering Department
Auburn University

David W. Smith
Research Engineer
Technical Center
Deere & Company

Makoto Hoki
Professor and Chair
Department of Bioproduction and Machinery
Mie University
Tsu Japan

CBS Publishers & Distributors Pvt. Ltd.

New Delhi • Bengaluru • Chennai • Kochi • Kolkata • Mumbai
Hyderabad • Nagpur • Patna • Pune • Vijayawada

CBS Pubs ISBN: 81-239-0501-7
ASAE   ISBN: 0-929355-72-5

First Indian Edition: 1997
Reprint: 2002, 2004

Published by **Satish Kumar Jain** and produced by **Varun Jain** for
CBS Publishers & Distributors Pvt. Ltd.,
4819/XI Prahlad Street, 24 Ansari Road, Daryaganj, New Delhi - 110002
delhi@cbspd.com, cbspubs@airtelmail.in • www.cbspd.com
Ph.: 23289259, 23266861, 23266867 • Fax: 011-23243014

Corporate Office: 204 FIE, Industrial Area, Patparganj, Delhi - 110 092
Ph: 49344934 • Fax: 011-49344935
E-mail: publishing@cbspd.com • publicity@cbspd.com

Branches:
• *Bengaluru:* 2975, 17th Cross, K.R. Road, Bansankari 2nd Stage,
  Bengaluru - 70 • Ph: +91-80-26771678/79 • Fax: +91-80-26771680
  E-mail: cbsbng@gmail.com, bangalore@cbspd.com
• *Chennai:* No. 7, Subbaraya Street, Shenoy Nagar, Chennai - 600030
  Ph: +91-44-26681266, 26680620 • Fax: +91-44-42032115
  E-mail: chennai@cbspd.com
• *Kochi:* Ashana House, 39/1904, A.M. Thomas Road, Valanjambalam,
  Ernakulum, Kochi • Ph: +91-484-4059061-65
  Fax: +91-484-4059065 • E-mail: cochin@cbspd.com
• *Kolkata:* 6-B, Ground Floor, Rameshwar Shaw Road, Kolkata - 700014
  Ph: +91-33-22891126/7/8 • E-mail: kolkata@cbspd.com
• *Mumbai:* 83-C, Dr. E. Moses Road, Worli, Mumbai - 400018
  Ph: +91-9833017933, 022-24902340/41 • E-mail: mumbai@cbspd.com

Representatives:
• Hyderabad: 0-9885175004   • Nagpur: 0-9021734563
• Patna: 0-9334159340   • Pune: 0-9623451994
• Vijayawada: 0-9000660880

Printed at:
Neekunj Print Process, Delhi

CONTENTS

PREFACE   xi

ACKNOWLEDGMENTS   xiii

1 DEVELOPMENT OF THE TRACTOR   1

Tractor Types   4
Some World Variations in Tractors   8
Functions of a Farm Tractor   8
Sources of Energy   13
Energy Conversion Devices   14
The Future Engine   16
Trends in Tractor Design   16

2 THERMODYNAMIC PRINCIPLES OF INTERNAL-COMBUSTION ENGINES   20

Specific Heat   22
Entropy   23
Energy Changes   23
Constant-Volume Changes   24
Constant-Pressure Changes   24
Isothermal Changes   25
Adiabatic or Constant-Entropy Changes   26
General or Polytropic Changes   28
Changes in an Adiabatic Expansion of a Perfect Gas   28
Determination of the Value of n from an Actual Compression Line   31
Equations for Nonflow Processes   33

3 INTERNAL-COMBUSTION ENGINE CYCLES   37

Simplifying Assumptions   37
The Ideal Air-Standard Otto Cycle   38
The Ideal Air-Standard Diesel Cycle   41
Actual Cycles and Causes for Deviation from the Ideal   43

4 FUELS AND COMBUSTION   48

Sources of Fuels   48
Chemical Composition of Petroleum   48
Petroleum Refining   51
Combustion   54
Gasoline Tests and Their Significance   56
Antiknock Quality   57

Factors Affecting the Octane Number Requirement   58
Volatility   59
Distillation   61
Reid Vapor Pressure Test   62
Sulfur Content   64
Corrosion Test   65
Gum Content   65
Gravity   65
Heating Values   66
Gasoline Additives   67
Diesel Fuel Tests and Their Significance   67
Cetane Number   69
Heating Value of Diesel Fuel   72
Viscosity   72
Carbon Residue   72
Flash Point   72
Pour Point and Cloud Point   73
Ash Content   73
Diesel Fuel Additives   73
Alternate Fuels   74

5 ENGINE DESIGN   77
Engine Design—General   78
Stroke-to-Bore Ratio   78
Crankshafts and Firing Orders   80
Tractor Engine   82
Valve Design   83
Valve Materials, Design, and Application   85
Valve Timing   87
Valve Clearance Adjustment   90
Valve Seats   91
Valve-Opening Area   91
Cams   93
Combustion Chamber Design   94
Effect of Compression Ratio   94
Piston Crank Kinematics   95
Inertia Force of Connecting Rod   98
Inertia Force of Single-Cylinder Engines   100
Crank Effort   101
Flywheels   102
Balance of Single-Cylinder Engines   104
Balance of Multicylinder Engines   106
Principles of CI Engine Operation   110
Construction of Diesel Engines   111
Combustion Chamber Design   112
Fuel-Injection Systems   116

Fuel Injectors   119
Turbochargers   120
Matching of Turbocharger to Engine   122
Aftercooling   129
Engine Noise   129

6 ELECTRICAL SYSTEMS   134
Battery   134
Starting Motor   137
Electrical Charging   139
Battery Ignition System with Breaker Points   140
Magneto Ignition System   145
Ignition Timing   145
Spark Plugs   148
Sensors   149
Environmental Problems   152

7 ENGINE ACCESSORIES   156
Speed Control   156
Principles of Centrifugal Governor Action   157
Spark Arresters   160
Mufflers   162
Air Cleaners   163
Air Inlet Location   163
Precleaners   164
Dry-Type Air Cleaner   164
Dust Composition   165
Air Cleaner Tests   166
Engine Cooling Systems   166
Cooling Load   167
Heat Transfer   170
Air Cooling   170
Radiator Design   171
Radiator Construction   174
Temperature Control   174
Antifreeze Materials and Coolants   175
Corrosion and Radiator Deposits   177
Pressure Cooling   178
Quantity of Air   178
Fans   179
Engine Cooling Summary   181

8 LUBRICATION   183
Types of Lubricants   183
Properties of Lubricants   183
Viscosity   184
Classification of Oil by Viscosity   188

Classification by Service 193
Universal Tractor Oils 194
Oil Additives 194
Journal Bearing Design 195
Oil Contamination 198
Oil Filters 199
Lubricating Oil Systems 201

9 HUMAN FACTORS IN TRACTOR DESIGN 203
Operator Exposure to Environmental Factors 203
Operator Exposure to Noise 206
Operator Exposure to Vibration 210
The Operator-Machine Interface 214
Noise and Vibration Control 216
Operator Seating 217
Sound Control in Operator Enclosures 224
Spatial, Visual, and Control Requirements of the Operator 226
Rollover Protection for Wheeled Agricultural Tractors 231
Thermal Comfort in Operator Enclosures 233
Safety 234

10 TRACTION 240
Traction Mechanics 240
Mohr-Coulomb Failure Criteria 240
Traction Performance Equations 243
Traction Prediction from Dimensional Analysis 247
Performance of Four-Wheel, Tandem, and Dual Tires 254
Tire Size, Load, and Air Pressure Relationship 257
Radial-Ply Construction 258
Tread Design 259
Effect of Lug Spacing 263
Traction Improvement 263
Tracks 265
Tire Testing 265
Traction Devices for Paddy Fields 265

11 MECHANICS OF THE TRACTOR CHASSIS 272
Simplifying Assumptions 272
Equations of Motion 273
Static Equilibrium Analysis—Force Analysis 278
Static Equilibrium Analysis—Maximum Achievable Drawbar Pull 280
Static Equilibrium Analysis— Four-Wheel-Drive 282
Longitudinal Stability 283
The Tractor as a 2-Degree-of-Freedom Vibratory System 285
Transient and Steady-State Handling 289
Lateral Stability in a Steady-State Turn 295
Three-Dimensional Static Analysis 297

Center of Gravity Determination   303
Moment of Inertia Determination   306

12 HYDRAULIC SYSTEMS AND CONTROLS   314
Hydraulic Component Symbols   314
Hydraulic Components   314
Motor Performance   315
Accumulators   319
Valves   321
Hydraulic Fluids   323
Classification of Hydraulic Controls   328
Draft Sensing   329
Automatic Control   330
Complete Hydraulic System   332
Power Steering   333
Noise in Hydraulic Systems   350
Hitches   350
Tractor Kinetic Energy   351
Integral Hitch Systems   353
Three-Point Hitches   353
Quick-Attaching Coupler for Three-Point Hitches   355

13 TRANSMISSIONS AND DRIVE TRAINS   360
Complete Drive Train   362
Transmission Types   364
Friction Brakes and Clutches   371
Spur and Helical Gears   377
Bevel Gears   377
Gear Design   378
Planetary Gear Systems   383
Differentials   389
Transmission Drive Shafts   390
Antifriction Bearings   391
Drive Train Speeds and Loads   392
Electronic Transmission Controls   392
Torsional Vibration   393
Computer Simulation   397

14 TRACTOR TESTS AND PERFORMANCE   403
Tractor Performance Criteria   403
Power Measurement Methods   404
Absorption Dynamometers   406
Electric Direct Current Dynamometers   409
Shop-Type Dynamometers   410
Drawbar Dynamometers   411
Torsion Dynamometers   413
Chassis Dynamometer   413

Power Estimating—Field Method   415
Air-Supply Measurement   416
Engine Pressure Indicators   418
Fuel Flow Meter   420
The Nebraska Tractor Tests   420
Globalization of the Tractor Industry   432
Correction for Atmospheric Conditions   433
Torque Curves   435
Engine Performance   435
Efficiency of Tractor Engines   435
Actual Power Output and Fuel Consumption   436
Tractor Reliability   438

APPENDIXES

A Standards for Agricultural Tractors   447
B Standard Graphical Symbols   451
C Agricultural Tractor Tire Loadings, Torque Factors, and Inflation
   Pressures—SAE J709d   453
D Conversion Factors   457

INDEX   459

PREFACE

At the time of the writing of the fourth edition of this textbook, the agricultural economy in the United States and Canada was depressed. The prices paid to farmers for their grain crops were very low, and consequently most farmers in North America could not afford to buy a new tractor when needed; therefore, the sales of tractors and other farm machines were much below normal. The farmer who was the victim of the depressed economy was forced to "make do." Instead of purchasing a new tractor when the old one needed to be replaced, the farmer usually purchased a used or second-hand tractor or repaired the old one. In a strict sense, tractors usually do not wear out; instead, they become obsolete. The farmer who owns an obsolete tractor would prefer to replace it with one having more power, more speeds, more conveniences, a better hydraulic system, lower operating cost, or all of the above.

But farmers in the United States, Canada, and other industrial nations will continue to want to purchase tractors that have all of the features, including microprocessors, found on other vehicles.

The authors have tried to make this textbook more international in scope, and it is partly for that reason that a new co-author from Japan has been added. Makoto Hoki, through his experience in the Philippines, Malasia, and other Third World countries, brings to this textbook broad experience in the mechanization of agriculture. To a farmer in a Third World country, a tractor is likely to have less than 10 kW of power, only three forward speeds, and a single cylinder engine, and the farmer is fortunate indeed if the tractor has a seat on which to ride.

The primary purpose of this book is as a text for engineering seniors and graduate students. A secondary intended use for this book is as a reference.

Measurements in this edition, as in the previous edition, are in SI units. In a few instances (e.g., tire sizes) customary units remain because of their continued international use. Conversion factors are included in Appendix D. We give credit to Dr. Hussam Al-Deen of the John Deere Werke, Mannheim, West Germany, for most of the conversion factors from customary to SI units.

ACKNOWLEDGMENTS

We pay tribute here to three previous authors. Each made significant contributions to the farm power literature both in farm power research and teaching. Dr. E. G. McKibben, now deceased, was a pioneer in farm power research and teaching, especially in matters relating to tractor stability and traction. E. L. Barger, now retired, is widely recognized for his contributions to transport and traction literature and for his leadership in farm power and machinery economics. With others, they also played a significant part in establishing the Ferguson Foundation Agricultural Engineering Series in cooperation with Harold Pinches of the original Ferguson Foundation, without which this book and others in the original Ferguson Foundation Series would likely not have been written. Dr. W. M. Carleton, now retired, also had a distinguished career in teaching, research, and service. Dr. Carleton's attention to details insured that the first edition was properly organized and written.

Credit is due Purdue University for providing the time and space to write this edition.

Special thanks go to the Society of Automotive Engineers and the American Society of Agricultural Engineers from whose literature came much of the new information in this edition.

Many educators have been helpful, especially those who participated in the Power and Machinery Textbook Conference at the ASAE Meeting at Bozeman, Montana, in June 1982. Their report helped in the reorganization of this book.

Many engineers in industry contributed information. Special thanks must go to Richard A. Michael and William H. Lipkea of the John Deere Product Engineering Center in Waterloo, Iowa, for their editorial work on Chapters 2, 3, and 13. Other engineers who contributed represented Massey-Ferguson Ltd.; Case IH; Ford Tractor Operations; Duetz Allis; Perkins Engines Group, England; Cummins Engine; Caterpillar; Kubota Ltd., Japan; Fiat Trattori, Italy; Versatile, Canada; and Yanmar, Japan.

Special thanks to Harriet Liljedahl for patiently proofreading this edition.

TRACTORS AND THEIR POWER UNITS

1
DEVELOPMENT
OF THE TRACTOR

Research is a high-hat word that scares a lot of people. It needn't.
Research is nothing but a state of mind—a friendly welcoming
attitude toward change.

CHARLES F. KETTERING

Mechanization of agriculture has several objectives:

1. To increase the productivity per agricultural worker
2. To change the character of farm work, making it less arduous and more attractive
3. To improve the quality of field operations, providing better soil environment for seed germination and plant growth

Modern tractors play an important role in achieving these goals. Human beings as power units or engines are limited to less than 0.1 kW continuous output and are therefore worth almost nothing as a primary source of power. If agricultural workers are to receive an adequate return for their labor, they must be efficient producers by controlling power rather than being the source of power.

Although tractors have been in existence for more than a century, they got their first impetus during World War I and really came into their own during World War II, in each case because of the enormous increase in demand for food and fiber, with less available agricultural labor.

Evolution of the tractor has accompanied changes in farm technology and sizes of farms. The tractor has progressed from its original primary use as a substitute for animal power to the present units designed for multiple uses. Traction power, belt power, power takeoff drives, mounted tools, and hydraulic remote control units, as well as climate-controlled cabs and power

1

steering, all serve to extend the usefulness and efficiency of the modern tractor.

The word *tractor* has been attributed to various sources (*Farm Implement News*, 1952, 1953); however, according to the Oxford Dictionary, the word was first used in 1856 in England as a synonym for *traction engine*. The term *tractor* appears in an 1890 U.S. patent for a track-laying steam traction engine. The rapid increase in numbers and versatility of tractors has been accompanied by a similar decrease in numbers of horses and mules on farms in the United States (USDA, 1960). Depending upon local and domestic conditions, various types and sizes of tractors have been developed and used worldwide.

The early steam engines furnished belt power, but they had to be pulled from place to place by horses or oxen. The next step in the evolution in farm power was the conversion of the steam engine into a self-propelled traction engine. Successful steam plows were developed in the decade of the 1850s, and continuous development occurred until 1900.

Inadequate traction plagued the inventors of the large heavy tractors, who tried to solve the problem by making the driving wheels wider and wider. One big-wheeled tractor made in 1900 for use in California had two wood-covered drive wheels, each 5 m wide and 3 m in diameter. The tractor weighed about 36 tonnes. Other attempts to solve the problem of traction resulted in the development of track-type agricultural tractors about 1900.

Early attempts to develop gasoline tractors were stimulated by the need for reduction in the number of workers required to attend the steam tractors, both when plowing and when operating threshing machines. Early gasoline tractors resembled steam tractors. Considerable development was needed before a reasonably successful internal-combustion engine was available for a tractor. The internal-combustion engine did not assume much importance until after the expiration of the Otto patents in 1890.

The first Winnipeg tractor trials were held in 1908, giving the public an opportunity to compare field operations of steam and gas tractors. Succeeding trials were held each year through 1912, when they were discontinued. The first U.S. tractor demonstration, held at Omaha, Nebraska, in 1911, was conducted as an exhibition and not as a competition between machines.

The Nebraska Tractor Test Law, passed in 1919, specified that every tractor sold in the state of Nebraska should be tested and the results published. In addition, the manufacturer was required to maintain an adequate supply of parts for repairs. The tests, which attained worldwide recognition, have provided standards for rating tractors, have speeded improvements, and have eliminated many types that were inferior in design and performance.

Some of the important highlights in tractor development are as follows:

1858	Steam plowing engine by J. W. Fawkes drew eight plows at 4.8 km/h in virgin sod.
1873	The Parvins steamer was probably the first U.S. attempt at a track-laying device, although the U.S. Patent Office recorded crawler tractor developments in the early 1850s.
1876	Otto patents were issued for an internal-combustion engine.
1889	At least one company built a tractor with an internal-combustion engine.
1908	The first Winnipeg tractor trials were held.
1910–1914	1. The first tractor demonstration was held in the United States at Omaha, Nebraska, in 1911.
	2. Smaller, lightweight tractors were introduced.
	3. The frameless-type tractor was introduced.
1915–1919	1. The power takeoff was introduced.
	2. The Nebraska tractor test law was passed.
1920–1924	A highly successful all-purpose farm tractor was developed.
1925–1929	The power takeoff was gradually adopted.
1930–1937	1. The diesel engine was applied to larger tractors.
	2. Pneumatic tires and higher speeds were introduced.
	3. Full electric equipment was adopted.
	4. Interest in high-compression engines increased.
	5. The all-purpose tractor was generally accepted.
1937–1941	1. Standardized ASAE and SAE power takeoff and hitch locations were generally accepted.
	2. Pressurized cooling systems were introduced.
	3. Liquid fill was widely used in tires to add ballast for traction.
	4. Three-point implement hitch and linkage were introduced.
	5. Automatic hydraulic draft control was introduced.
1941–1949	1. Live power takeoff was introduced.
	2. Hydraulic controls for drawn implements were adopted.
	3. Tractors for burning liquefied petroleum gases were introduced.
	4. Number of lawn and garden tractors expanded rapidly.
1950–1960	1. Power of tractors increased rapidly.
	2. Percentage of diesel tractors increased.
	3. Refinements such as power steering, automatic transmissions, and transmissions with greater speed selections became widely available.
1961–1970	1. Power of tractors continued to increase.
	2. Except for the smallest tractors in the United States, all tractors now have diesel engines.
	3. Much more emphasis was placed on operator comfort and safety.
	4. Full power-shift transmissions became available.
	5. Radial-ply tractor tires became available.
	6. The number of power tillers started to expand in Japan.
1970–1978	1. Turbocharger and intercoolers were added to diesel engines.
	2. Rollover protective structures (ROPS) became available and, begin-

ning October 15, 1976, were required on all new tractors sold for use by employees in the United States.

3. Most large tractors were equipped with cabs.
4. Nebraska tractor tests included sound level measurements.
5. Four-wheel drive increased in popularity.
6. Percentage of tractors over 75 kW continued to increase in North America.

1979–1985   1. Tractors equipped with electronic sensing and control systems became popular.
2. Tractor size and power have appeared to reach upper limits.

Tractor Types

The early farm tractor, essentially a substitute for the draft horse, was first powered by a steam engine. The internal-combustion engine came later. The first tractors were large and cumbersome and were suited mainly for plowing and threshing. As time passed, the tractor was developed for many more purposes. Adaptations were made for its use as a motor cultivator, and later came the general-purpose tractor to perform major farm operations.

The many uses, adaptations, and refinements of the tractor have resulted in the gradual evolution of several recognized classifications. One descriptive classification, listed below, is based on steering method, the arrangement of the frame, and traction members (see figs. 1-1 to 1-11).

1. Crawler	8 Power tiller
2. Standard row-crop	9. Tree skidder
3. High-clearance	10. Skid-steer loader
4. Utility	11. Four-wheel drive with smaller front steering wheels
5. Orchard	
6. Multipurpose	12. Four-wheel drive with equal-sized wheels and articulated frame steering
7. Lawn and garden	

The *crawler* tractor (fig. 1-1) is less important today in agriculture than it once was. It is usually used on very soft soil or where stability of a wheel tractor is a problem. At one time crawler tractors were larger than wheel tractors, and thus they were used on some large farms.

The *standard* tractor (fig. 1-2) was developed primarily for traction. It is characterized by a drive through the two rear wheels, with center of gravity located at approximately one-third the wheelbase ahead of the rear axles. *Row-crop* tractors, in addition to being suitable for traction work, are especially adapted for use on row crops, allowing quick attachment of various tillage and cultivation tools as well as implements for such tasks as mowing and handling forages.

FIGURE 1-1   Crawler tractor. When used in agriculture, this type of tractor will range in size from 20 to 100 kW. (Courtesy Caterpillar Tractor Co.)

High-clearance tractors (not shown) are usually modified row-crop tractors designed especially for crops that require extra clearance, such as cane.

Utility tractors (fig. 1-3) generally have less clearance than standard or row-crop tractors. They are used for many jobs on the farm or ranch and are often equipped with a front loader. A typical use for such a tractor is cleaning a feedlot.

An *orchard* tractor (not shown) differs little from a utility tractor except that it is a few centimeters lower. The ROPS standard by the Occupational Safety and Health Administration (OSHA) does not apply, and therefore the tractor is not equipped with a rollover protective structure.

A *multipurpose* tractor is designed to operate in either direction. It not only carries the implement but also furnishes the power. One such tractor is shown in figure 1-4.

Lawn and garden tractors (fig. 1-5) have a power output of less than 15

FIGURE 1-2   Standard, or row-crop, tractor. Power ranges from 15 to 150 kW.
(Courtesy Ford Motor Co.)

kW and are primarily designed for the care of large lawns. They can tow or
carry a lawn mower, a sweeper, a snowblower, a bulldozer blade, and many
other attachments. Because these tractors have very low clearance and are
designed for light-duty use, they are not generally used for agricultural production.

A *power tiller* (fig. 1-6) is a two-wheeled tractor guided by hand and used
commonly in the rice-growing regions of Japan and Southeast Asia. The
power tiller is equipped with a rotary tiller and powered by a horizontal single-
cylinder engine of 5 to 12 kW.

The *tree skidder* (fig. 1-7) is a four-wheel-drive tractor developed especially
for moving tree trunks out of the forest to an area where they can be loaded
onto trucks. The tree skidder is a combination of an agricultural and an
industrial tractor.

The *skid-steer* tractor (fig. 1-8) was designed especially for industrial use,
but it is often used in agriculture when turning room is restricted, such as in
a dairy or a fruit storage building. The name "skid-steer" comes from the

FIGURE 1-3 Utility or light industrial tractor. Power ranges from 15 to 100 kW. (Courtesy Ford Motor Co.)

method of turning, which is exactly like that of a crawler. The wheels on one side are braked or reversed, causing the tractor to skid in order to turn.

A *four-wheel-drive tractor with smaller front wheels* (fig. 1-9) is simply a standard or a row-crop tractor with the front wheels also being driven. In regard to price and traction, this type of tractor falls between the standard and the four-wheel-drive tractor with equal-sized wheels (fig. 1-10). This tractor has become particularly popular in Japan because of its excellent steering and traction characteristics in soft, wet rice fields.

Four-wheel-drive, or simply 4WD, tractors have been developed so as to be able to produce more drawbar power. The size of 4WD tractors varies in the United States and Canada from 100 kW to more than 300 kW. In Europe, 4WD tractors may be as small as 15 kW and are used especially in vineyards. Four-wheel-drive tractors can be steered by pivoting the tractor in the center (frame steer) or steering the wheels (as in fig. 1-10).

FIGURE 1-4   Reversible or "bidirectional" tractor. Designed for either front- or rear-attached implements. Has four-wheel drive and articulated-frame steering. (Courtesy Versatile Farm Equipment Co.)

The complexity of a modern tractor is illustrated by figure 1-11, which shows such features as power steering, epicyclic gear reduction in the front wheels, wet disk brakes, power shift transmission and an ROPS type cab.

Some World Variations in Tractors

1.   Tractors made and used in Japan are usually equipped with *riceland* tires.
2.   The power-to-mass ratio is greater for Japanese tractors than for others.
3.   European tractors more commonly use radial-ply tires for traction.
4.   Tractors made outside of North America may have as many as four different speeds on the power takeoff.
5.   In North America, 4WD is used mainly on tractors over 100 kW; in Japan, however, tractors as small as 10 kW may have front-wheel-assisted 4WD.
6.   In Europe, the United Kingdom, and Japan, 4WD with front wheels smaller than the rear is common, whereas almost all 4WD tractors in North America have equal-sized wheels.
7.   Crawler tractors are more popular in Europe and the United Kingdom.
8.   Power tillers are quite popular in Japan and Southeast Asia, primarily because of the small rice farms and soft, wet land conditions found there.

Functions of a Farm Tractor

Today's agricultural tractor is a complex vehicle used to propel and power a large variety of implements for agricultural production. Implement applications, therefore, have considerable effect on tractor design, and the tractor en-

FIGURE 1-5 Lawn and garden tractor. Power ranges from 5 to 15 kW. (Courtesy Wheel Horse.)

gineer must be aware of implement requirements to weigh properly such design considerations as crop clearance, wheelbase, width of tractor, power-to-mass ratio, and mass distribution, as well as to provide placement of controls to satisfy the majority of operations. Every design will necessarily include some compromises.

Implements are generally attached to and operated by tractors in one of four ways:

1. Towed, single-point hitch
2. Mounted, three-point hitch
3. Semimounted, three-point hitch
4. Frame mounted

Power may be transmitted to the implement by a power takeoff shaft or by oil-hydraulic hoses. Each new tractor provides many engineering problems related to proper drawbar height, provision for sufficient horizontal and vertical angular movement, and implement clearance.

The widespread adoption of hitch-mounted and frame-mounted equip-

FIGURE 1-6 Power tiller. Power ranges from 5 to 15 kW. (Courtesy Yanmar Agricultural Machinery Co.)

FIGURE 1-7 Tree skidder. This tractor usually has four-wheel drive. Power ranges from 75 to 150 kW. (Courtesy Caterpillar Tractor Co.)

FIGURE 1-8   Skid-steer tractor. This type of tractor usually has four-wheel drive. Power ranges from 15 to 60 kW. (Courtesy J.I. Case Co.)

FIGURE 1-9   Four-wheel drive tractor with smaller steering wheels in front. Power ranges from 7 to 120 kW. (Courtesy Fiat Trattori. Also see fig. 1-11.)

FIGURE 1-10 Four-wheel-drive tractor with equal-sized wheels and wheel steering. Power ranges from 20 to 300 kW. (Courtesy J.I. Case Co.)

FIGURE 1-11 Phantom view of a modern tractor showing the design complexity. (Courtesy Ford Motor Co.)

ment is probably the result of one or both of two primary factors: decreased investment cost to the farmer and increased versatility. In addition, the extra weight transfer of mounted implements adds significantly to the tractive ability of the tractor.

Sources of Energy

One of the characteristics of a technically advancing society, at least before 1978, is that the rate of growth of energy use is considerably higher than the rate of population growth and closely parallels the growth of economic output (Sporn 1957). Possession of surplus energy is a requisite for any kind of advanced civilization, for if human beings must depend strictly on their own muscles, it will take all of their mental and physical strength to supply the bare necessities of life. Humans become beasts of burden, and their civilization declines. The difference in per capita energy consumption is closely related to differences in standards of living. The important sources of energy are as follows:

1. Solar energy, direct
2. Solar energy, indirect
   a. Fossil fuels
      (1) Oil
      (2) Natural gas
      (3) Coal
      (4) Peat
      (5) Shale and tar sands
   b. Biomass (wood, corncobs, etc.)
   c. Wind
   d. Tides
   e. Water
3. Nuclear energy
4. Geothermal energy

Our present age may sometime be called the fossil fuel age since most of our present power use is coming from stored fossil fuels—oil, gas, and coal. How long these supplies will last is debatable, but they are nonrenewable. Modern farming in the world today depends on petroleum fuels, so it is of interest to know when the supply will begin to diminish. The date is uncertain, but in the United States, for example, production of petroleum has not met demand since 1947 (Scarlott 1957). In 1973 the United States imported about 40 percent of its oil needs (Ray 1973), and by 1978, over 50 percent was imported. It has also been predicted that we may have nearly reached the year of peak production, after which crude oil production will begin to decline.

Oil shales, from which it is possible to obtain nearly a barrel of oil per tonne, are available in considerable but indeterminate amounts. For power plant use, some attention is being given to recovery of the oil by methods other than mining.

Solar energy is our one inexhaustible energy source. A rough measure of energy falling on the earth is a maximum of approximately 940 W/m^2 and an average of 630 W/m^2. Although this energy source will likely become very useful for space heating and for agricultural operations such as crop drying, at present it is doubtful that it will be useful for operating tractors because of the low concentration of energy and a lack of suitable means of collecting and concentrating it. Another problem is the large difference in the amount of available solar energy at different times of year or day and at different locations.

Nuclear energy will ease the load on fossil fuels, as it is well adapted to large power plant operations where radiation shielding can be employed. It is a concentrated, clean, and easily controlled energy source, the energy in 4 ml of uranium 235 being equivalent to the energy stored in 140 m^3 of coal ("Nuclear Energy" 1961).

Controlled fusion, the process of combining the lighter elements rather than breaking up the heavy atoms as is done in the fission process, may offer some hope for future power plants, even those as small as a farm tractor, since fusion atomic energy does not require heavy shielding because it is not radioactive.

Energy Conversion Devices

Common techniques for the conversion of energy include the following:

1. Human and animal energy
2. Piston engines
   a. Diesel
   b. Spark ignition
3. Gas turbine
4. Rotary (Wankel) engine
5. Free-piston engine
6. Sterling engine
7. Steam piston engine
8. Steam turbine engine
9. Thermoelectricity
10. Fuel cell
11. Solar cell
12. Electric motor and electric generator

13. Storage battery
14. MHD (magnetohydrodynamic) devices

The second law of thermodynamics states that conversion of heat to work is limited by the temperatures at which conversion occurs. For a cycle with a high temperature T_1 and a low temperature T_2 the maximum efficiency, usually called the *Carnot cycle efficiency*, is

$$e = \frac{T_1 - T_2}{T_1}$$

Thus the Carnot cycle establishes an inherent limit to the maximum useful conversion of energy that may be obtained from a fuel used in a power plant, such as in a tractor engine. In the best tractor engines only about one-third of the heating value of the fuel is converted into useful work.

Conventional piston engines have hundreds of parts, and the reciprocating units must be stopped and started thousands of times per minute. Nevertheless, they are highly successful units.

Rotary engines are of interest because of their simple design. Inside the combustion chamber is a three-lobed rotor with curved sides. The shape of the rotor and the chamber causes the rotor to move in an eccentric orbit as it rotates. All three corners of the rotor stay in contact with the walls of the chamber at all times. Vanes, which correspond to piston rings in the conventional engine, seal the chambers from loss of gas. As the rotor turns, its corners form three moving cavities, which in turn go through the usual cycle of intake, compression, power, and exhaust. The output shaft turns at three times the speed of the rotor in this example.

The power-to-mass ratio is greater than 1.0, which is not an advantage on a tractor. The major mechanical problem, the rotating seals, has been solved well enough to include the Wankel in production line automobiles but not yet in industrial-type diesel engines. The Wankel engine so far is less efficient than conventional spark ignition engines used in tractors.

The simplicity and reliability of a gas turbine engine plus its high power-to-weight ratio makes it a desirable power unit for aircraft. Its high power-to-weight ratio is, however, not an asset on a tractor used for the usual agricultural or industrial purposes. The efficiency of gas turbine engines, when the power is transmitted through a shaft, is roughly one-half to two-thirds that of diesel engines. The efficiency compares more closely with gasoline-burning spark ignition engines.

The principle of operation of a gas turbine engine has much in common with the conventional piston-type internal-combustion engine. The events that take place in a piston engine (intake, compression, ignition, power, and exhaust) also take place in a gas turbine engine. Gas turbine engines run at a relatively constant rpm, which also restricts their use on a tractor in which variations in both speed and torque are required.

The Future Engine

The piston engine, especially the diesel, seems well established (see Amann 1974) and is not likely to be replaced immediately with other types of energy conversion systems. However, development of new engine materials has progressed considerably. Ceramic materials, which have high heat resistance, as well as high anticorrosion and antiwear properties, have indicated their potential applications to piston engines, permitting operations at higher temperatures and thus higher thermal efficiency compared to conventional metal-based engines. Although their brittleness has been a limitation to their wide application for engine structures, recent development of ultraplastic fine ceramics may open a new era for wide usage in piston engines. Complete replacement of engine parts with ceramics is not likely to occur in the near future, but applications of ceramics to specific engine parts including turbochargers will likely occur soon. Tractor and engine manufacturers continue to use better engine materials and to develop new types of energy conversion devices.

Trends in Tractor Design

The design and performance of tractors have changed considerably since the first Nebraska tractor tests. The future tractor design greatly depends on technological advancement and agricultural and economic development, including energy and environmental systems (Cragle 1983). From figure 1-12 it is clear that tractors continue to have a greater power/mass ratio, which results in the tractors traveling faster, since improvements in traction have not kept pace with the increase in power. It is also clear that the power output of tractors has continued to increase. The growth, however, in power and weight of tractors will probably not continue at the same rate as in the past. One reason is the failure of farm sizes to grow because of the decrease in the availability of land, as well as market limitations and the instability of export opportunities. The trend of farm size stabilization is now clearly noted in most European countries (Gohlich 1984). Another reason for limiting the increase in size of future tractors is increasing soil compaction affecting soil conservation as well as plant growth. Detailed discussion on the future trends in tractor development will not be made here since it is beyond the scope of this

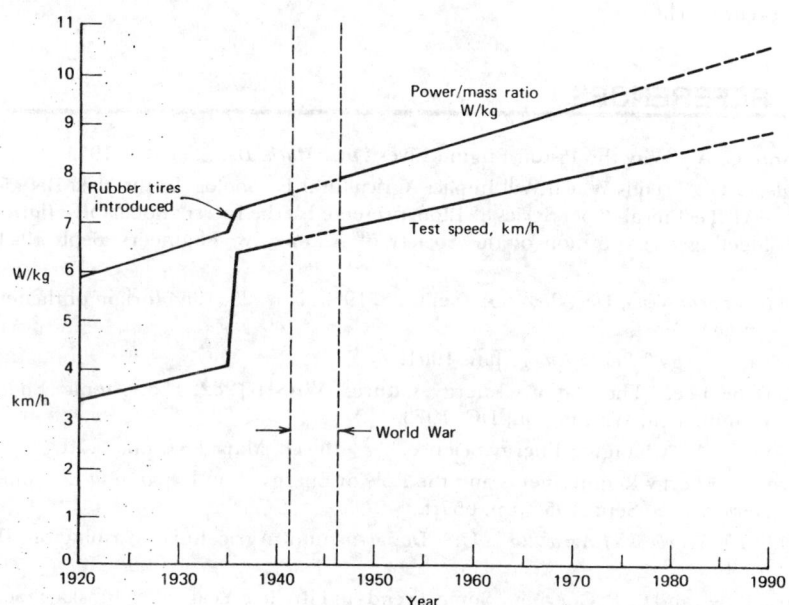

FIGURE 1-12 Trend of changes in tractor speed and power. (From Yahya and Goering 1977.)

book. However, Gohlich (1984) presents an excellent discussion on this subject. He concludes that:

1. More emphasis will be concentrated on reducing the total weight of the tractor-implement system in order to minimize the ground pressure and soil compaction, particularly during seed bed preparation.
2. Front- and rear-mounted implements will contribute increasingly to better, more convenient, and more economical field operations.
3. Instead of the standard heavy tractors today, for most of the required field operations a lighter, second tractor power unit with low-pressure tires, which would transfer the majority of its power through the power take off (pto) rather than through the ground drive, may be a desirable approach on the future farm.
4. The optimization of power transfer will be facilitated by driver information displays as well as by control circuits that may eventually take over some present manual functions.
5. Components for the driver's comfort and safety will be further improved,

thereby helping to ensure that farming remains a satisfying job for the
farm worker.

REFERENCES

Amann, C. A. "Why the Piston Engine Lives On." *Mach. Des.*, Feb. 21, 1974.

Cragle, R. G. "Trends Which Will Impact Agricultural Technology in the Next Decade."
SAE Technical Paper Series 831268, presented at the International Off-Highway
Meeting & Exposition of the Society of Automotive Engineers, Sept. 12–15,
1983.

Farm Implement News, Dec. 25, 1952; Sept. 25, 1953; Nov. 25, 1953 (origin of the term
tractor).

"Nuclear Energy." *Power Eng'g*, Jan. 1961.

Ray, Dixie Lee. "The Nations Energy Future," WASH-1282. U.S. Atomic Energy
Commission, Washington, DC, 1973.

Scarlott, C. A. "Changing Energy Science." *Sci. Month.*, May 1987, pp. 221ff.

Sporn, P. "Energy Requirements and the Role of Energy in an Expanding Economy."
Agric. Eng'g, Sept. 1957, pp. 657ff.

USDA. *The Yearbook of Agriculture.* U.S. Department of Agriculture, Washington, DC,
1960.

Yahya, R. K., and C. E. Goering. "Some Trends in Fifty-five Years of Nebraska Tractor
Test Data." Paper No. MC 77-503, presented at the 1977 Mid-central Regional
Meeting of the American Society of Agricultural Engineers, Mar. 25–26, 1977.

SUGGESTED READINGS

Angrist, S. W. *Direct Energy Conversion,* 3d ed. Allyn and Bacon, Boston, 1976.

Brinkworth, B. J. *Solar Energy for Man.* John Wiley & Sons, New York, 1972.

Cheney, E. S. *Energy Conversion.* Prentice-Hall, Englewood Cliffs, NJ, 1963.

———. "U.S. Energy: Limits and Future Outlook." *Am. Scientist.* Jan.–Feb. 1974.

Corliss, W. R. "Direct Conversion of Energy." U.S. Atomic Energy Commission, Washington, DC, Mar. 1964.

Csorba, Julius J. "Farm Tractors: Trends in Type, Size, Age and Use." *Agric. Info.
Bull.*, No. 231. Agricultural Research Service, USDA, Washington, DC.

"Fifty Years of the Farmall." *Implement & Tractor*, May 21, 1972.

"How Big Can Tractors and Equipment Get?" *Implement & Tractor*, Jan. 7, 1977.

Kawamura, Noboru. "Besouderheiten der Land Technik in Japan." *Grundl. Landtech.*,
vol. 25, no. 4, 1975.

Kisu, M. "Special Requirements for Tractors in Japan." *Proc. Inst. Mech. Eng.*, vol. 184,
1969–1970.

"The Lamp: Understanding Energy." Exxon Corporation, Spring 1975.

Menrad, Halger, W. Lee, and W. Bernhardt. "Development of a Pure Methanol Fuel Car." SAE Paper 770790, Sept. 1977.

Othmer, Donald F. "Energy Prospects for the Rest of the Century." *Mech. Eng.*, Aug. 1974.

"Power to Produce." *The Yearbook of Agriculture.* USDA, Washington, DC, 1960.

"Purdue Energy Conference of 1977." Proceedings published by Purdue University, Apr. 29–30, 1977.

"Report on Unconventional Power Sources." *Power Eng'g,* Jan. 1961.

Roberts, R. "Fuel and Continuous Feed-cells." In G. Seise and U. C. Calhoun (eds), *Primary Batteries.* John Wiley & Sons, New York, 1971.

———. "Energy Sources and Conversion Techniques." *Am. Sci.,* Jan.–Feb. 1973.

Winger, John G., et al. "Outlook for Energy in the United States to 1985." Chase Manhattan Bank, New York, June 1972.

World Energy Conference. "Survey of Energy Resources." World Energy Conference, New York, 1974.

2
THERMODYNAMIC PRINCIPLES OF INTERNAL-COMBUSTION ENGINES

Certain concepts of heat and work are necessary before the operation of the internal-combustion engine can be understood. It is assumed that the student at this point has had an introductory course in heat physics or thermodynamics. This chapter is intended only as a review of pressure-volume-temperature relationships for systems of fixed mass that explain the internal-combustion engine cycle or working process.

Thermodynamics is the study of changes in which energy is involved. The *first law of thermodynamics,* which is a statement of the conservation of energy, says:

> All forms of energy are mutually convertible. The energy of a closed and isolated system remains constant.

The *second law of thermodynamics* cannot be stated simply, but it deals with the fact that in real systems heat cannot be fully converted to work. The second law is based on experiment and indicates the practical limitations on engine performance. It means that no engine can convert all the energy supplied to it, inasmuch as a large part of the energy supplied is rejected in the form of unused heat.

20

In heat engines, work is done by changes in the volume of the gas. Certain laws have been established concerning these pressure, volume, and temperature relationships. For perfect gases *Boyle's law* states:

If a given mass of gas is compressed or expanded at constant temperature, the pressure varies inversely as the volume, or the product of pressure and volume remains constant.

Thus, if P_1 and V_1 are the initial pressure and volume and P_2 and V_2 are the final pressure and volume, then

$$P_1V_1 = P_2V_2 = P_nV_n = \text{constant} \tag{1}$$

Charles' laws state:

1. Under constant pressure the volume of a given mass of gas varies directly as the absolute temperature.
2. Under constant volume the absolute pressure of a given mass of gas varies directly as the absolute temperature.

If V_1 and T_1 are the initial volume and temperature and V_2 and T_2 the final volume and temperature, then

$$\text{With constant pressure} \quad \frac{V_1}{V_2} = \frac{T_1}{T_2} \tag{2}$$

$$\text{With constant volume} \quad \frac{P_1}{P_2} = \frac{T_1}{T_2} \tag{3}$$

Equations 1, 2, and 3 may be combined to give the equation of state:

$$\frac{P_1V_1}{T_1} = \frac{P_2V_2}{T_2} \cdots = \frac{PV}{T} = R \tag{4}$$

in which R is a constant, from which

$$PV = RT \tag{5}$$

Equation 5 is based on a unit mass of gas, and thus V refers to the specific volume. If a unit mass is not used, V refers to the volume of the entire amount, and the expression becomes

$$PV = MRT \tag{6}$$

where M = the mass of gas considered
P = the absolute pressure
V = the volume of the mass M
T = the absolute temperature

Note that absolute temperatures and pressures must be used. V can be volume (m^3) or specific volume (m^3/kg).

Avogadro's law states:

Equal volumes of gases at the same temperature and pressure contain the same number of molecules.

Thus for any given volume, the densities of two gases at the same temperature and pressure are proportional to their molecular weight. A "mole" is the mass of a definite number of molecules of gas and is equal numerically to the molecular weight of that gas. For example,

Oxygen	molecular weight = 32	1 kg mole = 32 kg
Carbon dioxide	molecular weight = 44	1 kg mole = 44 kg
Nitrogen	molecular weight = 28	1 kg mole = 28 kg
Air	apparent molecular weight = 29	1 kg mole = 29 kg
Methane	molecular weight = 16	1 kg mole = 16 kg

Further, the molar volume is the volume occupied by a mole of material, and its molar density is the reciprocal of the molar volume.

In like manner, from Avogadro's law, the volume of any gas at 0°C and 1 atm is 22.414 L/g mole (liters per gram mole).

If, as in equation 5, $PV = RT$, V is expressed as volume per unit mass, then the value of the constant R will be different for different gases. If V is expressed as the volume of 1 mole of gas, then the gas constant will be the same for all perfect gases. This constant is known as the universal gas constant and is equal to the value of R for any gas multiplied by the molar mass of that gas. The value of the universal gas constant is

$$R = 8.314 \frac{J}{g \text{ mole K}}$$

or

$$R = 1.986 \frac{cal}{g \text{ mole K}} \tag{7}$$

Specific Heat

Specific heat is defined as the amount of heat necessary to change the temperature of a unit mass of a substance by 1 degree. The specific heats of solids or liquids are nearly the same at any given temperature, regardless of pressure, but this is not true of gases. If the gas container increases in volume, the gas does work and not all the heat energy appears as an increase in temperature.

For a system of fixed mass, this situation may be represented by the following statement of the first law.

$$Q = Mc(T_2 - T_1) = (U_2 - U_1) + W \tag{8}$$

where Q = heat gained or rejected
M = mass of the gas
c = specific heat
$(U_2 - U_1)$ = change in internal energy
W = work done

From equation 8 it may be seen that the quantity of heat necessary to change the temperature of a given mass of gas 1 degree may have any number of values, depending on how much work the gas does during the operation. It will be sufficient for the purpose of this book to state that

c_v = specific heat at constant volume
c_p = specific heat at constant pressure
$k = c_p/c_v$

Entropy

Entropy as used in engineering thermodynamics has two principal applications when considering ideal or "reversible" processes: first, as a coordinate on a diagram in which heat transfer is shown graphically; second, as an index of the unavailability of heat energy for conversion into work.

When heat energy Q is represented graphically as an area and temperature T is the ordinate, then entropy S is used as the abscissa (fig. 2-1). The scales are such that $T\,dS = dQ$.

For ideal processes, a positive entropy indicates heat energy added during a process; a negative entropy indicates heat rejected; and zero entropy change signifies no heat energy added or rejected. For real processes, $T\,dS$ is generally greater than dQ.

Energy Changes

The general equation that expresses the relation among the three energy quantities involved in a given change for permanent gases is

$$Q = (U_2 - U_1) + W \tag{9}$$

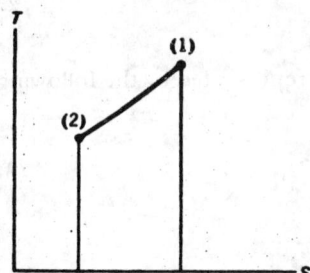

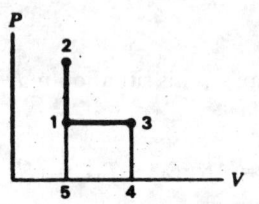

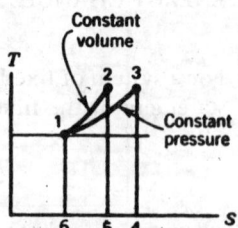

FIGURE 2-1 Heat energy represented as an area on temperature-entropy plane.

FIGURE 2-2 Constant-volume and constant-pressure curves on pressure-volume plane.

FIGURE 2-3 Constant-volume and constant-pressure curves on temperature-entropy plane.

It is the purpose of the following sections to develop some of these working expressions and to fit them into the general expression.

Constant-Volume Changes

Constant-volume heating is represented by the lines 1-2 in figures 2-2 and 2-3. The heat supplied $= c_v(T_2 - T_1) =$ area 1-2-5-6 (fig. 2-3). There is no work done by or upon the gas and hence $W = 0$, and there is no area covered on the PV plane.

Substitution in the general energy equation gives

$$c_v(T_2 - T_1) = (U_2 - U_1) + 0 \tag{10}$$

All the heat supplied appears as an increase in internal energy, $c_v(T_2 - T_1)$. It is always true that, for any perfect gas, the increase in internal energy is $c_v(T_2 - T_1)$, which can be written in place of the general quantity $(U_2 - U_1)$.

Constant-Pressure Changes

Constant-pressure heating is represented by the lines 1-3 in figures 2-2 and 2-3. The heat supplied is $Q = c_p(T_3 - T_1) =$ area 1-3-4-6 in figure 2-3. The work done is $W = P(V_3 - V_1) =$ area 1-3-4-5 in figure 2-2. Substituting in the general energy equation gives

$$c_p(T_3 - T_1) = c_v(T_3 - T_1) + P(V_3 - V_1) \tag{11}$$

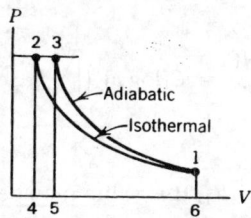

FIGURE 2-4 Isothermal and adiabatic compression lines on pressure-volume plane.

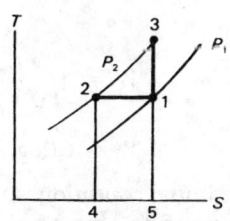

FIGURE 2-5 Isothermal and adiabatic compression lines on temperature-entropy plane.

but

$$P_3V_3 = RT_3 \quad \text{and} \quad P_1V_1 = RT_1$$

Substituting in equation 11, since $P_1 = P_3$,

$$c_p(T_3 - T_1) = c_v(T_3 - T_1) + R(T_3 - T_1) \tag{12}$$

$$c_p = c_v + R$$

$$c_p - c_v = R \tag{13}$$

Isothermal Changes

For a perfect gas, Boyle's law applies:

$$PV = C = \text{a constant} \tag{14}$$

Equation 14 is a special case of the general equation $PV^n = $ a constant. Here the exponent $n = 1$, and the process may be represented by line 1-2 in figures 2-4 and 2-5.

The external work is represented by the area under the curve between 1 and 2. The work done during any change in volume from V_1 to V_2 is

$$W = \int_{V_1}^{V_2} P \, dV \tag{15}$$

Assume that P and V are related by equation 14. Substituting $C/V = P$ in equation 15,

$$W = \int_{V_1}^{V_2} \frac{C}{V}\, dV = C \int_{V_1}^{V_2} \frac{dV}{V} = C[\log_e V]\Big|_{V_1}^{V_2}$$

$$W = C(\log_e V_2 - \log_e V_1) \tag{16}$$

Since the initial conditions are $PV = C = P_1 V_1$, this value may be substituted in equation 16 to obtain

$$W = P_1 V_1 (\log_e V_2 - \log_e V_1)$$

or

$$W = P_1 V_1 \log_e \frac{V_2}{V_1} \tag{17}$$

For any mass of gas other than unity there may be substituted

$$P_1 V_1 = MRT$$

and

$$\frac{V_2}{V_1} = \frac{P_1}{P_2}$$

The work is

$$W = MRT \log_e \frac{V_2}{V_1} = MRT \log_e \frac{P_1}{P_2} \tag{18}$$

Letting $r = V_1/V_2$ be the compression ratio or ratio of expansion, we have

$$W = MRT \log_e \frac{1}{r} \tag{19}$$

There is no change in internal energy during an isothermal expansion or contraction of a perfect gas since the temperature is constant. The gas receives or rejects an amount of heat equal to the work done on it or by it. From equation 18 the amount of heat may be calculated as follows:

$$Q = MRT \log_e \frac{V_2}{V_1} = MRT \log_e \frac{1}{r} \tag{20}$$

Adiabatic or Constant-Entropy Changes

In adiabatic expansion or compression, the gas neither receives nor rejects heat. Therefore, any work done by the gas during expansion is done at the expense of internal energy and, conversely, any work done upon the gas during compression increases the internal energy of the gas. For ideal or reversible processes, the compression of a gas in a cylinder approaches adiabatic when the process is extremely rapid, because then there is little time for heat to be lost by conduction and radiation.

The line 1-3 in figures 2-4 and 2-5 represents an adiabatic compression from a pressure of P_1 to a pressure of $P_3 = P_2$. The work on the gas is represented in figure 2-4 by the area 1-3-5-6. Since this is an adiabatic change, no heat is received from or rejected to an outside body. Therefore the line 1-3 in figure 2-5 covers no area.

Equation 9 now becomes

$$0 = c_v(T_3 - T_1) + W$$

$$W = -c_v(T_3 - T_1)$$

for a unit mass

$$W = -c_v \left(\frac{P_3 V_3}{R} - \frac{P_1 V_1}{R} \right) = -\frac{1}{R} c_v (P_3 V_3 - P_1 V_1) \qquad (21)$$

Since

$$R = c_p - c_v$$

then

$$W = -\frac{c_v}{c_p - c_v} (P_3 V_3 - P_1 V_1)$$

$$W = \frac{P_3 V_3 - P_1 V_1}{1 - k} = \frac{P_1 V_1 - P_3 V_3}{k - 1} \qquad (22)$$

for a unit mass of gas.

For M mass units of gas,

$$W = \frac{MR(T_1 - T_3)}{k - 1}$$

which is a convenient expression for determining the work during an adiabatic change. The equation will give a negative arithmetical answer for compression of the gas (work on the gas) and a positive answer for expansion.

An expression for the work under an adiabatic curve can be developed by integration in a manner similar to that used for the isothermal process. The procedure is shown in the next section for the general expression for work under any polytropic change.

General or Polytropic Changes

Consider a polytropic change (PV^n = a constant) from the condition of P_1 and V_1 to P_2 and V_2:

$$W = \int P \, dV$$

$$P_1 V_1^n = PV^n$$

$$P = P_1 V_1^n \frac{1}{V^n}$$

Substituting,

$$W = P_1 V_1^n \int_{V_1}^{V_2} \frac{dV}{V^n} = P_1 V_1^n \int_{V_1}^{V_2} V^{-n} \, dV$$

$$= \frac{P_1 V_1^n}{1 - n} (V_2^{1-n} - V_1^{1-n})$$

$$= \frac{P_1 V_1^n V_2^{1-n} - P_1 V_1^n V_1^{1-n}}{1 - n}$$

Substituting $P_1 V_1^n = P_2 V_2^n$,

$$W = \frac{P_2 V_2^n V_2^{1-n} - P_1 V_1^n V_1^{1-n}}{1 - n} = \frac{P_2 V_2 - P_1 V_1}{1 - n} \tag{23}$$

If $n = k$, then

$$W = \frac{P_2 V_2 - P_1 V_1}{1 - k} \tag{24}$$

Equations 23 and 24 assume that work is done *by* the gas. If the solution gives a positive value, the gas has done work by expansion. If a negative result is obtained, work has been done on the gas by compression. The sign of the result then becomes an index of the nature or direction of the operation.

Changes in an Adiabatic Expansion of a Perfect Gas

For adiabatic expansion of a perfect gas it can be shown that PV^k = a constant, or

$$P_1 V_1^k = P_2 V_2^k \tag{25}$$

Combining equation 25 with the equation of state

$$\frac{P_1 V_1}{T_1} = \frac{P_2 V_2}{T_2} \tag{26}$$

allows the determination of final conditions of pressure, volume, and temperature if two initial conditions and one final condition are known.

$$T_2 = T_1\left(\frac{V_2}{V_1}\right)^{1-k} = T_1\left(\frac{V_1}{V_2}\right)^{k-1} = T_1\left(\frac{P_2}{P_1}\right)^{(k-1)/k}$$

$$= T_1\left(\frac{P_2}{P_1}\right)^{1-(1/k)} \tag{27}$$

$$P_2 = P_1\left(\frac{V_1}{V_2}\right)^{k} = P_1\left(\frac{T_2}{T_1}\right)^{k/(k-1)} \tag{28}$$

$$V_2 = V_1\left(\frac{P_1}{P_2}\right)^{1/k} = V_1\left(\frac{T_1}{T_2}\right)^{1/(k-1)} \tag{29}$$

It should be noted that the above formulas can be used for any expansion of a perfect gas following PV^n = a constant, provided that k is replaced by n. (It is assumed that both k and n are constants. This assumption is, in general, not true and will lead to error if the above equations are used for real gases.) The equations may be used for any system of units as long as the same system of units is employed throughout and both pressure and temperature values are absolute.

Polytropic expansion differs from adiabatic in that heat may be added or removed during the process. Ordinarily the value of n in a polytropic expansion is greater than unity and less than $k = c_p/c_v$.

For example, in a cylinder of volume $V_1 = 0.5$ L, an ideal gas is compressed from $P_1 = 10$ N/cm^2 to $P_2 = 40$ N/cm^2. The mass M of the gas is 0.001 kg. Consider the compression governed by $Pv^{1.4} = P_1 v_1^{1.4}$ (see fig. 2-6), where $v_1 = V_1/M$ and $v_2 = V_2/M$ are the specific volumes at conditions 1 and 2 respectively. The specific internal energy, $u = U/M$, is given by $u = 1.5\ vP + u_0$. Find the heat generated by the compression.

The heat $_1Q_2$ is given by the following equation from the first law:

$$_1Q_2 = U_2 - U_1 + W = M(u_2 - u_1) + Mw$$

The change of the specific internal energy is

$$u_2 - u_1 = 1.5\ (P_2 v_2 - P_1 v_1)$$

The specific volume $v_1 = V_1/M = 0.5$ L/0.001 kg $= 500$ L/kg, and v_2 is found from $Pv^{1.4} = P_1 v_1^{1.4}$ as follows:

$$v_2 = v_1\left(\frac{P_1}{P_2}\right)^{1/1.4} = 500\ \frac{L}{kg}\left(\frac{1}{4}\right)^{1/1.4} = 186\ \frac{L}{kg}$$

Then

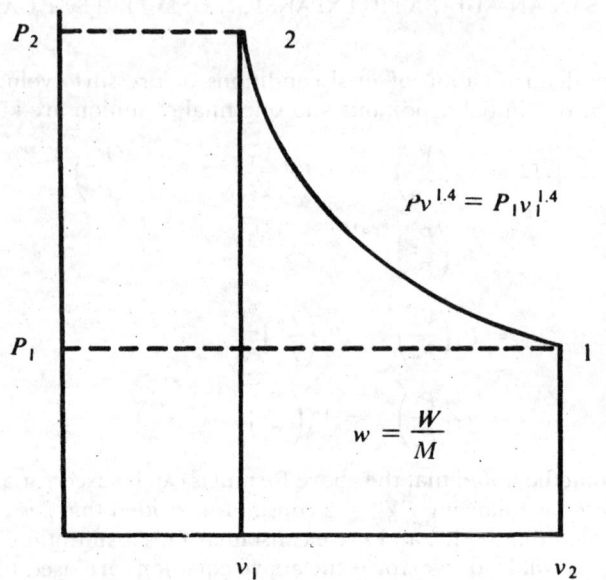

FIGURE 2-6   Ideal gas.

$$u_2 - u_1 = 1.5\left(40\ \frac{N}{cm^2} \cdot 186\ \frac{L}{kg} \cdot \frac{1000\ cm^3}{L} - \frac{10\ N}{cm^2} \cdot 500\ \frac{L}{kg} \cdot \frac{1000\ cm^3}{L}\right)$$

$$= 36.6 \times 10^5\ \frac{N\ cm}{kg} = 36.6 \times 10^3\ \frac{N\ m}{kg} = 36.6\ \frac{kJ}{kg}$$

The specific work $w = W/M$ is given by the area under the curve and is

$$_1w_2 = \int_1^2 P\ dv = \int_{v_1}^{v_2} P_1 v_1^{1.4}\ \frac{dv}{v^{1.4}}$$

$$= P_1 v_1^{1.4}\left(\frac{1}{-0.4}\right)\left[\frac{1}{v_2^{0.4}} - \frac{1}{v_1^{0.4}}\right]$$

$$= \frac{-P_1 v_1}{0.4}\left[\left(\frac{v_1}{v_2}\right)^{0.4} - 1\right] = -\left(\frac{10\ N}{0.4\ cm^2}\right)\left(500\ \frac{L}{kg}\right)\left[\left(\frac{500}{186}\right)^{0.4} - 1\right]$$

$$= -6065\ \frac{N\ L}{cm^2\ kg} \cdot \frac{1000\ cm^3}{L} = -(60.65) \cdot 10^5\ \frac{N\ cm}{kg}$$

$$= -60.65 \cdot 10^3\ \frac{N\ m}{kg} = -60.65\ \frac{kJ}{kg}$$

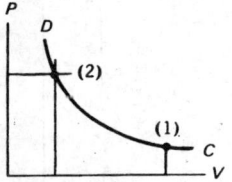

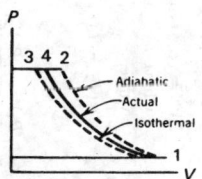

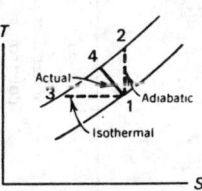

FIGURE 2-7
A polytropic curve.

FIGURE 2-8
Pressure-volume
diagrams.

FIGURE 2-9
Temperature-
entropy diagrams.

The heat is then

$$_1Q_2 = 0.001 \text{ kg}\left(36.6 \; \frac{\text{kJ}}{\text{kg}}\right) - 0.001 \text{ kg} \left(60.65 \; \frac{\text{kJ}}{\text{kg}}\right)$$

$$_1Q_2 = -0.024 \text{ kJ} = -24 \text{ J}$$

The minus sign indicates that this amount of heat has gone out of the system.

Determination of the Value of n from an Actual Compression Line

In figure 2-7 let CD represent a polytropic line. Let the values of the ordinates be measured at the two points 1 and 2 and introduced into the equation

$$P_1V_1^n = P_2V_2^n$$

Since n is the only unknown, it may be calculated.

Another method for determining n is to lay off on log-log graph paper successive values of P and V measured at chosen points in the curve; or lay off values of $\log P$ and $\log V$ on ordinary cross-section paper. If n is a constant, the points will lie in a straight line in either event, and the slope of the line gives the value of n.

If two points are chosen, then

$$n = \frac{\log(P_1/P_2)}{\log(V_2/V_1)} \tag{30}$$

Several pairs of points should be used to test the constancy of n.

TABLE 2.1 Summary of Nonflow Thermodynamic Processes for Ideal Gases with Constant Specific Heats

	Constant P (absolute)	Constant V	Constant T (isothermal)	Isentropic or Reversible Adiabatic	General or Polytropic
Ideal gas law	$\dfrac{V_1}{T_1} = \dfrac{V_2}{T_2}$	$\dfrac{P_1}{T_1} = \dfrac{P_2}{T_2}$	$P_1 V_1 = P_2 V_2$	$\dfrac{P_1 V_1}{T_1} = \dfrac{P_2 V_2}{T_2}$	$\dfrac{P_1 V_1}{T_1} = \dfrac{P_2 V_2}{T_2}$ or $PV = MRT$
Exponential law	$PV^n =$ constant, where $n = 0$	$PV^n =$ constant, where $n = \infty$	$PV^n =$ constant, where $n = 1$	$PV^k =$ constant, where $k = c_p/c_v$	$PV^n =$ constant, where $n =$ any value
Change in internal energy $(U_2 - U_1)$	$Mc_v(T_2 - T_1)$	$Mc_v(T_2 - T_1)$	0	$Mc_v(T_2 - T_1)$	$Mc_v(T_2 - T_1)$
External work $({}_1W_2)$	$P(V_2 - V_1)$	0	$P_1 V_1 \log_e \dfrac{V_2}{V_1}$ or $MRT \log_e \dfrac{V_2}{V_1}$ or $MRT \log_e \dfrac{P_1}{P_2}$	$\dfrac{P_1 V_1 - P_2 V_2}{k - 1}$ or $Mc_v(T_1 - T_2)$ or $\dfrac{MR(T_1 - T_2)}{k - 1}$	$\dfrac{P_1 V_1 - P_2 V_2}{n - 1}$
Total heat added $({}_1Q_2)$	$Mc_v(T_2 - T_1) + P(V_2 - V_1)$ or $Mc_p(T_2 - T_1)$	$Mc_v(T_2 - T_1)$	$P_1 V_1 \log_e \dfrac{V_2}{V_1}$	0	$Mc_v(T_2 - T_1) + \dfrac{P_1 V_1 - P_2 V_2}{n - 1}$

Equations for Nonflow Processes

In table 2-1 is a set of equations for nonflow thermodynamic processes for gases. These have been assembled in one location for convenience.

Figures 2-8 and 2-9 illustrate on pressure-volume and temperature-entropy diagrams the various thermodynamics processes shown in table 2-1, where 1 and 2 refer to initial and final conditions, respectively. No attempt has been made to plot any curve to scale.

PROBLEMS

1. Sketch P-V and T-S diagrams for
   (a) a constant-volume change
   (b) a constant-pressure change
   (c) a polytropic change.
Draw the P-V and T-S diagrams sufficiently large that all three changes may be placed on one diagram. Show the work accomplished in each process and indicate it on the proper diagram.

2. One kilogram of air is compressed from a volume of 14 m³ and a pressure of 101 kPa to a volume of 1 m³ and a pressure of 1825 kPa. What is the value of the exponent n? (Pressures given are always absolute unless gage is specifically indicated.)

3. What external work would be required to make possible the compression of problem 2?

4. A tractor tire contains 28.3 L of air at 184 kPa and a temperature of 27°C. How many kilograms of air does the tire contain?

5. The specific heat of air at constant volume is 0.767 kJ/(kg K). How much heat has been added to the air in the tire of problem 4 when the temperature of the air reaches 50°C if the process is assumed to be constant volume?

6. If the tire in problem 4 is punctured and all the air escapes, what will be the new volume of the air that was in the tire? How much heat will be exchanged between the outside air and the escaped air? Assume that the expansion is polytropic, the outside air temperature is 16°C, and the barometric pressure is 101 kPa.

7. A sample of tractor exhaust was found to have the following composition by volume at a temperature of 38°C: $N_2 = 73\%$; $H_2O = 13\%$; $CO_2 = 12\%$; $O_2 = 2\%$.
   (a) Determine the apparent molecular weight.
   (b) Determine the composition by weight.
   (c) Determine the value of R for the gas.
   (d) Determine the volume of 1 kg of the mixture (barometer = 101 kPa).

8. Water is pumped into a 10 m³ tank that initially contains only air at 100 kPa and 300 K. The air is compressed as water enters the tank. The tank is sealed so that the air cannot escape. The pump is stopped when the tank is half full by volume. If the air and water stay at 300 K, what is the work output by the pump?

9. Air at 100 kPa and 300 K is compressed to 500 kPa. Find the work (kg basis) if the process is
   (a) adiabatic ($k = 1.4$)
   (b) isothermal
   (c) polytropic ($k = 1.3$)
Also sketch all three processes on the same P-V and T-S diagrams.

10. Air is cooled in the following manner. Process 1-2 is an isothermal compression beginning at 100 kPa and 300 K and ending at 1000 kPa, whereas process 2-3 is an adiabatic expansion back to 100 kPa. Show the processes on P-V and T-S diagrams. Find the net work and the final temperature on a kg basis.

11. A system (fixed mass of air) undergoes a series of three processes and starts and returns to the same state (100 kPa and 300 K). Process 1-2 is an isothermal compression ($P_2 = 1000$ kPa), process 2-3 is a constant-pressure heat addition, and process 3-1 is an adiabatic expansion. Show all three processes on P-V and T-S diagrams and find the work and heat transfer for each process. What is the net work and net heat transfer? If efficiency is defined as the ratio of net work to heat transfer into the system, what is the efficiency of the cycle made up of these three processes? Calculate heat and work on a kg basis.

12. Repeat problem 11 with process 1-2 changed to adiabatic compression, and process 3-1 changed to constant-volume heat rejection.

13. The expansion of exhaust gases in an internal combustion engine is close to a polytropic expansion with $n = 1.3$. If air at 10 MPa and 2000 K is expanded through a volume ratio of 8, find the work and heat transfer on a kg basis and show the process on P-V and T-S diagrams.

14. A compression process starts at 101 kPa and ends at 930 kPa. What compression ratio is required to give this compression pressure if a general process is considered with $n = 1.25$?

15. Write a computer program to calculate the average polytropic exponent for the pressure-volume data given on p. 35. Calculate two values: (a) for the compression (52 to 172 crank angle degrees), and (b) for the expansion (242 to 322 crank angle degrees). The data were taken from a typical diesel engine. Note that 180 crank angle degrees is the point of minimum volume. Plot the data (P vs. V), and on the same graph also plot the values computed with your average values for the polytropic exponent.

Crank Angle (degrees)	Pressure (kPa)	Volume (m³)
Compression		
0.5200E + 02	0.3151E + 03	0.1176E − 02
0.6200E + 02	0.3475E + 03	0.1099E − 02
0.7200E + 02	0.3922E + 03	0.1009E − 02
0.8200E + 02	0.4547E + 03	0.9089E − 03
0.9200E + 02	0.5436E + 03	0.8006E − 03
0.1020E + 03	0.6731E + 03	0.6873E − 03
0.1120E + 03	0.8679E + 03	0.5727E − 03
0.1220E + 03	0.1172E + 04	0.4610E − 03
0.1320E + 03	0.1667E + 04	0.3569E − 03
0.1420E + 03	0.2506E + 04	0.2647E − 03
0.1520E + 03	0.3954E + 04	0.1888E − 03
0.1620E + 03	0.6304E + 04	0.1326E − 03
0.1720E + 03	0.9201E + 04	0.9875E − 04
Expansion		
0.2420E + 03	0.2869E + 04	0.4829E − 03
0.2520E + 03	0.2206E + 04	0.5955E − 03
0.2620E + 03	0.1761E + 04	0.7102E − 03
0.2720E + 03	0.1455E + 04	0.8228E − 03
0.2820E + 03	0.1240E + 04	0.9296E − 03
0.2920E + 03	0.1086E + 04	0.1028E − 02
0.3020E + 03	0.9742E + 03	0.1115E − 02
0.3120E + 03	0.8922E + 03	0.1190E − 02
0.3220E + 03	0.8326E + 03	0.1252E − 02

16. A long, insulated cylinder is divided into three compartments by frictionless, nonconducting pistons. The pistons are initially locked. For the conditions shown below, write a computer program to find the final pressure and temperatures after the pistons are unlocked. Assume all processes are ideal and adiabatic. Note that the three gases are different.

	Compartment		
	A (air)	B (CO_2)	C (H_2)
M (kg)	0.1	1	10
P_1 (kPa)	100	1000	10,000

(continued)

	Compartment		
	A (air)	B (CO_2)	C (H_2)
T (K)	300	300	300
MW (kg/mole)	29	44	2
c_p (kJ/kg K)	1.004	0.842	14.2
c_v (kJ/kg K)	0.717	0.653	10.1
R (kJ/kg K)	0.287	0.189	4.12
k	1.40	1.29	1.41

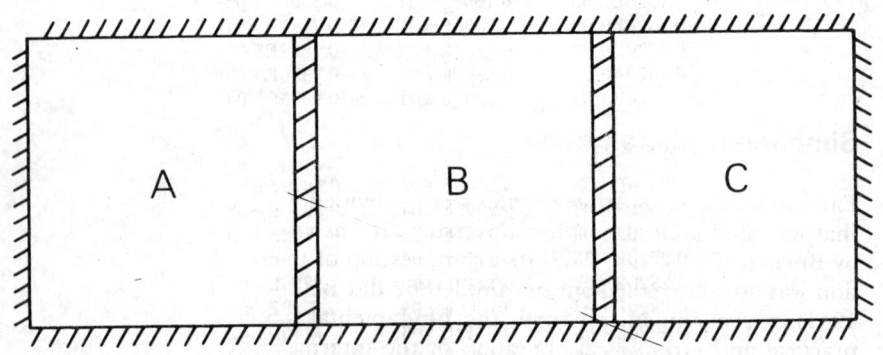

SUGGESTED READINGS

Lichty, L. C. *Combustion Engine Processes.* McGraw-Hill, New York, 1967.

Moyer, J. A., J. P. Calderwood, and A. A. Potter. *Elements of Engineering Thermodynamics.* Wiley, New York, 1941.

Obert, E. F. *Internal Combustion Engines and Air Pollution.* Harper & Row, New York, 1973.

Sneenden, J. B. and S. V. Kerr. *Applied Heat for Engineers.* International Ideas, Philadelphia, 1976.

Taylor, C. F. *The Internal Combustion Engine in Theory and Practice*, vol. 1: *Thermodynamics, Fluid Flow, and Performance*, 2d ed.; vol. 2: *Combustion, Materials, Fuel, and Design*, MIT Press, Cambridge, MA, 1985.

3

INTERNAL-
COMBUSTION
ENGINE CYCLES

Simplifying Assumptions

Early internal-combustion engines were provided with a combustible charge that was ignited at atmospheric pressure. It was recognized as early as 1838 by Burnett (see Lichty 1967) that compression of the charge before combustion was advantageous, but not until 1862 did Beau de Rochas (Moyer et al. 1941), a Frenchman, set forth the fundamental principles underlying the practical and economical operation of the internal-combustion engine. His requirements for maximum economy in the cycle were:

1. Smallest possible surface-volume ratio for the cylinder.
2. Most rapid expansion process possible.
3. Maximum possible expansion.
4. Maximum possible pressure at the beginning of the expansion process.

The first two requirements were designed to reduce heat loss through the cylinder walls to a minimum, thus maintaining the greatest possible amount of heat to be converted into work. The third requirement recognized that more work was produced by greater expansion. The fourth permitted higher pressures throughout the stroke for a given expansion ratio or permitted greater expansion ratios, either of which resulted in more work.

The first successful engine utilizing the principles of Beau de Rochas was built in 1878 and is credited to Nikolaus Otto, a German, from whom came the term "Otto cycle." The ordinary spark-ignition four-stroke-cycle engine draws in a charge of air, with which the fuel is usually mixed, on the first stroke; it compresses the charge on the second stroke, after which the fuel and air mixture is ignited; expansion takes place on the third stroke, and the burned products are exhausted on the fourth stroke. In any theoretical consideration of the engine cycle, attention is given to processes of compression, expansion, heating, and cooling.

Figure 3-1 illustrates the ideal indicator diagram for a four-stroke-cycle engine. MA represents the introduction of the charge; AB is adiabatic compression; BC is instantaneous combustion of the fuel and heating of the charge at constant volume; CD is adiabatic expansion of the hot gas; DA is an instantaneous drop of pressure following release; and AM is the ejection of the remainder of the charge.

Chapter 2 dealt with thermodynamic principles but did not discuss the complete cycle. A brief discussion of ideal Otto and diesel cycles in this chapter will be followed by a discussion of the reasons why the actual cycles do not attain the efficiency of the ideal cycles.

The Ideal Air-Standard Otto Cycle

In considering the ideal cycle certain assumptions will be made. In the ideal Otto cycle the four events are assumed to occur in two strokes, as in the actual case with a two-stroke-cycle engine. We may also assume for the ideal cycle that:

1. The piston has zero friction in the cylinder.
2. Air is used in the cylinder as the working fluid.
3. No heat transfer takes place through the engine walls.
4. The crank starts at the bottom of the stroke under conditions of P_1, V_1, and T_1.
5. Adiabatic compression occurs along AB and adiabatic expansion occurs along CD (fig. 3-1).
6. Constant-volume addition of heat occurs along BC, and a constant-volume rejection occurs along DA.
7. The working fluid (air) is treated as a perfect gas with constant specific heats. All thermodynamic processes are assumed ideal.

The efficiency of the cycle may be calculated as follows: Heat added at constant volume during the cycle is

$$Q_{in} = Mc_v(T_3 - T_2)$$

The heat rejected from D to A is

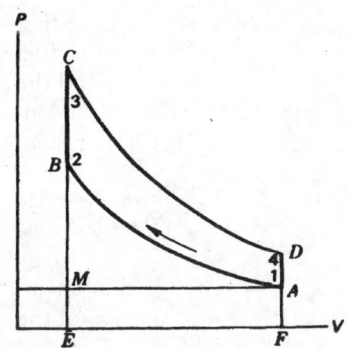

FIGURE 3-1 Ideal indicator diagram for a four-stroke-cycle engine.

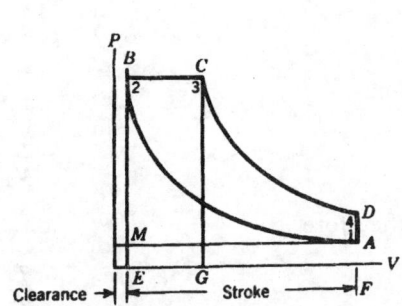

FIGURE 3-2 Ideal diesel cycle.

$$Q_{out} = Mc_v(T_4 - T_1)$$

where
Q = heat in or out of the system
M = total mass of cylinder contents (assumed constant)
c_p = the specific heat of air at constant pressure
c_v = the specific heat of air at constant volume
T = absolute temperature
$c_p/c_v = k = 1.4$ for air

The cycle efficiency is

$$e = \frac{Q_{in} - Q_{out}}{Q_{in}} = \frac{Mc_v(T_3 - T_2) - Mc_v(T_4 - T_1)}{Mc_v(T_3 - T_2)}$$

$$= 1 - \frac{T_4 - T_1}{T_3 - T_2}$$

$$= 1 - \frac{T_1\left(\dfrac{T_4}{T_1} - 1\right)}{T_2\left(\dfrac{T_3}{T_2} - 1\right)} \qquad (1)$$

since

$$\frac{T_2}{T_1} = \left(\frac{V_1}{V_o}\right)^{k-1}$$

and

$$\frac{T_3}{T_4} = \left(\frac{V_4}{V_3}\right)^{k-1}$$

and

$$\frac{V_1}{V_2} = \frac{V_4}{V_3}$$

therefore

$$\frac{T_4}{T_1} = \frac{T_3}{T_2}$$

and equation 1 becomes

$$e = 1 - \frac{T_1}{T_2} = 1 - \left(\frac{V_2}{V_1}\right)^{k-1}$$

$$= 1 - \frac{1}{(V_1/V_2)^{k-1}} \tag{2}$$

$$= 1 - \frac{1}{r^{k-1}}$$

where r = compression ratio. Also,

$$e = 1 - \left(\frac{P_1}{P_2}\right)^{(k-1)/k} \tag{3}$$

Hence it is apparent that the efficiency of an ideal Otto cycle increases as the compression ratio increases and is independent of the heat liberated by the fuel. It should be remembered that factors other than the compression ratio influence the efficiency of the actual cycle. These factors are discussed later. The differences between the ideal and actual cycles are so great that close quantitative agreement should not be expected.

It is the work done during each cycle that enables the engine to do work and develop power. For the ideal Otto cycle illustrated in figure 3-1 the work of the cycle is equal to the algebraic sum of the work under the compression and expansion curves, where compression work *FABE* is negative and expansion work *FDCE* is positive. From equation 24 in chapter 2, the cycle work may be calculated as

$$W = \frac{P_1V_1 - P_2V_2}{k - 1} + \frac{P_3V_3 - P_4V_4}{k - 1} \tag{4}$$

or, from conservation of energy,

$$W = Q_{in} - Q_{out}$$

The Ideal Air-Standard Diesel Cycle

The diesel cycle is usually completed in four strokes as is the Otto cycle. The ideal cycle may also be considered as a two-stroke cycle since the exhaust and inlet strokes are considered frictionless. In this cycle the gas is also assumed to be atmospheric air throughout the entire cycle. This cycle is represented in figure 3-2. In addition to the difference between the diesel and Otto cycles in the nature of the second operation (the combustion process), the diesel is characterized by the high compression pressures employed. In the diesel engine, the compression operation AB is performed upon a charge of air only. Consequently, there is no danger of preignition during compression. The only limits to pressure and temperatures are mechanical.

The heat quantities and efficiency of the diesel cycle can be worked out in a manner similar to that used for the Otto cycle. It will be noted that the efficiency of the diesel cycle is influenced not only by the compression ratio but also by the length of time during which the heat is being added to the compressed air. It is simpler to state merely that heat is added than to consider the addition of fuel, since the weight of the charge thereby remains constant.

Referring to figure 3-2, the following expressions will be obtained: No heat is added in adiabatic compression from 1 to 2. Heat is added from 2 to 3.

$$Q_{in} = Mc_p(T_3 - T_2)$$

No heat is rejected in adiabatic expansion from 3 to 4. Heat is rejected from 4 to 1.

$$Q_{out} = Mc_v(T_4 - T_1)$$

The efficiency of the cycle is

$$e = \frac{Q_{in} - Q_{out}}{Q_{in}} = 1 - \frac{Q_{out}}{Q_{in}}$$

$$= 1 - \frac{Mc_v(T_4 - T_1)}{Mc_p(T_3 - T_2)} = 1 - \frac{(T_4 - T_1)}{k(T_3 - T_2)}$$

$$= 1 - \frac{T_1}{T_2}\left[\frac{\left(\frac{T_4}{T_1} - 1\right)}{k\left(\frac{T_3}{T_2} - 1\right)}\right] \qquad (5)$$

From T, V relations, we have

$$T_3 = T_2 \frac{V_3}{V_2} \text{ at constant pressure}$$

also

$$T_4 = T_3 \left(\frac{V_3}{V_4}\right)^{k-1} \text{ adiabatically}$$

Thus

$$T_4 = T_2 \frac{V_3}{V_2} \left(\frac{V_3}{V_4}\right)^{k-1} = T_2 \frac{(V_3/V_2)^k}{(V_4/V_2)^{k-1}}$$

Since the compression from 1 to 2 is adiabatic and $V_4 = V_1$, we have

$$T_2 = T_1 \left(\frac{V_4}{V_2}\right)^{k-1}$$

Solving for T_1 and then dividing T_4 by T_1 we obtain

$$\frac{T_4}{T_1} = \frac{(V_4/V_2)^{k-1}(V_3/V_2)^k}{(V_4/V_2)^{k-1}} = \left(\frac{V_3}{V_2}\right)^k$$

Substituting in equation 5, we have

$$e = 1 - \frac{T_1[(V_3/V_2)^k - 1]}{kT_2[(V_3/V_2) - 1]}$$

$$T_1 = T_2 \left(\frac{V_2}{V_1}\right)^{k-1}$$

$$e = 1 - \left\{\frac{(V_2/V_1)^{k-1}[(V_3/V_2)^k - 1]}{k[(V_3/V_2) - 1]}\right\}$$

$$= 1 - \frac{1}{(V_1/V_2)^{k-1}} \left[\frac{(V_3/V_2)^k - 1}{k[(V_3/V_2) - 1]}\right]$$

$$e = 1 - \frac{1}{r^{k-1}} \left[\frac{(V_3/V_2)^k - 1}{k[(V_3/V_2) - 1]}\right] \tag{6}$$

From the equations it may be seen that the efficiency increases as the compression ratio V_1/V_2 increases and also as the fuel cutoff ratio V_3/V_2 is diminished. In other words, maximum efficiency will be obtained with high compression and early fuel cutoff. As the fuel cutoff approaches zero the efficiency of the diesel cycle approaches that of the Otto cycle for the same compression ratio. For the same compression ratio and the same heat input, the diesel cycle is less efficient than the Otto.

The external work per cycle in the case of the ideal diesel cycle consists also of the algebraic sum of the work under the compression and the expansion curves. With the ideal diesel cycle as illustrated in figure 3-2 there is negative work $ABEF$ under the adiabatic compression curve and positive compression work under constant pressure line BC and adiabatic curve CD.

$$W = \frac{P_1V_1 - P_2V_2}{k - 1} + \frac{P_3V_3 - P_4V_4}{k - 1} + P_{2,3}(V_3 - V_2) \qquad (7)$$

Actual Cycles and Causes for Deviation from the Ideal

The actual process in an engine results in a performance or efficiency considerably less than that indicated by the ideal analysis. The deviations are due to the fact that the assumptions made for the air-standard cycle analysis do not hold true for the actual cycle. The entire gas is not air, the specific heat is not constant, chemical reaction does occur during the cycle, and heat transfer does occur through the cylinder walls.

The efficiency of an internal-combustion engine is frequently given as the ratio of the work equivalent of the area on the actual indicator diagram to the work equivalent of the lower heat of combustion of the fuel used. This ratio expressed as a percent is called the *indicated thermal efficiency*. When viewed in this light, the engine is a relatively inefficient energy converter, with efficiencies of perhaps 25 to 35 percent. A fairer comparison would be to consider the relative efficiency as a measure of engine performance. The *relative efficiency* is defined as the ratio of the indicated thermal efficiency of an engine to the efficiency of the ideal process upon which the design of the engine is based, the ratio being a measure of the perfection of design and performance of the engine.

The air-to-fuel ratio supplied to the engine is not uniform among cylinders of a multicylinder engine. The intake stroke takes place at less than atmospheric pressure, whereas the exhaust stroke takes place above atmospheric pressure. Because of inertia forces in the intake mixture, the intake valve is not necessarily closed at the end of the intake stroke or opened at the beginning of the intake stroke. The exhaust valve opens considerably before the end of the expansion stroke. A more complete discussion of valve timing

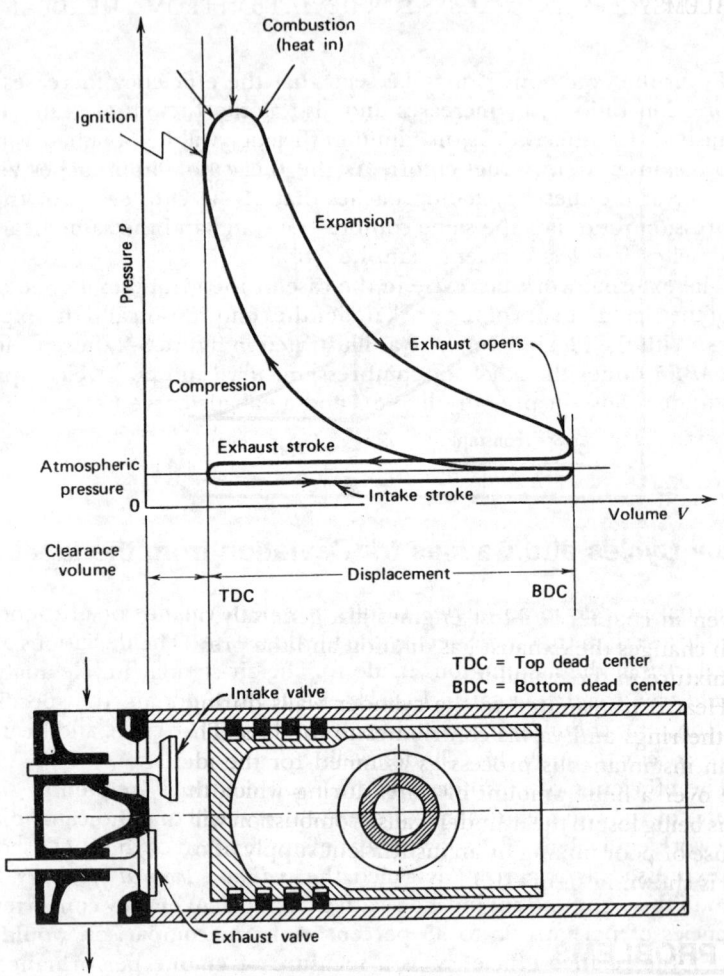

FIGURE 3-3 Actual *P-V* diagram showing the deviation from the ideal cycle and the loss from pumping for a four-stroke-cycle engine.

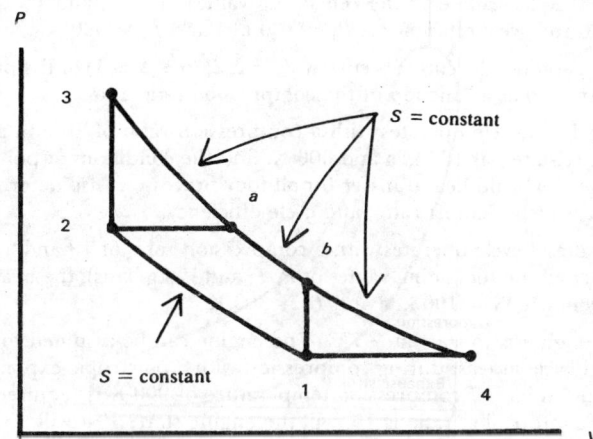

FIGURE 3-4 Ideal gas processes.

is given in chapter 5. Most engines are generally operated at part throttle, which changes the exhaust gas dilution and the pressure and temperature of the mixture in the cylinder.

Heat is transmitted to the cylinder walls during the cycle, and leakage past the rings and valves contributes to the loss. The combustion process is not an instantaneous process as assumed for the ideal Otto cycle but takes place over a finite amount of time, during which the piston is moving and heat is being lost to the cylinder walls. Combustion will usually not be complete because of poor mixing or an insufficient supply of oxygen. An actual engine cycle is shown in figure 3-3.

PROBLEMS

1. An ideal gas can undergo any of the processes shown in figure 3-4. Process 1-2 and process 3-a-b-4 are adiabatic. All processes are ideal. Four possible cycles beginning and ending at stage 1 can be devised. Describe these cycles and determine which cycle has the highest and which cycle has the lowest thermal efficiency. Draw the corresponding T-S diagram.

2. An ideal Otto cycle operates with a compression ratio of 8 and a heat input of 2000 kJ/kg. If state 1 is 100 kPa and 300 K, find the conditions at points 2, 3, and 4 as well as the work and heat transfer for all four processes. Also determine the net work, net heat transfer, and cycle efficiency.

3. An ideal Otto cycle operates with a compression ratio of 6 and a temperature at point 3 of 3500 K. Calculate all the remaining values of P, T, and V (kg basis), the heat input, and the cycle efficiency if $P_1 = 100$ kPa and $T_1 = 300$ K.

4. In a diesel engine, the cutoff occurs at $V_3 = XV_1$ $(0 \leq X \leq 1)$ of the piston stroke. Find the cutoff ratio as a function of the compression ratio and X.

5. An ideal diesel cycle operates with a compression ratio of 16 and a heat input of 2000 kJ/kg. If state 1 is 100 kPa and 300 K, find the conditions at points 2, 3, and 4, as well as the work and heat transfer for all four processes. Also determine the net work, net heat transfer, cutoff ratio, and cycle efficiency.

6. An ideal diesel cycle operates with a compression ratio of 18 and a cutoff ratio of 4.0. Calculate all the remaining values of P, T, and V (kg basis), the heat input, and the cycle efficiency if $P_1 = 100$ kPa and $T_1 = 300$ K.

7. A diesel engine is to start at $-7°C$. The engine can be assumed to operate on the ideal diesel cycle at least during compression with a polytropic exponent of 1.25. It is known that an end of compression temperature of 900 K is required to start an engine. If the compression ratio is 16, will the engine start? If it will not start, what compression ratio will be required?

8. For the same temperature limits and equal amount of heat input, which ideal cycle (Otto or diesel) has the highest thermal efficiency? Use P-V and T-S diagrams to illustrate your conclusion.

9. If the ideal Otto cycle is allowed to expand to the initial pressure ($P_4 = P_1$) so that the heat rejection is at constant pressure instead of at constant volume, how is the thermal efficiency affected? If the diesel cycle is also changed in this way, how is the thermal efficiency affected? Explain your results.

10. Write a computer program to calculate the efficiency of an ideal Otto cycle and an ideal diesel cycle for compression ratios of 10 to 20 if $P_1 = 100$ kPa and $T_1 = 300$ K. Plot the results on the same graph using efficiency for the vertical axis and compression ratio for the horizontal axis. Use at least five different heat inputs between 1000 and 2000 kJ/kg. Explain your results.

11. The dual cycle is identical to the ideal Otto cycle except that it involves two heat addition processes (one at constant volume and a second at constant pressure). Write a computer program to compute the thermal efficiency and compression ratio of the dual cycle as a function of the split in heat addition between the two processes. Define split as the ratio of heat added at constant volume to the total heat added. Use five values of split including 0 and 1. Plot the results on the same graph using efficiency as the vertical axis and compression ratio as the horizontal axis. Explain your results. Let $P_1 = 100$ kPa and $T_1 = 300$ K, and assume a total heat input of 2000 kJ/kg.

REFERENCES

Lichty, L. C. *Combustion Engine Processes*. McGraw-Hill, New York, 1967.

Moyer, J. A., J. P. Calderwood, and A. A. Potter. *Elements of Engineering Thermodynamics*. Wiley, New York, 1941.

SUGGESTED READINGS

Obert, E. F. *Internal Combustion Engines and Air Pollution*. Harper & Row, New York, 1973.

Sneenden, J. B. and S. V. Kerr. *Applied Heat for Engineers*. International Ideas, Philadelphia, 1976.

Taylor, C. F. *The Internal Combustion Engine in Theory and Practice*, vol. 1: *Thermodynamics, Fluid Flow, and Performance*, 2nd ed.; vol. 2: *Combustion, Materials, Fuel, and Design*. MIT Press, Cambridge, MA, 1985.

4
FUELS AND COMBUSTION

Sources of Fuels

Almost all of the fuels commonly used in farm tractors are products of crude petroleum. As fossil fuels become scarce, other fuels will have to be used. There are many factors that must be considered in the selection of an alternate fuel. Some of these factors are cost per unit of work done, availability, compatibility with the engine, safety, storage, management, and convenience. A section at the end of this chapter describes the properties of some of the alternate fuels most likely to be used in the future.

Chemical Composition of Petroleum

Crude petroleum is made up of combined carbon and hydrogen in approximately the proportion of 86 percent carbon to 14 percent hydrogen. The atoms of carbon and hydrogen may be combined in many ways to form many different hydrocarbon compounds in crude oil. The fuels made from crude oil are never one single hydrocarbon. The engineer is concerned chiefly with the physical properties and operating characteristics of a fuel; therefore, only a very brief space will be devoted to fuel chemistry in this book.

The hydrocarbons making up most of the refined fuels belong to the paraffin (C_nH_{2n+2}), olefin (C_nH_{2n}), diolefin (C_nH_{2n-2}), naphthene (C_nH_{2n}), and aromatic (C_nH_{2n-6}) families. As an example, for the paraffin series when n equals 1, the first would be CH_4 or methane gas. When n equals 4, the compound would be butane, with the formula C_4H_{10}. When n equals 1 to 4, the hydrocarbons are gases at normal temperatures and pressures. Butane boils at 0°C at atmospheric pressure. The normal compounds of the paraffin series have straight-chain molecular structures as follows:

$$
\begin{array}{c}
\quad\ \text{H}\ \ \text{II}\ \ \text{H}\ \ \text{H}\ \ \text{H}\ \ \text{H}\ \ \text{H}\ \ \text{H} \\
\quad\ \ |\quad |\quad |\quad |\quad |\quad |\quad |\quad | \\
\text{H---C---C---C---C---C---C---C---C---H} \\
\quad\ \ |\quad |\quad |\quad |\quad |\quad |\quad |\quad | \\
\quad\ \text{H}\ \ \text{H}\ \ \text{H}\ \ \text{H}\ \ \text{H}\ \ \text{H}\ \ \text{H}\ \ \text{H}
\end{array}
$$

C_8H_{18} or n-octane

Paraffins may have the same number of atoms of carbon and hydrogen but have different molecular structure. These are called *isomers*. An example is iso-octane (C_8H_{18}), structured as follows:

C_8H_{18} or iso-octane

The paraffins are saturated, quite stable, and low in gum-forming properties. Kerosene is a good example of a commercial fuel made up largely of paraffins. The straight-chain hydrocarbons generally detonate badly in an engine. On the other hand, the isomers are highly knock resistant. Normal heptane (n-

$$
\begin{array}{c}
\quad\ \text{H}\ \ \text{H}\ \ \text{H}\ \ \text{H}\ \ \text{H}\ \ \text{H}\ \ \text{H} \\
\quad\ \ |\quad |\quad |\quad |\quad |\quad |\quad | \\
\text{H---C---C---C---C---C---C---C---H} \\
\quad\ \ |\quad |\quad |\quad |\quad |\quad |\quad | \\
\quad\ \text{H}\ \ \text{H}\ \ \text{H}\ \ \text{H}\ \ \text{H}\ \ \text{H}\ \ \text{H}
\end{array}
$$

C_7H_{16} n-heptane

C_7H_{16} triptane

heptane (n-heptane), for example, is the low antiknock fuel (zero octane) used as a reference fuel in knock testing (see page 57). Triptane, an isomer of heptane, is one of the most knock-free hydrocarbons known. These fuels have the same chemical formula but are different structurally and vary widely in burning characteristics.

The *olefins* are important because they occur in fuels made by the cracking process. They have the general formula C_nH_{2n}, are more resistant to detonation than the paraffins, and are unsaturated. They are capable of taking up additional hydrogen, becoming paraffinic as in the following example:

$$
\begin{array}{ccc}
 & & \text{H—H} \\
 & & |\quad| \\
\text{H—C}{=}\text{C—H} + 2\text{H} \rightarrow & \text{H—C}-\text{C—H} \\
|\quad| & & |\quad| \\
\text{H}\ \ \text{H} & & \text{H}\ \ \text{H}
\end{array}
$$

The *diolefins* are unsaturated, but have a straight-chain structure. They have the general equation C_nH_{2n-2}. This family of hydrocarbons is the principal cause of instability of cracked gasoline in storage, causing gum and off-color and lower antiknock properties.

Naphthenes have the same chemical formula as the olefins, but they are saturated compounds, and have high antiknock values. They have molecules made up of carbon and hydrogen with a ring structure, and the prefix cyclo is used in their name. Cyclohexane C_6H_{12} can be illustrated as follows:

C_6H_{12} cyclohexane

The *aromatics* are ring-structured. They are unsaturated but stable chemically. The general formula is C_nH_{2n-6}. The formula C_6H_6 represents benzene having a structure as follows:

C_6H_6 benzene

Alpha-methyl naphthalene, $C_{11}H_{16}$, is another aromatic of importance because it is used as a reference fuel in testing diesel fuel. It has a high antiknock value as a fuel for spark ignition engines, but poor ignition quality as a diesel fuel.

Petroleum Refining

Each hydrocarbon has a definite specific gravity and boiling point, and in any series a higher value of n generally increases the specific gravity and thus enhances the possibility of refining or fractionating by distillation. Figure 4-1 is a simplified diagram of the refining process for a mixed-base crude for the manufacture of gasoline, kerosene, distillate, lubricating oil, and asphalt. Crude oil from the well or storage is fed into the still, where the temperature is raised to a higher temperature than the high-boiling-point constituents. The vapors pass up through the bubble or fractionating tower and "cuts" or fractions of specific boiling-point ranges are separated and subsequently condensed.

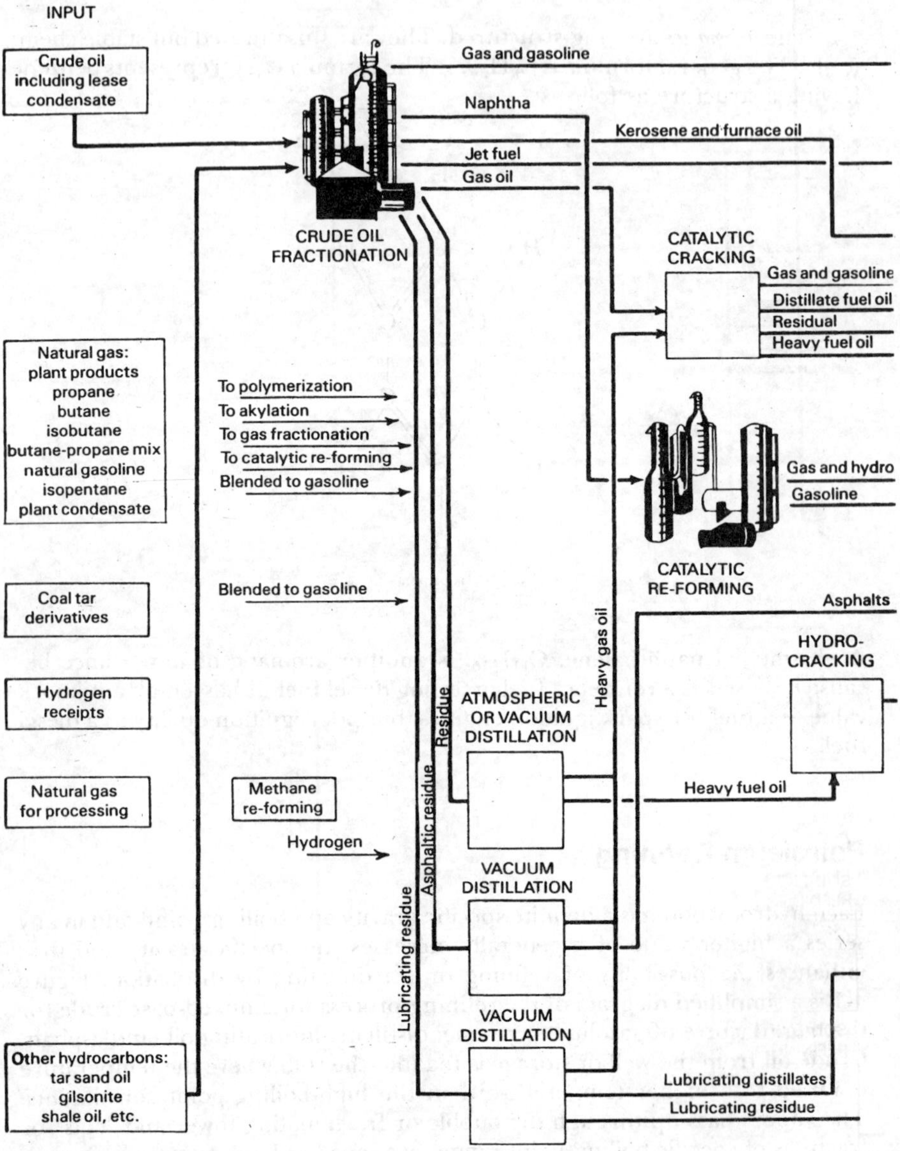

INPUT

Crude oil
including lease
condensate

Gas and gasoline

Naphtha

Kerosene and furnace oil

Jet fuel

Gas oil

CRUDE OIL
FRACTIONATION

CATALYTIC
CRACKING

Gas and gasoline

Distillate fuel oil

Residual

Heavy fuel oil

Natural gas:
plant products
propane
butane
isobutane
butane-propane mix
natural gasoline
isopentane
plant condensate

To polymerization

To akylation

To gas fractionation

To catalytic re-forming

Blended to gasoline

Gas and hydro

Gasoline

CATALYTIC
RE-FORMING

Coal tar
derivatives

Blended to gasoline

Asphalts

HYDRO-
CRACKING

Hydrogen
receipts

Heavy gas oil

Residue

ATMOSPHERIC
OR VACUUM
DISTILLATION

Asphaltic residue

Natural gas
for processing

Methane
re-forming

Hydrogen

Heavy fuel oil

VACUUM
DISTILLATION

Lubricating residue

VACUUM
DISTILLATION

Other hydrocarbons:
tar sand oil
gilsonite
shale oil, etc.

Lubricating distillates

Lubricating residue

FIGURE 4-1 Simplified flow diagram of a modern petroleum refinery. (Courtesy American Petroleum Institute.)

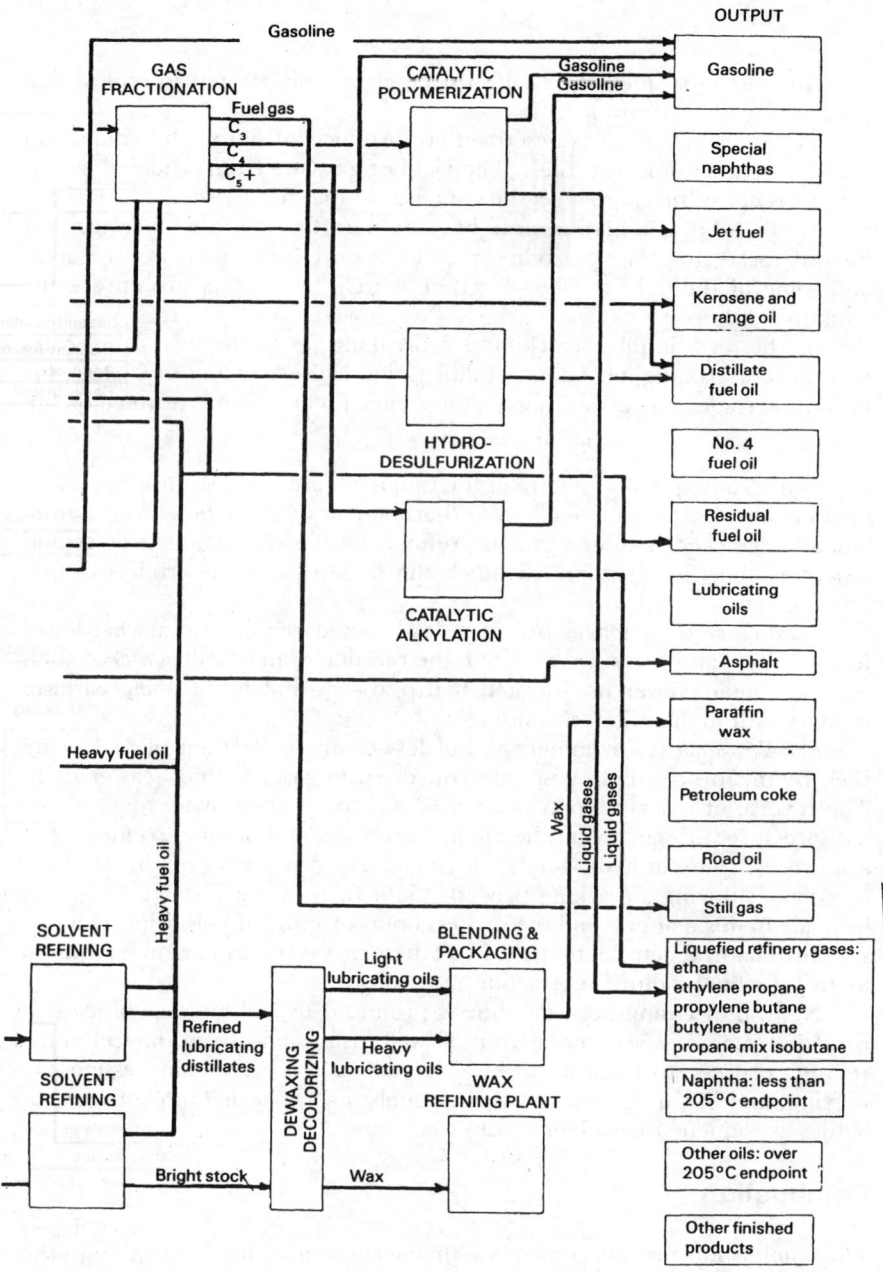

FIGURE 4-1 Continued

53

After its separation the fuel must be chemically treated to remove impurities, chiefly sulfur and gum.

The fuels produced as described are products of simple distillation and are known as straight-run fuels. The yield of gasoline from crude oil by this process is not sufficient to meet the demand for that fuel. Also, many straight-run fuels tend to run high in straight-chain paraffins that are inherently low in antiknock value. The prevailing practice is to re-form the heavy fractions, and some of the light gasoline fraction as well, by *cracking* to improve the antiknock value.

As the term implies, cracking is a breaking up of the heavy molecules into more useful, lighter ones. An illustration is the cracking of $C_{14}H_{30}$, the paraffin tetradecane, producing a lighter paraffin and an unsaturated olefin.

$$C_{14}H_{30} \rightarrow C_7H_{16} + C_7H_{14}$$

The two resulting molecules are in the range of gasoline. Pressure, heat, and time are the factors essential to the thermal-cracking reaction. After an oil has been cracked it must again be refined by the distillation process and separated into fuel fractions in much the same way as the crude was first fractionated.

Catalytic cracking makes use of a catalyst and requires somewhat lower temperatures and pressures to obtain the reaction. Catalytically cracked fuels tend to contain fewer unsaturated hydrocarbons and more stable paraffin isomers with high antiknock values.

Polymerization is a commercial, but less common, method of producing fuel. By this process light gases are converted to gasoline and heavier fuels. The reverse of cracking, polymerization is accomplished by a catalyst, heat, and pressure, principally on the olefins, propane, and butane. *Hydrogenation* is a process wherein hydrogen is added chemically to heavy unsaturated hydrocarbons during a cracking process. Light fuels in the gasoline range can be made in this manner, and the process holds promise of bolstering a limited supply of natural petroleum fuels. In all these processes distillation is required to give the final motor-fuel fractions.

Natural or casing-head gasoline is produced by still another process. It is made from gases recovered from oil well casings and from the still in the refining process. Its manufacture is essentially a process of compressing, liquefying, and blending with less volatile fuels; or it is sealed in containers or bottles as "liquefied petroleum" (LP) gas.

Combustion

The combustion process consists of chemically combining oxygen from the air with carbon and hydrogen in the fuel. Heat is liberated in the process, and a pressure increase results. The final or desired result occurs during the power or expansion stroke when a volume change takes place and work is done.

Air is the source of oxygen. By weight, air is 23.1 percent oxygen and 76.9 percent nitrogen. Actually there are minor amounts of other elements present in air, but they need not be considered in combustion calculations. Air by volume consists of 20.8 percent oxygen and 79.2 percent nitrogen. Thus, for 1 kg of oxygen there is 3.33 kg of nitrogen, and for each cubic meter of oxygen in the air there is 3.8 m^3 of nitrogen.

According to Avogadro's law, which states the relationship between moles and volumes, at the same temperature and pressure, equal volumes of ideal gases contain the same number of molecules. The mole mass of a gas is numerically equal to its molecular weight. At standard conditions of 760 mm Hg and 0°C the molar volume of ideal gases is 22.41 m^3/kmol = 22.41 L/mole. Atomic and molecular weights of some of the elements entering into the combustion process are given in table 4-1.

When combustion is complete, C and H burn to CO_2 and H_2O, respectively. Engine fuels seldom are one chemical compound, and combustion calculations serve only to illustrate general principles. Hydrocarbon fuels, as pointed out earlier in this chapter, are made up of a multitude of different carbon and hydrogen compounds. Methane gas (CH_4), the simplest of the paraffin series, can be used to illustrate the combustion process. The calculation of the masses of the compounds is as follows:

$$CH_4 + 2O_2 + 7.6N_2 = CO_2 + 2H_2O + 7.6N_2$$

$$(12 + 4) + (2 \times 32) + (7.6 \times 28) = (12 + 32) + 2(2 + 16) + (7.6 \times 28)$$

$$16 \quad + \quad 64 \quad + \quad 212.8 \quad = \quad 44 \quad + \quad 36 \quad + \quad 212.8$$

$$292.8 = 292.8$$

From the above calculation it can be seen that 64 kg of oxygen is required

TABLE 4-1 Atomic and Molecular Weights of Common Combustion Substances

Substance	Symbol for Element	Atomic Weight	Symbol for Molecule	Molecular Weight
Carbon	C	12	C	12
Hydrogen	H	1	H_2	2
Oxygen	O	16	O_2	32
Nitrogen	N	14	N_2	28
Sulfur	S	32	S	32
Air (apparent)	—	—	—	29.0

for 16 kg of fuel. Since air is represented by the oxygen plus the nitrogen in the process, 276.8 kg of air is required for 16 kg of fuel. The theoretically correct air-fuel ratio then becomes

$$\frac{Air}{Fuel} = \frac{276.8}{16} = \frac{17.3}{1}$$

Calculating on a volume basis,

1 mole CH_4 + 2 moles O_2 + 7.6 moles N_2
$$= 1 \text{ mole } CO_2 + 2 \text{ moles } H_2O + 7.6 \text{ moles } N_2$$

or

$$1 \text{ m}^3 \text{ } CH_4 + 2 \text{ m}^3 \text{ } O_2 + 7.6 \text{ m}^3 \text{ } N_2 = 1 \text{ m}^3 \text{ } CO_2 + 2 \text{ m}^3 \text{ } H_2O + 7.6 \text{ m}^3 \text{ } N_2$$

The volume of air needed for each unit of CH_4 gas is 9.6 units. Since nitrogen is inert and does not enter into the combustion process, it may be eliminated and air requirements may be calculated after the oxygen requirement is known, by means of the percentages of oxygen in air mentioned above.

Even under ideal conditions or an excess of air, combustion in an engine is not complete. Carbon monoxide is always present in the by-product, as may be illustrated as follows:

$$2CH_4 + 3.5O_2 \rightarrow CO_2 + CO + 4H_2O$$

The heating value of CO when oxidized to CO_2 is 1095 kJ/kg or approximately 21 percent of the heating value of gasoline. Certain hydrocarbons are usually taken as representative of the various liquid fuels for purposes of combustion calculations. Octane (C_8H_{18}) is usually used for gasoline, and cetane ($C_{16}H_{34}$) for diesel oil. The combustion of gasoline is expressed as

$$C_8H_{18} + 12.5O_2 = 8CO_2 + 9H_2O$$

$$(12 \times 8) + (1 \times 18) + (12.5 \times 32) = 8(12 + 32) + 9(2 + 16)$$

$$114 \quad + \quad 400 \quad = \quad 352 \quad + \quad 162$$

$$1 \quad + \quad 3.50 \quad = \quad 3.08 \quad + \quad 1.42$$

Since air is 23.1 percent oxygen, 3.50/0.231 or 15.1 units of air are required for 100 percent air requirement. There are 1.42 units of water produced per unit of fuel. The water appears in the engine exhaust products as a vapor.

Gasoline Tests and Their Significance

Most of the gasoline and diesel fuel tests are those specified by the American Society for Testing and Materials (ASTM), Society of Automotive Engineers

(SAE), and the American Petroleum Institute (API). A fuel is tested in order
to estimate its (1) volatility, (2) burning characteristics, (3) freedom from
excessive impurities, and (4) storage stability. Inspection tests have been developed to measure or determine all these properties. These tests can be found
in the many standards published by the ASTM. Only the most important are
discussed here.

Antiknock Quality

The fuel-air mixture in the cylinder of a spark ignition engine will, under
certain conditions, burn spontaneously in localized areas instead of progress-
ing from the spark. This usually causes an audible "ping," knock, or detonation. The actual loss of power and damage to an engine because of detonation is generally not significant until the intensity becomes very severe.
However, heavy and prolonged detonation may have an adverse effect in
terms of power loss and possible damage to the engine.

The octane number of a gasoline is a measure of its tendency to resist
detonation during combustion in an engine. The octane number of the fuel
is an arbitrary scale based on two hydrocarbons that define the ends of the
scale. Normal heptane, by definition, has an octane number of zero and iso-
octane has a value of 100. The test is made with a special engine similar to
that in figure 4-2. Briefly, the test procedure is as follows: A sample of the
fuel to be tested is burned in a test engine, and the compression ratio is
adjusted until detonation occurs. The intensity of detonation is measured by
special instruments and is compared with a fuel made up of varying amounts
of two reference fuels. The octane number of the fuel being tested is the
percentage of iso-octane in the mixture of the two reference fuels that has
the same knock intensity as the fuel being tested. The octane number is a
measure of the suitability of a fuel for use in an engine of given compression
ratio. High octane rating is necessary for high-compression, high-performance engines. It is generally the most important single quality characteristic
for fuels used in spark ignition engines.

There are two tests in general use for measuring the octane number of
gasoline, the Motor Method (ASTM D2700) and the Research Method (ASTM
D2699). The tests are similar in all respects except for the engine operating
conditions. In the Motor Method the operating conditions are more severe
than in the Research Method. The laboratory ratings obtained by these tests
on a gasoline sample are nearly always different. Since the Research Method
is less severe, ratings by this method are usually higher than those obtained
by the Motor Method.

It is not possible to use field tests to determine the octane number of fuel

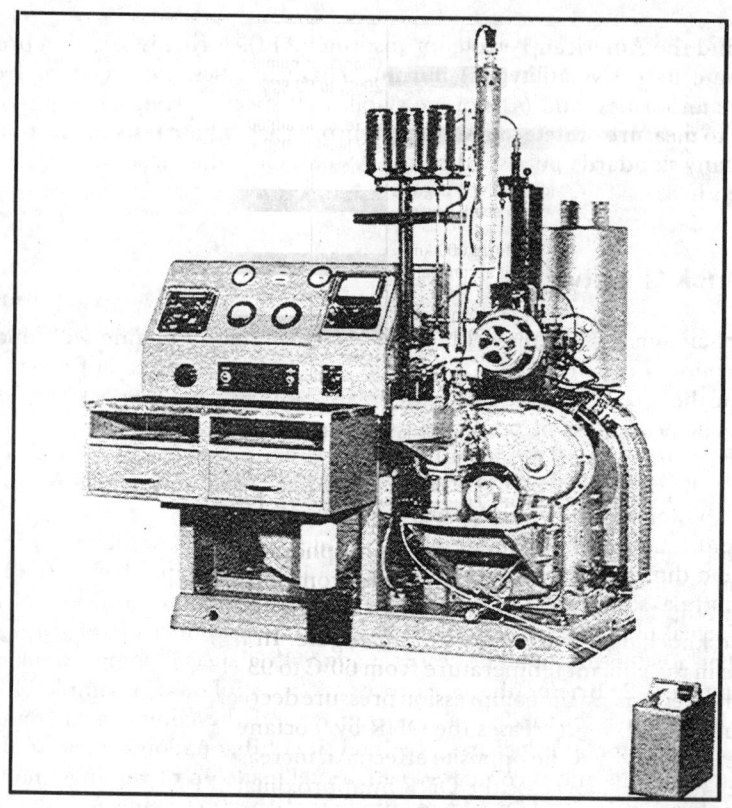

FIGURE 4-2   Test engine for determining diesel cetane rating using ASTM D613 method. (Courtesy Waukesha Engine Division.)

because of the many variables that affect the octane number requirement of the engine.

Factors Affecting the Octane Number Requirement

The effect of several engine operating variables on the octane number requirement (ONR) is shown in figure 4-3. ONR is also affected by other factors such as compression ratio (fig. 4-4), air-fuel ratio, combustion chamber design, and operating conditions such as speed.

Ignition timing has a decided effect. As the ignition timing of several typical automobile engines was advanced 4 degrees ahead of the basic timing, the ONR

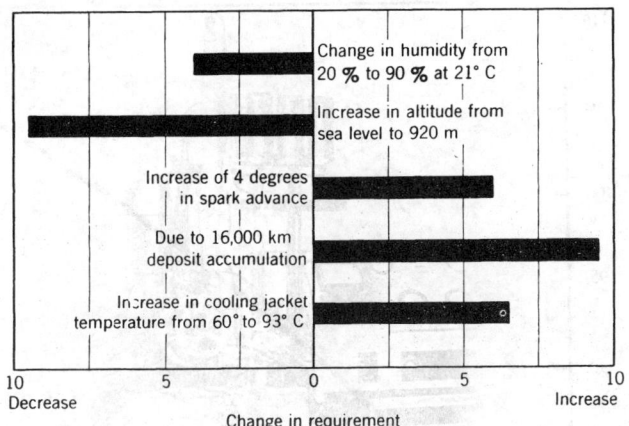

FIGURE 4-3 Effect of several factors on the octane number requirement. (From H. J. Gibson, *SAE J.*, July 1949.)

increased approximately 6 numbers. Atmospheric conditions influence knocking in an engine. An increase in humidity from 4 to 18 g/kg of air reduced the ONR by 3 numbers. The temperature of the engine or its coolant seriously affects the knocking characteristics of an engine. In figure 4-3 we note that an increase in the coolant temperature from 60°C to 93°C increased the ONR by 7 numbers. Decreasing the compression pressure decreases the ONR. An increase in altitude of 920 m decreases the ONR by 9 octane numbers. Increasing the compression ratio has the opposite effect. An increase in the compression ratio from 6:1 to 8:1 will increase the ONR by approximately 12.

Combustion deposits, mostly carbon, result in a reduction in the heat transfer rate and also a slight increase in the compression ratio. As a result, the ONR increases. It was found that 16 000 km of use cause the ONR of an automobile engine to increase by 9. The study indicated that the carbon deposits became relatively stable at 16 000 km of use. This study clearly indicates that the ONR of an engine cannot be determined until the carbon deposits have become stabilized.

The configuration and size of the combustion chamber markedly affect the ONR. In general, the smaller the combustion chamber the lower will be the ONR.

Volatility

In spark ignition engines that burn gasoline the volatility is an extremely important characteristic. Gasolines that vaporize too readily may boil in fuel

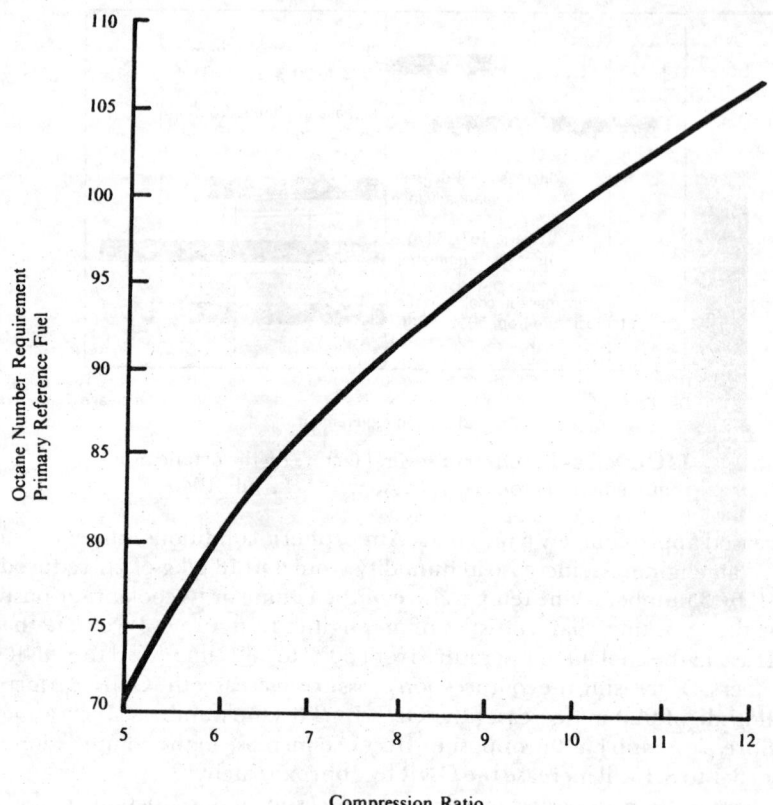

FIGURE 4-4   Fuel antiknock quality required as compression ratio is increased. (From SAE J312 JUN82.)

lines or in carburetors and cause a decrease in the fuel flow to the engine, resulting in rough engine operation or stoppage. On the other hand, gasolines that do not vaporize readily enough may cause hard starting and slow warm-up, as well as unequal distribution of fuel to the individual cylinders. These two requirements are sometimes in conflict. For example, during spring and fall a gasoline of volatility suitable for satisfactory starting at the frequent low temperatures encountered may be susceptible to vapor lock during the warmer time of the day.

There are two methods in general use for evaluating the volatility of fuels, the distillation test and the Reid vapor pressure test.

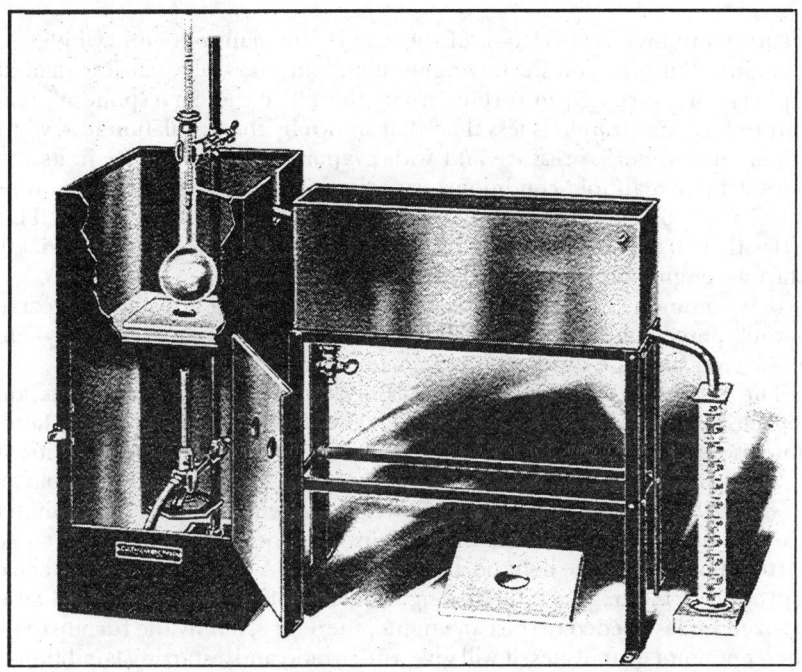

FIGURE 4-5   Apparatus for making distillation tests of liquid fuels. (Courtesy C. J. Tagliaboe Manufacturing Co.)

Distillation

The distillation procedure is described in ASTM D86 (*Distillation of Petroleum Products*). A sample (100 ml) of the fuel to be tested is placed in a distilling flask (fig. 4-5), and heat is applied at a specified uniform rate until the first drop falls from the condenser. The thermometer is read, and this reading is called the initial boiling point. Heating is continued, and the temperature is read as each one-tenth part of the sample is distilled over. The end point is the highest temperature observed on the thermometer; it will be reached just as the last of the volatile portion of the fuel is boiled away. As pointed out earlier, a petroleum fuel is made up of hydrocarbons of many different boiling temperatures. The temperatures observed in the test are the boiling temperatures of the various fractions.

A distillation curve is an indirect measure of the temperature that will be required to give vaporization of the fuel in the manifold and cylinders of the engine. Vaporization in the engine manifold takes place at less than atmospheric pressure and in turbulent air; therefore the corresponding temperature for vaporization is less than that shown by the distillation test, which is run at atmospheric pressure and with evaporation taking place in its own vapors. Under manifold conditions, vaporization may occur at temperatures from 57°C to 70°C lower than those shown by the distillation test results. Thus a fuel with a 10 percent point of 52°C may be expected to vaporize sufficiently to start an engine at temperatures near −18°C.

It is common practice to plot the temperature readings against corresponding percentages distilled to show the distillation range graphically. Figure 4-6 gives average distillation curves for several fuels.

The 10 percent point of the curve is associated with starting, vapor lock, and carburetor icing characteristics. The 10 percent evaporated temperature should be low enough to assure ready starting under normal temperature conditions, but not so low that vapor lock may be experienced. In temperate and cold climates the 10 percent point should be lower in the winter than in the summer. Since only fuel vapors enter into the combustion process, it is possible to create a starting mixture on the light portions of the fuel. A carburetor on full choke can produce a 1:1 air-fuel ratio by weight. A ratio of approximately 10:1 air to vaporized fuel is needed to start an engine; therefore, if only the fuel up to the 10 percent point evaporates, it will give a 10:1 ratio and a starting condition.

The 50 percent point is important as an index of the engine warm-up characteristics of gasoline; the lower the 50 percent temperature, the faster the warm-up.

The 90 percent evaporated temperature is associated with engine acceleration, crankcase dilution, and fuel economy. The presence of very heavy ends in the gasoline is likely to cause poor mixture distribution in the intake manifold of an engine and, therefore, may affect engine performance during acceleration. The presence of a large portion of fuel that has not evaporated can cause crankcase dilution. The unburned portion may enter the crankcase past the piston rings and dilute the oil.

The initial boiling point is difficult to measure accurately and although it has some importance in cold-weather starting, it is generally of less importance as a specification for gasoline. The final boiling point has little significance and is not usually included in gasoline specifications.

Reid Vapor Pressure Test

The Reid vapor pressure test is another means of estimating the volatility of the fuel. The vapor pressure of the fuel is measured with a special vapor

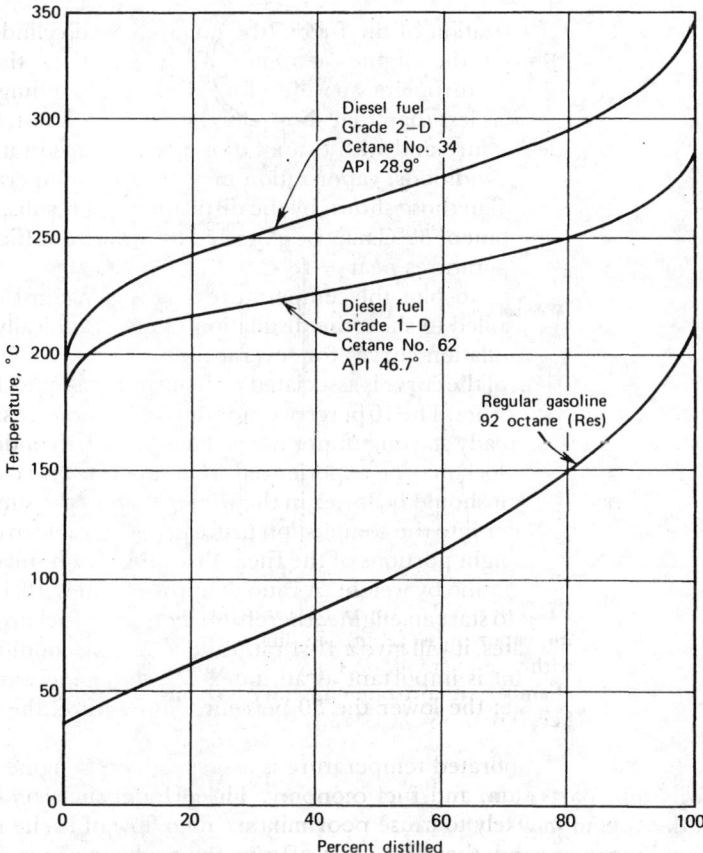

FIGURE 4-6 Characteristic distillation curves of fuels.

pressure bomb illustrated in figure 4-7. Fuel is placed in the bottom section of the bomb and saturated air-water vapors occupy the upper portion. The bomb is immersed in a water bath and heated to a temperature of 37.8 ± 0.1°C. Summer gasolines will have a vapor pressure of 48 to 62 kPa with the winter grade gasolines being 14 to 21 kPa higher than the summer grades. The Reid vapor pressure is a good indication of the tendency of the fuel to vaporize, but the actual amount of vapor formed also depends on the front-end volatility of the gasoline. It has been shown that for easy engine starting a quantity of low-boiling front-end fractions are required. Unfortunately, the boiling temperatures of these fractions may sometimes be lower than the gasoline tem-

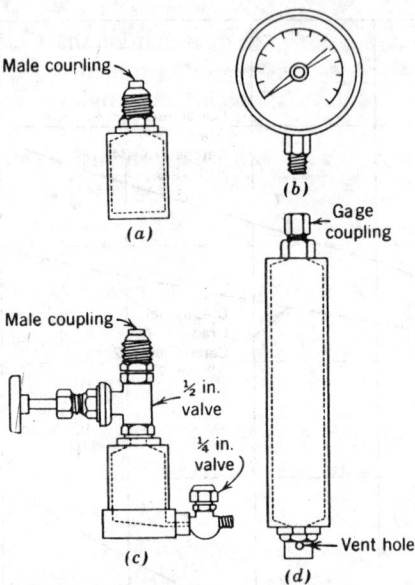

FIGURE 4-7   Apparatus for determining vapor pressure. (*a*) Gasoline chamber with one opening. (*b*) Gage. (*c*) Gasoline chamber with two openings. (*d*) Air chamber. (From ASTM D323.)

perature in some parts of the fuel system in the vehicle. Under such conditions vapor in the system may completely or partially stop the flow of fuel, causing a condition known as vapor lock. On farm tractors the problem of vapor lock is not so serious since the fuel tanks are generally located above or in front of the engine, and therefore the fuel is under greater static pressure than in an automobile.

Sulfur Content

Two tests for sulfur are generally used. One is a test of corrosive sulfur compounds, and the other is a quantitative measure of the sulfur present. ASTM D439 specifications on gasolines place the limit at 0.1 percent; however, the technical data available do not afford a good basis for specifying maximum sulfur content. Diesel fuels contain varying amounts of sulfur, depending on the crude oil source, refining processes, and grade. The ASTM requirements for diesel fuels are shown in table 4-3 (see p. 70). This table specifies that for

No. 1-D and 2-D diesel fuel the sulfur content shall be a maximum of 0.5 percent. High sulfur content can become a problem in diesel engine operation at low temperatures and during intermittent engine operation. Under these conditions, where more moisture condensation can take place, cold corrosion and increased engine wear may accompany the use of diesel fuels containing excessive amounts of sulfur. Sulfur content is based on the method described in ASTM D1266.

Corrosion Test

Some sulfur compounds may be corrosive to certain metals, particularly copper. The presence of corrosive sulfur may be indicated by immersing a polished strip of copper in the fuel for 3 h at 50°C (ASTM D130). The sample strip is then compared with freshly polished copper. Any discoloration indicates the presence of corrosive sulfur.

Gum Content

Hydrocarbon fuels, particularly those made by cracking and containing unsaturated unstable compounds, have a tendency to form viscous liquids or solids called gums resulting from oxidation during storage. The presence of large quantities of gum can cause intake valve sticking and heavy intake manifold deposits. One gum test, ASTM D381, is performed as follows: A 50-ml sample of gasoline is poured into a beaker, which is then placed in a bath heated to 163°C. A stream of hot air is directed on the surface of the gasoline. After the gasoline has evaporated, the beaker is weighed and the residue is reported as gum. The ASTM specification for motor gasoline, D439, sets a maximum limit of 5 mg of gum per 100 ml of fuel.

Motor gasolines are often deliberately blended with nonvolatile oils, or additives, to minimize the problem of gum formation.

Gravity

The gravity or density of a fuel is usually expressed in degrees API, a scale devised by the American Petroleum Institute. The API gravity, although an easy test to perform, does not give direct information. The boiling temperature, the volatility, and the heat value are somewhat related to the gravity so that it is used as a means of estimating these values. Tables and equations of heat values of petroleum products can be based upon the gravity. A high API gravity fuel commonly contains more heat units per kilogram but, because of its lightness, will have a lower heat value per liter.

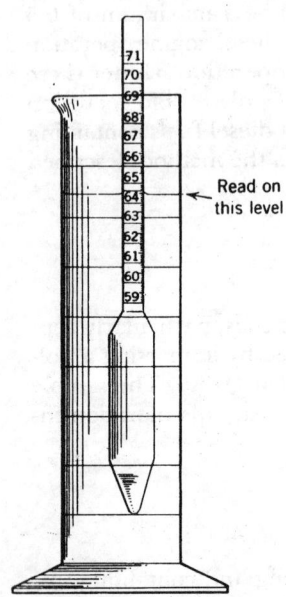

FIGURE 4-8 Hydrometer and cylinder used in API gravity determination.

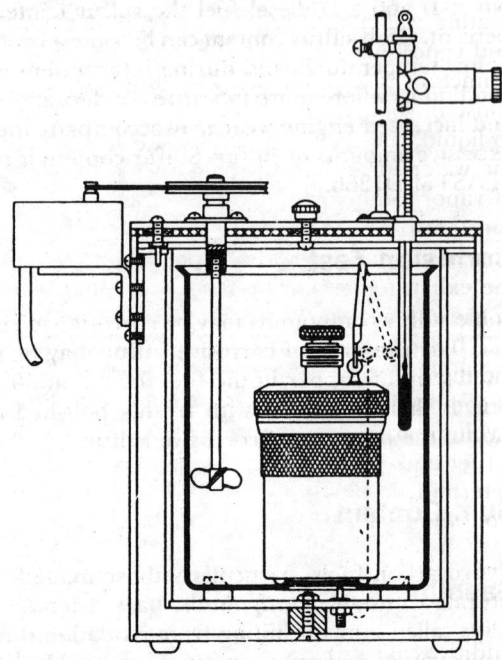

FIGURE 4-9 Oxygen bomb calorimeter. (Courtesy Parr Instrument Co., Inc.)

The method of determining the gravity of petroleum products is specified in ASTM D287. The relationship between the API gravity scale and the specific gravity scale is expressed as follows:

$$\text{API degrees} = \frac{141.5}{\text{specific gravity at } 15.6°C} - 131.5$$

Thus the API gravity of water is 10 and liquids lighter than water have values above 10. The hydrometer illustrated in figure 4-8 provides a simple means for measuring the gravity of fuels.

Heating Values

The heating value, or heat of combustion, of a fuel is an important measure of its worth, since it is the heat produced by the fuel within the engine cylinder that enables the engine to do work. Other factors being equal, it follows that the fuel with the highest heating value will be the most valuable and will give the greatest amount of work output when it is burned in an engine. The heating values may be reported as high or low, depending on the methods and conditions of the test. Where the results are obtained by the bomb calorimeter (fig. 4-9), in which a unit quantity of fuel is burned with oxygen at constant volume and the water formed is cooled from a gaseous state or steam to liquid, the total heating value is known as high or gross heating value. If the water formed in the combustion process remains as steam and the heat of vaporization contained therein is not accounted for, it is called the low or constant-pressure heating value. In the practical application of fuel combustion in engines the water vapor is not condensed; instead it passes out with the exhaust gases in the form of steam. Therefore, the combustion process in the engine is more nearly represented by the low heating value.

It is common practice to calculate engine thermal efficiencies in terms of the high heating value. Also, unless otherwise stated, printed values of heat of combustion are high values. Since fuels are generally purchased by volume, a volume basis of comparing heating values is logical. If the mass per liter of a fuel and the heat per kilogram are known, the heat per liter is readily determined. The gross or high heating value of diesel fuel can be estimated from figure 4-10.

Gasoline Additives

Additives are important ingredients of modern gasolines. Additives are used to raise octane number and to combat surface ignition, spark plug fouling, gum formation, rust, carburetor icing, deposits in the intake system, and intake valve sticking. Although in most cases, a chemical compound satisfies one of these functions, sometimes an additive may perform more than one function (see table 4-2).

Diesel Fuel Tests and Their Significance

Until recently, diesel fuels were not well standardized or regulated. Fuel oil for furnaces and diesel fuels for engines were commonly taken from the same storage tank. In an effort to standardize diesel fuels the American Society for Testing and Materials issued *Classification of Diesel Fuel Oils*, ASTM D975. The two grades of diesel fuels commonly used are Grades No. 1-D and No. 2-D.

Grade No. 1-D comprises the class of volatile fuel oils from kerosene to the intermediate distillates. Fuels within this classification are applicable for use in high-speed engines in services involving frequent and relatively wide variations in loads and speeds and also for use in cases where abnormally low fuel temperatures are encountered.

Grade No. 2-D includes the class of distillate oils of lower volatility. These fuels are applicable for use in high-speed engines in services involving rela-

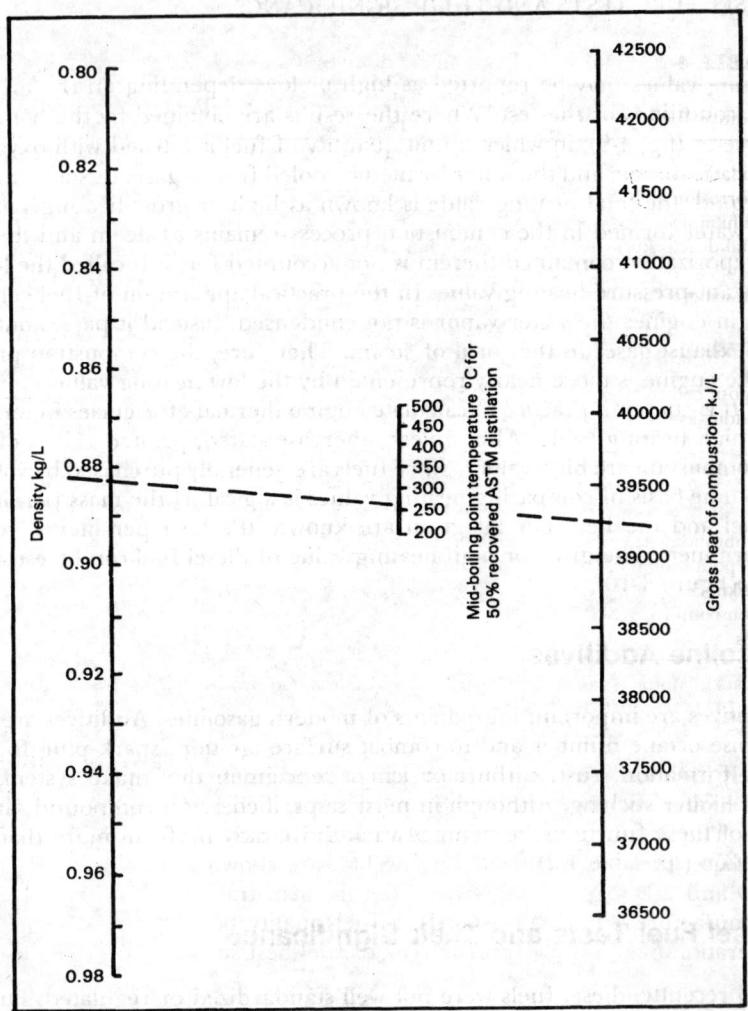

FIGURE 4-10 Nomograph for estimating gross heating value of diesel fuels. (From SAE, *Diesel Fuels,* Apr. 1982.)

TABLE 4-2 Summary of Gasoline Additives

Additive	Type	Function
Oxidation and corrosion inhibitors	Aromatic amines and phenols	Inhibit gum formation and corrosion
Metal deactivators	Chelating agent	Inhibit gum formation from catalytic action of certain metals
Anti-icing additives	Alcohols and surfactants	Prevent icing in carburetor and fuel system
Detergents	Amines and amine phosphates	Prevent deposits in carburetor
Deposit control additives	Polybutene amines	Removes and prevents deposits in carburetors, PCV systems, and intake ports and valves—controls varnish and sludge deposits on the pistons and in the crankcase
Combustion chamber deposit modifiers	Phosphorus compounds	Minimize surface ignition and spark plug fouling
Antiknock compounds	Tetraethyl or tetramethyl lead	Increase octane numbers
Dyes		Identification

SOURCE: "Motor Gasoline," *Chevron Research Bulletin,* Chevron Research Company, Richmond, CA, 1974.

tively high loads and uniform speeds or in engines not requiring fuels having the higher volatility or other properties specified for Grade No. 1-D.

The limiting requirements of various characteristics of two of the grades of diesel fuels, as established by ASTM, are shown in table 4-3. Grades No. 1-D and 2-D are the only ones used in farm tractors in the United States. Manufacturers normally specify a fuel quality because engine design and operation materially affect the type of fuel best suited for an engine.

Cetane Number

The ignition quality of diesel fuel is determined by a method somewhat similar to that used in determining the antiknock quality of gasoline. As in the case of the octane number scale, the scale of cetane number represents blends of two pure hydrocarbon reference fuels. The reference fuels are *n*-cetane with a cetane number of 100 and heptamethylnonane with a cetane number of 15. The cetane number for a blend is calculated by cetane number = (percentage of *n*-cetane) + 0.15 (percentage of heptamethylnonane). A cetane rating of

TABLE 4-3 Detailed Requirements for Diesel Fuel Oils[a]

Grade of Diesel Fuel Oil	Flash Point, °C Min	Cloud Point, °C Max	Water and Sediment, vol % Max	Carbon Residue on 10% Residuum, % Max	Ash, weight, % Max	Distillation Temperatures, °C 90% Point Min	Distillation Temperatures, °C 90% Point Max	Viscosity Kinematic cSt at 40°C[b] Min	Viscosity Kinematic cSt at 40°C[b] Max	Sulfur[c] weight % Max	Copper Strip Corrosion Max	Cetane Number Min
No. 1-D A volatile distillate fuel oil for engines in service requiring frequent speed and load changes	38	—[d]	0.05	0.15	0.01	—	288	1.3	2.4	0.50	No. 3	40
No. 2-D A distillate fuel oil of lower volatility for engines in industrial and heavy mobile service.	52	—[d]	0.05	0.35	0.01	282[e]	338	1.9	4.1	0.50	No. 3	40

SOURCE: Condensed from *Diesel Fuels*, SAE J313 Jun 86. Published by permission of the Society of Automotive Engineers.

[a] To meet special operating conditions, modifications of individual limiting requirements may be agreed upon between purchaser, seller, and manufacturer.

[b] 1 cSt = 1 mm²/s.

[c] In countries outside the United States, other sulfur limits may apply.

[d] It is unrealistic to specify low-temperature properties that will ensure satisfactory operation on a broad basis. Satisfactory operation should be achieved in most cases if the cloud point (or wax appearance point) is specified at 6°C above the tenth percentile minimum ambient temperature for the area in which the fuel will be used.

[e] When cloud point less than −12°C is specified, the minimum viscosity shall be 1.7 cSt (or mm²/s) and the 90% point shall be waived.

66 indicates that a diesel fuel has the same ignition delay in an ASTM-CFR engine as a mixture, by volume, of 60 parts n-cetane and 40 parts heptamethylnonane.

The cetane number is determined in a standard, single-cylinder, variable-compression-ratio engine (fig. 4-2) with special loading and accessory equipment and instrumentation. The engine, the operating conditions, and the test procedure are standardized in ASTM D613, *Test for Ignition Quality of Diesel Fuels by the Cetane Method.*

The cetane number of a diesel fuel depends primarily on its chemical composition. In general, the aromatic hydrocarbons are low in cetane number, the paraffins have high cetane numbers, and the naphthalenes fall somewhere in between. Thus it is apparent that the base material and refining processes used in making diesel fuels largely determine ignition quality. The ignition quality of a diesel fuel can be improved by the use of additives, such as hexyl nitrates.

The relationship between the cetane number of a diesel fuel and the performance of a diesel engine should not be confused with the relationship between the octane number of gasoline and the performance of a spark ignition engine. With a spark ignition engine, raising the octane number improves potential engine performance by allowing the compression ratio to be increased, thereby increasing the power and efficiency of the engine. In the diesel engine, the desirable level of cetane number is established by the requirements of good ignition quality at light loads and low temperatures.

In general, high-cetane fuels permit an engine to be started at lower air temperatures, provide faster engine warm-up without misfiring or producing white smoke, reduce the formation of varnish and carbon deposits, and eliminate diesel knock. Cetane numbers above the commercially available range may lead to incomplete combustion and exhaust smoke if the ignition delay period is too short to allow proper mixing of the fuel and air within the combustion space. Generally, diesel fuels marketed in the United States will range from 33 to 64 cetane number.

In contrast to gasoline, which cannot have its octane number predicted accurately, the cetane number of diesel fuel can be calculated with reasonable accuracy without actually using the test engine. The method of calculating the cetane number was developed by the American Petroleum Institute. It is now published under ASTM 976-80. The calculated cetane index is not an optional method for expressing cetane number. It is a supplementary tool for predicting cetane number with considerable accuracy when used with due regard for its limitation. It is not applicable to fuels containing additives for raising cetane number. Neither is it applicable to pure hydrocarbons, synthetic fuels, alkylates, or coal tar products. It should not be used to calculate the cetane number of fuels having a distillation endpoint below 260°C.

The equation for calculating the cetane number is

$$454.74 - 1641.416D + 774.74D^2 - 0.554B + 97.803(\log B)^2$$

where D = density at 15°C, g/ml (ASTM D1298) and B = mid-boiling temperature, °C (ASTM D86).

Heating Value of Diesel Fuel

The direct method of determining the heating value is to burn it in a bomb calorimeter (fig. 4-9). For most purposes, however, the heating value can be calculated (fig. 4-10). Two relatively simple tests are performed to measure (1) the API gravity and (2) the mid-boiling-point temperature from a distillation test.

Viscosity

The injector pumps perform best when the fuel has the proper viscosity. If the viscosity is too low, more frequent maintenance and repair of the injection system may result. If the viscosity is too high, excessively high pressures may result in the injection system. Therefore the proper viscosity, or resistance to internal flow, of the diesel fuel becomes a requirement. In the table of requirements for diesel fuel oils the desired viscosity is given in centistokes at 40°C. The method for determining the viscosity is specified in ASTM D445.

Carbon Residue

A diesel fuel often has a tendency to form carbon deposits in the engine. This tendency may be estimated by using the Ramsbottom coking method, described in ASTM D524, which specifies heating the fuel for 20 min at 549°C.

Carbon residue tests are very effective in predicting carbon formation of base fuels, but they may be in error when the fuels contain additives such as ignition improvers.

Flash Point

The flash point is not directly related to engine performance. It is, however, of importance in connection with legal requirements and safety precautions involved in fuel handling and storage, and is normally specified to meet insurance and fire regulations. The method commonly used is the Pensky-Martens closed tester for flash point determination, commonly referred to as ASTM D93.

Pour Point and Cloud Point

The lowest temperature at which the fuel ceases to flow is known as its pour point. It is obviously important that the diesel fuel flow at the lowest atmospheric temperature that may be encountered in operation. The pour point specification should not be any lower than necessary because a low pour point can often be obtained only at the expense of lower cetane numbers or higher volatility.

The cloud point of a petroleum oil is the temperature at which paraffin wax or other solid substances begin to crystallize out or separate from solution when the oil is chilled under prescribed conditions. The cloud point will generally occur about 5°C above the pour point.

The methods of determining the pour point and the cloud point have been standardized in ASTM D97 and ASTM D2500.

Ash Content

Diesel engine injectors are precision made and therefore are quite sensitive to any abrasive material in the fuel. Since the ash content is directly related to wear of the injection system, it must be kept quite low. For example, diesel fuel No. 1-D has a maximum allowable ash content of 0.01 percent. The method for determining the ash content has been standardized as ASTM D482. The ash-forming materials in fuel contribute both to wear in the injection system and to engine deposits.

Diesel Fuel Additives

The additives in diesel fuel are for engine protection and to improve engine performance. Antioxidants, metal deactivators, and corrosion inhibitors of the types used in gasoline and jet fuel are also used in diesel fuels. The ignition quality of diesel fuel, though primarily influenced by hydrocarbon composition and volatility, can be improved by the addition of amyl nitrate or hexyl nitrates. Dispersants such as petroleum sulfonates and alkyl succinimides may be added to diesel fuels. These additives keep residues in suspension and also help prevent lacquer-type deposits from forming on critical engine parts. Pour depressants are used quite extensively in European diesel fuels but relatively infrequently in the United States.

Alternate Fuels

Liquefied petroleum gas, commonly known as LP gas, LPG, or bottle gas, consists of propane (C_3H_8) or butane (C_4H_{10}), or both, in varying proportions depending partly on the ambient temperature at which the engine is to be used since there is a difference in their boiling temperatures. Some properties are shown in table 4-4. At summer temperatures the pressure in the fuel tank may be above 1.4 mPa (200 psi) in order to maintain the LPG in a liquid state.

Methane and *LPG* were the only gases available for fuel in tractors in 1986. Methane can also be liquefied, but at a much higher pressure than propane or butane. Because of the very high pressure required to liquefy methane, it is better suited for stationary engines such as those used on irrigation pumps, in which case it is transported and stored in the gaseous state. Although methane is found naturally in petroleum wells, it can also be produced from coal or in biogas digesters. A *biogas digester,* or generator, consists of a closed container into which water; waste materials, such as manure; and bacteria are placed. The decomposition produces a gas that is mostly methane, which when drawn off and cleaned up can be used as a fuel for engines. The design and use of biogas digesters are described by Stout (1979), the Office of Technical Assessment (1980), and Liljedahl (1983). Although biogas digesters are mostly experimental in the United States, they are being used extensively in China and India.

The most common alcohols used for motor fuel are ethanol (C_2H_5OH)

TABLE 4-4   Comparison of Physical Properties of Some Fuels

Fuel	Density, kg/L	High Heating Value, kJ/L	Octane Number,[a] Average	Nominal Compression Ratio	Stoichiometric Air/Fuel Ratio, by Mass
Propane	0.509	25 640	111	8.4:1	15.7
Butane	0.575	28 430	98	8.0:1	15.5
Ethanol	0.784	21 200	111	9.0	9.0
Regular gasoline	0.733	34 560	93	7.4:1	14.7
Diesel, No. 1-D	0.823	37 630	—	16.6:1	—
Diesel, No. 2-D	0.847	39 020	—	16.6:1	—

[a]Research method employed.

and methanol (CH_4O). A list of properties can be found in SAE J1297 (1982). Ethanol, which is commonly made from the grain of corn (maize) and cane sugar, is used both as a fuel and as an additive to raise the octane number of gasoline. Ethanol has an octane number (research) of about 111; however, because it is oxygenated it contains only 66 percent as much energy per unit of volume as gasoline.

Use of plant oils, such as sunflower oil, for internal combustion engines has been investigated since 1976. The technical problems often reported include excess carbon deposits on pistons and exhaust valves, fuel filter clogging, and engine oil dilution. These problems must be solved before plant oils can become a viable alternate fuel.

PROBLEMS

1. Gasoline can be represented by the hydrocarbon octane (C_8H_{18}). Assume that when gasoline is used in an engine, the air supply is 95% of that theoretically required for complete combustion (a rich mixture). Assume that all the hydrogen is burned and that the carbon burns to carbon monoxide and carbon dioxide, so that there is no free carbon left. Calculate the percentage analysis of the dry exhaust gases by volume. Air contains 23% oxygen by weight.

2. Assume that diesel fuel can be represented by the hydrocarbon cetane ($C_{16}H_{34}$). When used in an engine, the air supply is 5% greater than that theoretically required for complete combustion (a lean mixture). Assuming that all the hydrogen is burned and that no free carbon is left in the exhaust, calculate the percentage analysis of the dry exhaust gases by volume. Also, how much water is produced per kilogram of fuel burned?

3. Repeat problem 1 using methane (CH_4) for the fuel.

4. Repeat problem 2, except that the air supply is equal to that theoretically needed for complete combustion and the fuel is ethyl alcohol (C_2H_6O).

5. Repeat problem 4 using hydrogen (H_2) for fuel.

REFERENCES

SAE. *Alternate Automotive Fuels.* Information Report, SAE J1297. Society of Automotive Engineers, Warrendale, PA, Apr. 1982.

SAE. *Diesel Fuels.* Information Report, SAE J313. Society of Automotive Engineers, Warrendale, PA, Apr. 1982.

SAE. *Automotive Gasolines.* Information Report, SAE J312. Society of Automotive Engineers, Warrendale, PA, June 1982.

SUGGESTED READINGS

ASTM. *Petroleum Products, Lubricants, and Fossil Fuels.* American Society for Testing and Materials, Philadelphia, 1985.

Barger, E. L. "Power Alcohol in Tractors and Farm Engines." *Agric. Eng'g,* Feb. 1941.

Bauer, D. J., D. S. Marks, and J. B. Liljedahl. "A Method for Evaluating the Thickening of Lubricating Oil When Vegetable Oil Is Used as a Fuel in Diesel Engines." *Proceedings of the 1982 International Conference on Plant and Vegetable Oil as Fuels,* Fargo, ND, Aug. 2–4, 1982.

BOSTID. *Producer Gas: Another Fuel for Motor Transport.* National Academy Press, Washington, DC, 1983.

Cummings, W. M. "Fuel and Lubricant Additives, Part I." *Lubrication,* vol. 63, no. 1, 1977.

Deere & Company. *FOS 58-Fuels, Lubricants and Coolants,* 5th ed., John Deere Service Publications, Moline, IL, 1984.

Gibson, H. J. " Effect of Several Factors on the Octane Number Requirement." *SAE J.,* July 1949.

Hill, F. J., and C. G. Schleyerback. *Diesel Fuel Properties and Engine Performance,* SAE paper 770316. Society of Automotive Engineers, Warrendale, PA, 1977.

Liljedahl, J. B., W. E. Tyner, James Butler, and John Caldwell. *Biogas Digesters in the Peoples Republic of China.* Paper No. 83-4060, presented at the 1983 Annual Meeting of the American Society of Agricultural Engineers, Bozeman, MT, June 27–29, 1983.

Lodwich, J. R. "Chemical Additives in Petroleum Fuels: Some Uses and Action Mechanisms." *J. Inst. Petrol.,* col. 50, no. 491, Nov. 1964.

National Research Council. *Proceedings of the Workshop on Energy and Agriculture in Developing Countries.* National Academy Press, Washington, DC, 1981.

Obert, E. F. *Internal Combustion Engines,* 3d ed. International Textbook Co., Scranton, PA, 1968.

Office of Technical Assessment. *Energy from Biological Processes.* Congress of the United States, Washington, DC, Ballinger Publishing Co., Cambridge, MA, July 1980.

SAE. *Alternate Fuels for SI Engines,* SAE SP638. Society of Automotive Engineers, Warrendale, PA, 1985.

Stout, B. A. *Energy for World Agriculture.* Food and Agriculture Organization of the United Nations, Rome, 1979.

5
ENGINE DESIGN

Many sizes and types of engines have been used for tractors. Most tractors have a four-stroke-cycle engine, although some two-stroke-cycle diesels are in successful operation. Early tractor history saw the rise and fall of a large number of companies that were making tractors, utilizing automotive parts to a great extent. Automobile engines are generally unsuitable for tractors.

Tractor engines are designed for a high load factor, that is, the power output may be 85 to 90 percent of the maximum brake power at rated speed, and the engine is expected to be able to produce this power for long periods of time. Automobile engines, on the other hand, are given ratings corresponding to the maximum power the engine will produce at the best power speed.

For example, one small U.S.-made automobile with a four-cylinder gasoline engine and with 1.6-L displacement develops 49 kW. A European-made diesel automobile engine with 1.47-L displacement develops 36 kW at 5000 rpm. A Japanese-made diesel tractor with 1.5-L displacement engine develops 20 kW at 2400 rpm. Obviously, the engine speed affects the power output. Engine speed also affects specific fuel consumption (liters per kilowatthour) and engine life. The higher speeds decrease life and increase specific fuel consumption.

Early tractor engines were designed with one to four cylinders. Single-cylinder engines were only slight modifications of stationary units. Although simple in construction, they were much larger and heavier for a given amount of power than the two- and four-cylinder engines, and a large proportion of the power was necessary to propel the machine itself. Tractors with one-cylinder engines were popular as late as 1914. One-cylinder engines are now used in large numbers on power tillers in small rice fields and on lawn and garden tractors.

The tractor engine was an outgrowth of the horizontal stationary engine. These engines were built with one, two, or four cylinders; the cylinders were situated side by side or opposed. Two-cylinder horizontal-opposed engines resulted in good mechanical balance, but lack of compactness and difficulty in servicing were serious disadvantages. However, small one-cylinder horizontal diesel engines have been developed and are widely used for power tillers because of simplicity and low center of gravity (fig. 5-1).

Engine Design—General

The tractor manufacturer must determine, usually by survey, the percentage of spark ignition and compression ignition engines to build. The problem of choosing a gasoline or diesel engine is largely one of economics.

New engines correspond essentially to current models but contain changes or modifications made possible by improvement in design, fuels, lubricants, or materials. The first step of any design problem is to select the speed, type, number, and size of cylinders and the arrangement of the cylinders for the required output. The second step is to calculate the sizes and materials of the parts to withstand the stresses.

Table 5-1 shows some of the changes that have occurred since 1919 in the design of tractor engines.

Stroke-to-Bore Ratio

In 1982 the average stroke-to-bore ratio in tractor engines was about 1.01. This ratio for the tractors tested at Nebraska since 1920 has varied as follows: 1920, 1.29; 1948, 1.22; 1960, 1.16; 1975, 1.16; and 1982, 1.01. Obviously, a decrease in the length of the stroke necessitates an increase in the bore in order to maintain the same displacement.

The stroke-to-bore ratio is a design consideration because higher compression ratios are generally permissible with small bores, and higher thermal efficiencies may be attained with the higher compression ratios. However, assuming equal compression ratio and equal displacement per stroke, a cylinder with a large stroke-to-bore ratio will have a higher surface-to-volume ratio than a cylinder with a small stroke-to-bore ratio. This larger surface permits more heat transfer to the combustion chamber walls, resulting in a decreasing efficiency. The effect of stroke-to-bore ratio on the friction power is shown in figure 5-2. The short-stroke engine, having less piston travel, has less friction.

The breathing capacity of volumetric efficiency of high-speed engines can be improved by decreasing the stroke-to-bore ratio, which permits the use of larger or multiple valves.

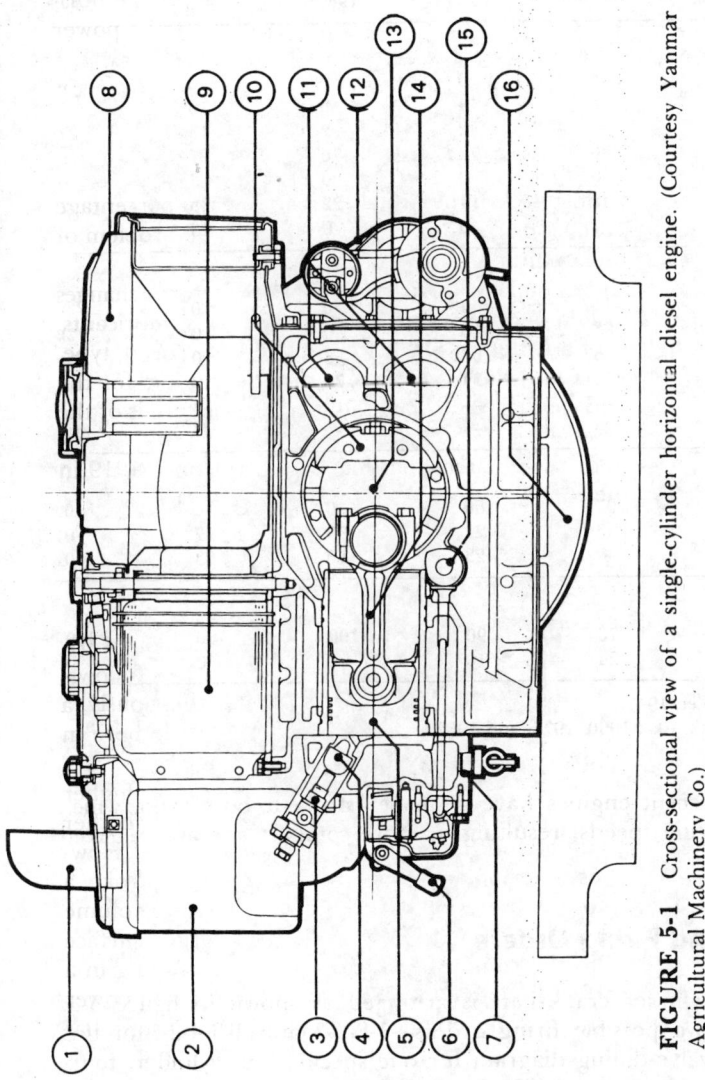

FIGURE 5-1 Cross-sectional view of a single-cylinder horizontal diesel engine. (Courtesy Yanmar Agricultural Machinery Co.)

Key:

1. Muffler
2. Air cleaner
3. Fuel injector
4. Swirl chamber

5. Aluminum alloy piston
6. Decompression lever
7. Cylinder head
8. Fuel tank

9. Radiator
10. Counterweight
11. Secondary balancer
12. Starting motor

13. Crankshaft
14. Connecting rod
15. Camshaft
16. Flywheel

TABLE 5-1   Trends in Tractor Engine Design (Percent in year shown)

	1919	1948[a]	1960[b]	1975[b]	1982[b]
Number of cylinders					
1	4	0	0	0	0
2	13	14	0	10	0
3	—	2	2	13	27
4	80	70	76	27	19
5	0	0	0	3	0
6	2	14	22	33	52
8	1	0	0	13	2
Engine speeds, rpm					
0–500	11	0	0	0	0
501–1100	83	10	0	0	0
1101–1500	5	56	8	0	0
1501–2000	1	34	76	0	0
2001–2500	—	—	16	83	90
2501–3000	—	—	—	17	10
Ignition					
Magneto	97	66	0	0	0
Battery	3	21	50	7	0
Diesel	—	23	50	93	100
Type of motor					
Vertical	76	90	100	100	100
Horizontal	24	10	0	0	0

[a]Tractors on the market, 1948.
[b]Tractors tested at Nebraska, 1960, 1975, and 1982.

In general, recent engines have a smaller stroke-to-bore ratio, which allows higher engine speeds, resulting in more compact size and reduced vibration.

Crankshafts and Firing Orders

The usual four-cylinder crankshaft is arranged as shown in figure 5-3. There are only two possible firing orders—1-2-4-3 and 1-3-4-2—for this arrangement. A valve-timing diagram for one specific four-cylinder, four-stroke-cycle engine is shown in figure 5-4. On this engine the inlet and exhaust do not overlap at the top of the stroke. The inlet valves of cylinders 1 and 3 are open at the same time during 35°. The late closing of

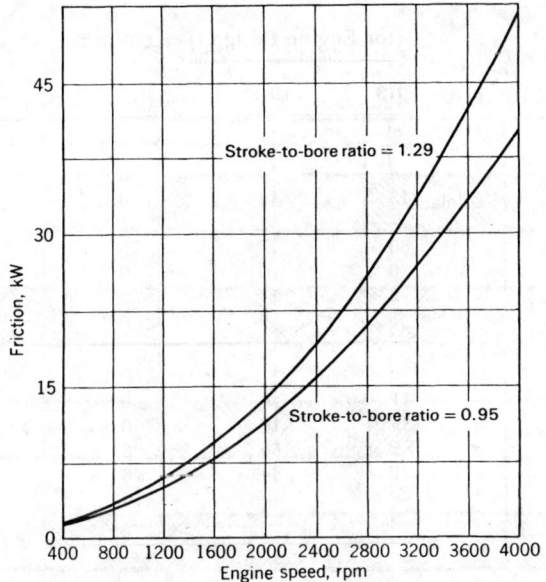

FIGURE 5-2 Effect of stroke-to-bore ratio on power to overcome friction. (From M. M. Roenseh, "Thermal Efficiency and Mechanical Losses of Automotive Engines," SAE Paper, May 1947.)

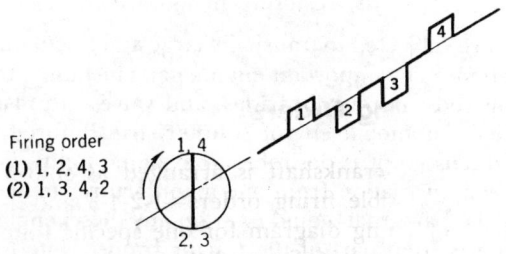

Firing order
(1) 1, 2, 4, 3
(2) 1, 3, 4, 2

FIGURE 5-3 Crankshaft arrangement for a four-cylinder engine.

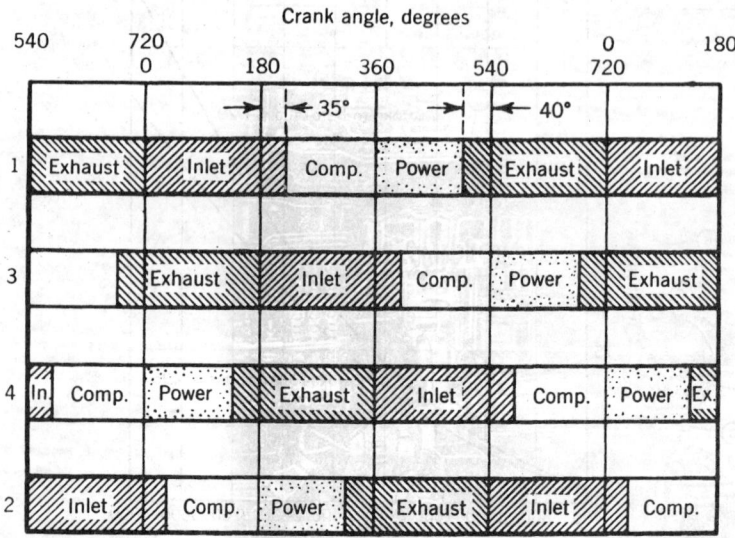

FIGURE 5-4 Valve-opening diagram for a four-cylinder tractor engine.

the inlet valves permits a ramming effect, which tends to raise the volumetric efficiency when the engine is running at high speeds. At low engine speed, with late inlet closing, some of the charge may be forced back out of cylinder 1 into the inlet manifold and may enter cylinder 3. As a result, volumetric efficiency and pressure are lowered and detonation for a given compression ratio is reduced.

Tractor Engine

Tractor engines are subjected to unusually large and fluctuating loads during most field operations. The important engine parts including the block, crankshaft, connecting rods, pistons, bearings, and valves are made heavier and stronger than their automobile engine counterparts. Before the design of the engine parts is discussed, it is desirable to examine figure 5-5 in detail. Although there is some variation from one manufacturer to another, figure 5-5 shows most of the parts that would exist on any diesel engine for a tractor. Tractor engines are often part of the tractor frame. Note the bolt holes in the side of the lower front part of the engine block. Some tractor manufacturers attach the front axle mounting brackets directly to the front of the block, thus eliminating the need for a frame. In other words, the tractor is

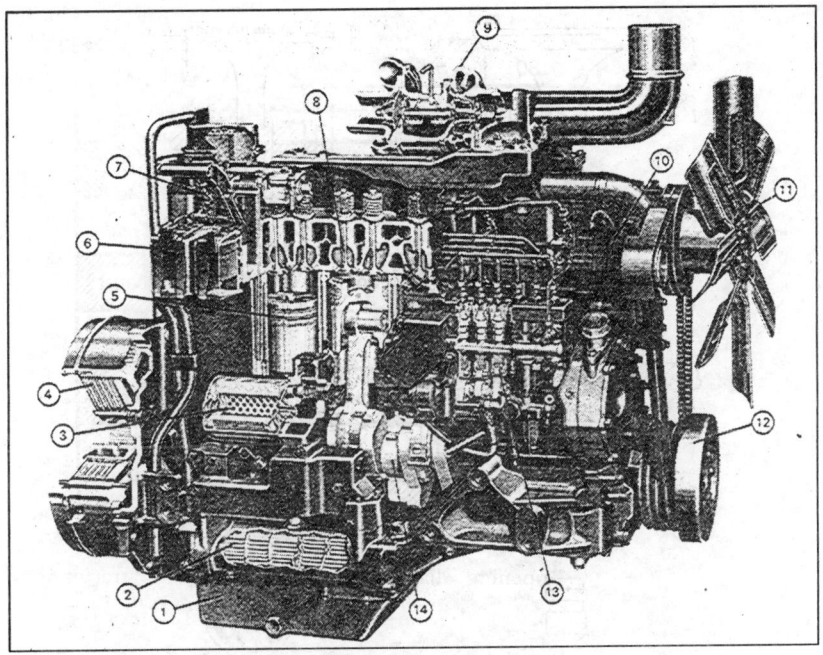

FIGURE 5-5   Six-cylinder diesel engine for a large tractor. (Courtesy Deere & Company.)

Key:

1. Oil pan
2. Oil cooler
3. Oil filter
4. Wet clutch
5. Piston and rings
6. Fuel filter
7. Fuel injector

8. Valve
9. Turbo charger
10. Alternator
11. Fan
12. Damper
13. Fuel pump
14. Crankshaft

frameless or partly without a frame, a situation that is more common with smaller tractors.

Valve Design

A valve may be defined as a device that is used for closing and opening a passage. The operation of an internal-combustion engine necessitates the admission, trapping, and exhausting of the working medium, all of which

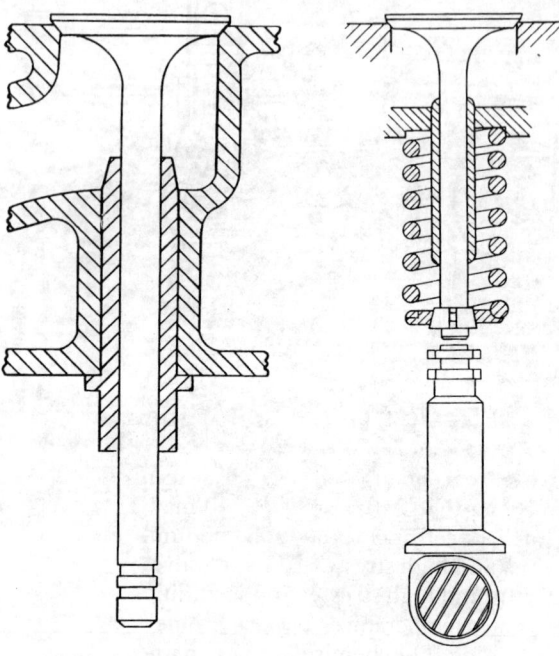

FIGURE 5-6 Poppet
valve.

FIGURE 5-7 Typi-
cal valve-lifting
mechanism. (Cour-
tesy Aluminum In-
dustries, Inc.)

are accomplished by valves. Their proper function, essential to both good
engine performance and fuel economy, largely depends upon design features.

The valve type, diameter, lift, and material are important design considerations affecting engine performance. The intake and exhaust valves are
required to allow smooth gas flow into and out of the cylinder. These valves
must close tightly to prevent leakage of gases during the compression and
combustion process.

The type of valve most commonly found in present-day internal-
combustion engines is the poppet valve (figs. 5-6 and 5-7). The common
valve arrangement found on tractor engines is the overhead (fig. 5-5). Two-
stroke engines (fig. 5-8) usually do not have conventional poppet valves.
Instead, the intake and exhaust processes are through ports in the cylinder walls.

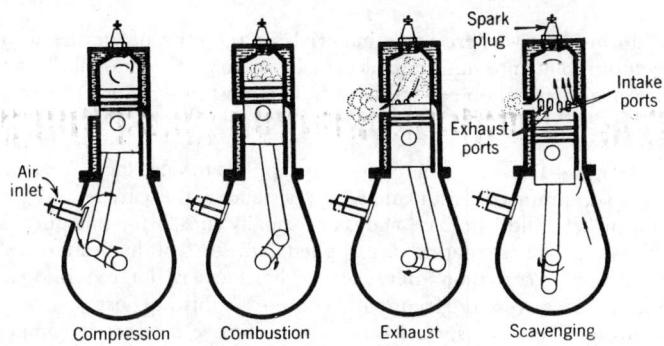

Compression   Combustion   Exhaust   Scavenging

FIGURE 5-8 Two-stroke cycle engine.

Valve Materials, Design, and Application

The materials used are frequently different for the inlet and the exhaust valves. Inlet valves run cooler than exhaust valves and may be made of materials such as carbon steel, nickel steel, or chrome-molybdenum alloys, which may be hardened and will withstand high stresses. Exhaust valves also must be made of materials that will withstand high stresses and in addition maintain their hardness and strength at elevated temperatures. Nickel-chrome steels provide great strength and resistance to etching. The chemical analyses of the alloying elements in valve materials commonly used in tractor engines can be found in SAE J775 (July 1980), Table 1 of which gives the chemical composition of different valve steels. A section of SAE J775 is reproduced as follows:

> The design of a valve and its application to the engine are as important as, and frequently much more important than, the selection of the valve material. Valve durability is limited by the operating temperature of the valve and by the stresses imposed upon it. It is sometimes, but not always, possible to select a stronger or more temperature-resistant material to overcome limitations in valve design and application.
>
> Good cooling is paramount to satisfactory valve durability, and this implies proper seating of the valve in the engine and proper cooling of the immediate area around the valve seat. There is no known material which will deliver satisfactory durability under conditions of blowby, excessive valve seat distortion, and/or inadequate seat cooling. In general, a reduction of temperature of 15°C will approximately double the burning durability of any given material. Again, valve stresses arise generally from the dynamics of the valve train and from the manner in which the valve closes against its seat. Thus careful analysis of the valve train kinematics is a necessity.

In certain instances, where valve materials or requisite properties to obtain the desired durability are not available or cannot be used economically, various means of valve fortification are employed. The most important of these are as follows:

1. Face Coatings—These are welded overlays applied to valve faces and intended to develop optimum corrosion and wear resistance at the valve seating surface. Cobalt-, nickel-, and iron-base alloys are usually chosen for this purpose.
2. Head Coating—These coatings are applied to the tops of the heads of exhaust valves to inhibit corrosion. Because hot hardness is not required, nickel-chromium alloys are most frequently chosen for this purpose.
3. Aluminizing—This is a special case of a protective coating. It comprises a thin layer of aluminum applied to the face and sometimes to the valve head. Aluminum, when diffused, alloys with the base material and provides a thin hard corrosion-resistant coating.
4. Internal Cooling—Hollow valves, partially filled with metallic sodium or sodium-potassium mixtures, transfer heat by convection from the hot head end of valve to the stem. Internal cooling reduces maximum valve temperatures to about 175°C.
5. Stem and Tip Welding—Occasionally, the valve tip or the entire stem is made of martensitic steel. The stem material is welded to the head section of the valve and heat-treated to provide better scuff resistance.
6. Valve Rotators—These are mechanical devices [see fig. 5-9] which cause the valve to turn in the opening portion of the cycle. Rotation tends to ensure good heat transfer from the valve face to the seat. Several types are in use. Valve rotators have been responsible for durability improvements of from two to ten times of the same material without rotation.
7. Hydraulic Valve Lifters—Hydraulic valve lifters [see fig. 5-10] compensate for wear and for thermal changes in the valve train. Their use tends to avoid excessive stresses resulting from too large a clearance, as well as excessive valve temperatures resulting from too small a clearance.

Valves are subjected to high acceleration forces. For example, in an engine operating at 2400 rpm, an exhaust valve must open, permit the burned mixture to escape, and then close, all within about 0.01 s. During this cycle the exhaust valves and seats are exposed to temperatures that may be as high as 2500°C during the combustion and may be about 650°C during the exhaust.

The temperature to which exhaust valves are subjected is shown clearly in figure 5-11. The exhaust temperature is decreased by increasing the compression ratio of the engine, although at the same time the maximum temperature will increase within the combustion chamber. Since the top of the valve is within the combustion chamber, it is obvious that a valve is subjected to increased thermal stresses as the compression ratio is increased. The

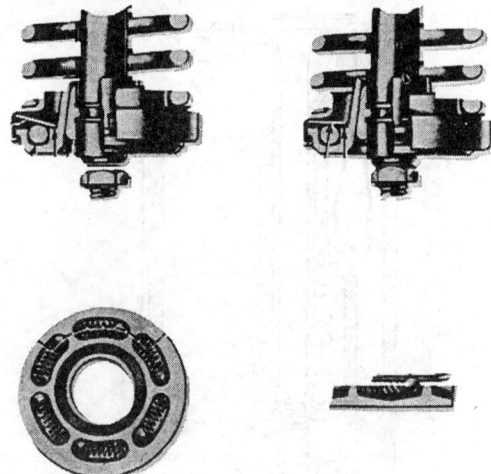

FIGURE 5-9   Valve rotator mechanism. (Courtesy Thompson Products.)

method by which a valve is cooled is shown in figure 5-12. Some heat flows through the valve seats, and the remainder is lost through the valve stem, which is cooled by lubricating oil and by contact with the valve stem guide.

Valve Timing

It would seem at first thought that the intake valve should open on head dead center and close on crank dead center, whereas the exhaust valve should open on crank dead center and close on head dead center. With a large and very slow-speed engine this procedure might operate satisfactorily. The proper valve timing is a function of several factors, including engine speed. With an increase in engine speed the intake valve must be closed later and the exhaust opened earlier. The best valve timing for any given engine can be determined only by actual tests because it depends greatly on the design of the intake and the exhaust passages. A given timing results in maximum compression pressure at some given speed. An ideal arrangement would be one in which the valve timing was automatically changed with the speed. Such an arrangement has recently been put into practical use, but it is limited to some gasoline engines and large diesel engines producing more than 1000 kW. With the rapid advancement of electronic control technology, the application of au-

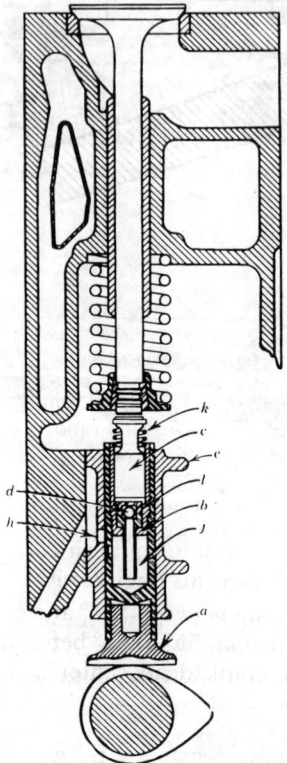

FIGURE 5-10 Zero-lash hydraulic valve lifter. (Courtesy Wilcox Rich Division, Eaton Manufacturing Co.)

tomatic control of valve timing and independent valve control system depending on load conditions may be put into practice for tractor engines in the near future.

Figure 5-13 shows the results of a test on an engine to determine the optimum valve timing at a given rpm. The data plotted are air consumption in kilograms per minute for varying inlet valve opening (ivo) and inlet valve closing (ivc). They are also a comparative measure of volumetric efficiency for different inlet valve timings. The curves represent combinations of inlet

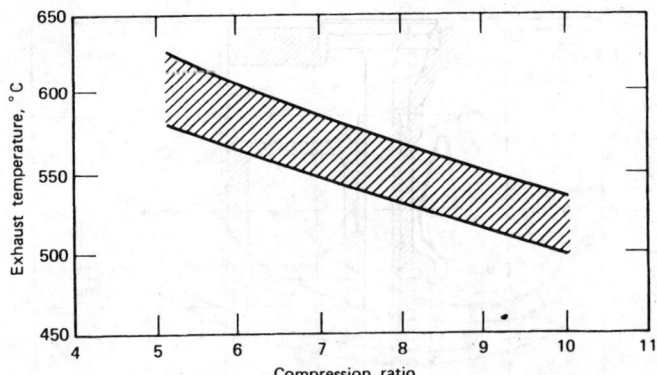

FIGURE 5-11 Range of exhaust-port temperatures at maximum power for different compression ratios. (From G. H. Larson, "Liquified Petroleum Gas for Tractors," *Kansas State Univ., Engr. Exp. Sta. Bull.*, no. 71, 1954.)

valve opening and closing that result in equal air consumption. The optimum of 1.27 kg/min was obtained with this engine with a valve opening between 295° and 305° and a valve closing between 225° and 240°. In terms of nearest dead center, this would be opening 55° to 65° before head dead center (hdc) and closing 45° to 60° after crank dead center (cdc). It should be clearly

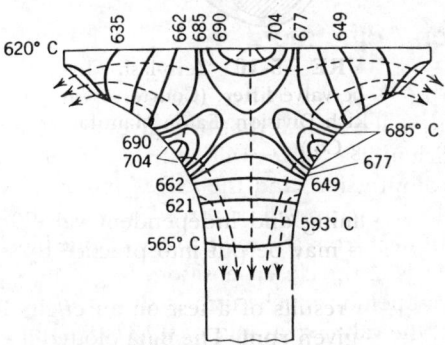

FIGURE 5-12 Heat flow in exhaust valve. (Courtesy TRW Valve Division.)

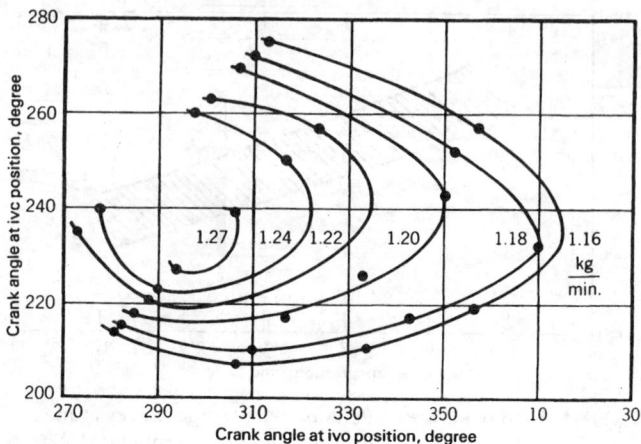

FIGURE 5-13   Selection of optimum inlet valve timing: ivo = intake valve open; ivc = intake valve closed.

understood that these results are only for one specific set of conditions. The speed was 1275 rpm, intake pipe was 372 mm, and the pipe diameter was 35 mm. To determine the most practical setting it would be necessary to run other tests at different engine speeds.

Valve Clearance Adjustment

A certain amount of clearance between the tappets and the valve stems is usually necessary to make certain that the valves are able to seat properly. If clearance is insufficient, the valves will be held off the seats, the engine will not have proper compression, and the valves and seats will be damaged by burning because the hot gases will be forced between the valve and the seat. Also, the transfer of heat from the valve to the engine block will be interfered with when the valve is in the closed position.

A typical valve-opening mechanism is shown in figure 5-7. To ensure complete closing of the valve, clearance of 0.15 to 0.41 mm must be provided at the end of the valve stem. This clearance provides for expansion of the valve stem and valve-lifting mechanism when it becomes hot. If the clearance is insufficient, the valve will be held open and the leaking valve will soon be burned. If the clearance is too large, the valve range will be decreased and the operation of the valve will be noisy.

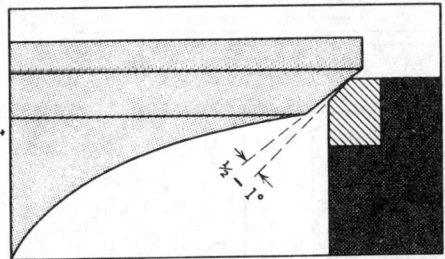

FIGURE 5-14 Recommended valve interference angle. (From A. T. Colwell, "Wear Reduction of Valves and Valve Gear," *SAE Trans.*, vol. 43, 1938, p. 366.)

Valve Seats

The valve and valve seat are the most critical parts, for it is the seal here that must prevent leakage of gases. Seat widths are usually specified by the manufacturer, but widths of 2.0 to 3.2 mm are recommended on many engines. Seats that are too wide tend to collect carbon and other deposits, whereas seats that are too narrow reduce the rate of heat dissipation into the coolant and may result in burned valves.

It has been found through experience that better sealing and longer valve life results from grinding the valve with a slight interference angle as shown in figure 5-14. This allows the valve to seat on the combustion chamber side and to minimize the accumulation of combustion products between the valve and the valve seat. Figure 5-14 also shows a valve seat insert of extra hard material. Such inserts are commonly used on exhaust valve seats and sometimes on the intake valve seat. Valves that are seated into the cast iron head of the cylinder block are only satisfactory for light-duty engines. Hard alloy materials are used for inserts. The insert must fit very tightly to ensure proper heat flow.

Valve-Opening Area

The port area (see fig. 5-15) is equal to

$$A_1 = \frac{\pi}{4} (D_1^2 - d^2)$$

where D_1 is the diameter of the port and d is the diameter of the valve stem.

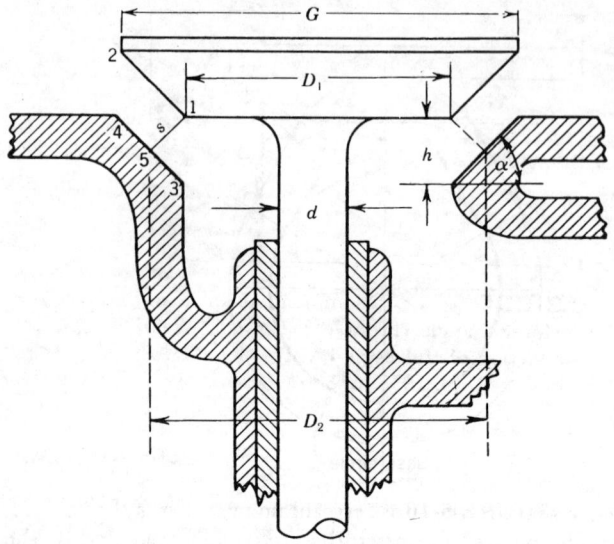

FIGURE 5-15 Poppet valve partly open.

The beveled poppet valve is almost universally used since it presents a conical seating surface that makes it self-centering. If the minimum diameter is D_1, and the slant height is s, then the valve opening area, A_3, is

$$A_3 = \pi \left(\frac{D_1 + D_2}{2}\right) s$$

since

$$s = h \cos \alpha$$

and

$$D_2 = D_1 + 2h \cos \alpha \sin \alpha$$

then

$$A_3 = \frac{\pi}{2} (D_1 + D_1 + 2h \cos \alpha \sin \alpha) h \cos \alpha$$
$$A_3 = \pi (D_1 + h \cos \alpha \sin \alpha) h \cos \alpha$$

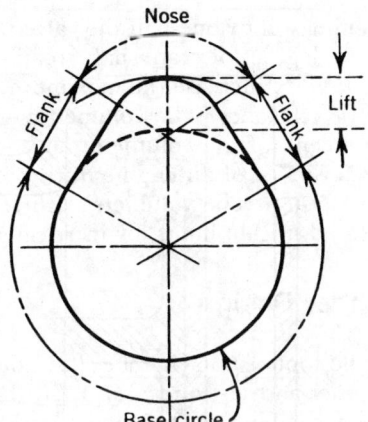

FIGURE 5-16 Typical cam outline.

When α is 45°,

$$\cos \alpha = 0.707, \sin \alpha = 0.707$$

$$\therefore A_3 = \pi(D_1 + 0.707 \times 0.707h)0.707h$$

$$A_3 = 2.22\left(D_1 + \frac{h}{2}\right)h \tag{1}$$

If the lifts are high, the line s will fall outside the point 4, in which event the line 1-4 becomes the slant height, from which the valve-opening area is determined.

The valve port area is commonly designed to produce a maximum gas velocity of 71 m/s in the exhaust port and 56 m/s in the intake valve port.

Cams

Valves are usually lifted by cams similar to that shown in figure 5-16. The cam contour is formed by two arcs joined by two flanks which may be either arcs or straight lines tangent to the arcs. The cam may be symmetrical or the leading flank may vary from the following flank. In a four-stroke-cycle engine the camshaft rotates at one-half the crankshaft speed and, therefore, the cam angle for the valve opening will be one-half the valve range of operation, the latter being expressed in degrees of crankshaft rotation.

One of the means used to increase the maximum power of an engine is to increase its volumetric efficiency. The volumetric efficiency is usually defined as the ratio, in percent, of the volume of air at standard atmospheric conditions entering the cylinder of an engine during its intake stroke to the volume displaced by the piston. The volumetric efficiency and the power of an engine can be increased, within limits, by increasing the valve lift.

Combustion Chamber Design

The size and shape of the combustion chamber has a decided effect upon the proper mixing of the fuel and air, and also upon the fuel detonation. In general, the combustion chamber is designed to create turbulence of the mixture for better combustion and greater flame propagation velocity, thereby decreasing the combustion time and improving antiknock characteristics. A short and compact combustion chamber generally requires less time for combustion and improves the antiknock characteristics.

Effect of Compression Ratio

The thermal efficiency of an engine does not approach the theoretical efficiency of an Otto cycle engine as defined by

$$\text{Eff.} = 1 - \frac{1}{r^{n-1}} \tag{2}$$

There are many reasons why the actual efficiency is substantially below the theoretical efficiency. Some of the discrepancy can be accounted for by the following factors:

1. The expansion and compression of gases is not adiabatic.
2. There are friction losses in the engine.
3. Work is required to draw in and exhaust gases.
4. For a free-breathing engine (i.e., one not supercharged) the volumetric efficiency will be less than 100 percent.
5. Complete combustion does not occur.

Nevertheless, there is a definite relationship between the compression ratio r, and the specific fuel consumption of an engine. For one LP gas engine the effect of the compression ratio upon efficiency is shown clearly in figure 5-17. Large changes in the compression ratio of diesel engines are not prac-

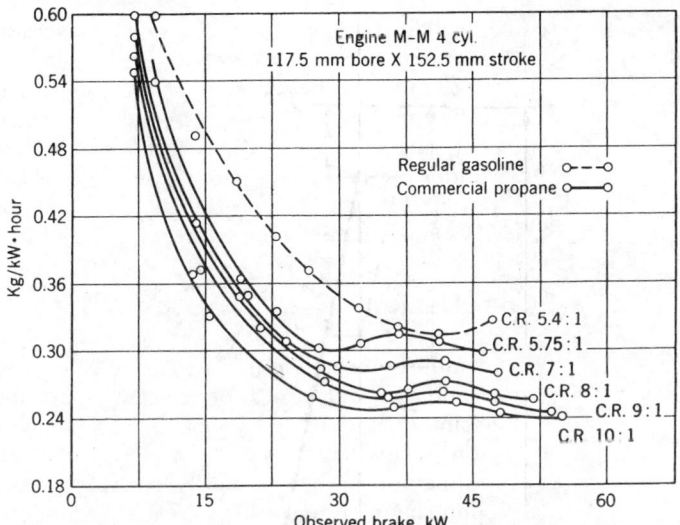

FIGURE 5-17 Specific fuel consumption for different compression ratios at various loads and constant speed of 1100 rpm. (From G. H. Larson, "Liquefied Petroleum Gas for Tractors," *Kansas State Univ., Engr. Exp. Sta. Bull.,* no. 71, 1954.)

tical. Most diesel engines have compression ratios that range from 16:1 to 22:1.

Piston Crank Kinematics

In figure 5-18 the piston displacement s after the crank has turned $\theta°$ from top dead center is expressed as

$$s = r + l - r \cos \theta - l \cos \phi \qquad (3)$$

where r and l are the crank radius and the connecting rod length, respectively. Since

$$l^2 = \overline{bc}^2 + \overline{ca}^2 = r^2 \sin^2 \theta + l^2 \cos^2 \phi$$

then

$$l \cos \phi = \sqrt{l^2 - r^2 \sin^2 \theta}$$

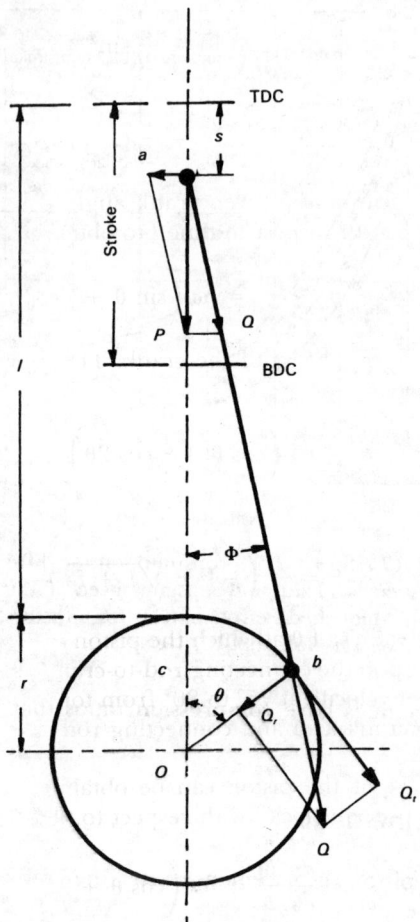

FIGURE 5-18 Piston and crank relation. TDC = top dead center, BDC = bottom dead center.

Substituting the latter in equation 3 and simplifying, we obtain

$$s = r\left[(1 - \cos \theta) + \frac{l}{r}(1 - \sqrt{1 - r^2/l^2 \sin^2 \theta})\right] \qquad (4)$$

Adding $r^4 \sin^4 \theta / 4l^4$ to the terms under the radical to complete the square results in the approximate relation

$$s = r\left[(1 - \cos \theta) + \frac{r}{2l} \sin^2 \theta\right]$$

$$= r\left[(1 - \cos \theta) + \frac{r}{4l}(1 - \cos 2\theta)\right] \tag{5}$$

The piston velocity v at a given crank angle can be derived by differentiating equation 5 with respect to time t to obtain

$$v = \frac{ds}{dt} = \frac{ds}{d\theta}\frac{d\theta}{dt} = r\omega\left(\sin \theta + \frac{r}{2l} \sin 2\theta\right) \tag{6}$$

where ω is the angular velocity of the crank. The maximum velocity of the piston is attained when

$$\frac{dv}{d\theta} = r\omega\left(\cos \theta + \frac{r}{l} \cos 2\theta\right) = 0$$

from which

$$\cos \theta = \frac{l}{4r}\left(-1 \pm \sqrt{1 + 8r^2/l^2}\right) \tag{7}$$

It is noted that the values of θ at which the piston velocities are maximum or minimum depend upon the connecting-rod-to-crank ratio. Usually the piston attains its maximum velocity at 75° to 80° from top dead center at which the angle between the crank arm and connecting rod is close to being perpendicular.

The acceleration of the piston can be obtained by differentiating the expression for the piston velocity with respect to time. Thus

$$a = \frac{dv}{dt} = \frac{dv}{d\theta}\frac{d\theta}{dt} = r\omega^2\left(\cos \theta + \frac{r}{l} \cos 2\theta\right) \tag{8}$$

The maximum and minimum values for acceleration are attained when

$$\frac{da}{d\theta} = 0 \quad \text{or} \quad \sin \theta + \frac{2r}{l} \sin 2\theta = 0$$

from which we obtain

$$\sin \theta = 0 \quad \text{and} \quad \cos \theta = -\frac{l}{4r}$$

Then at the angles of $\theta = 0$ and $\theta = 2\pi$, which correspond to top and bottom dead center respectively, the maximum accelerations are attained.

The piston and the upper part of the connecting rod are assumed to

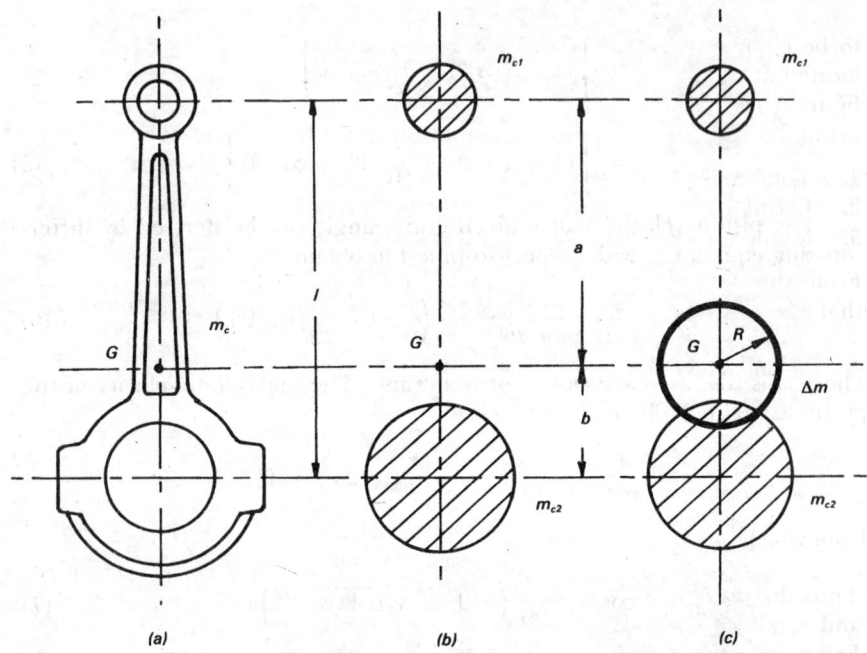

FIGURE 5-19   Equivalent dynamic system for a connecting rod.

have a reciprocating motion. If the mass of the reciprocating part is denoted as M, the inertia force F acting on the mass can be expressed as

$$F = Ma = Mr\omega^2 \left(\cos\theta + \frac{r}{l}\cos 2\theta\right) \tag{9}$$

In equation 9, the terms containing $\cos\theta$ and $\cos 2\theta$ are called the first and second harmonics of the inertia force respectively. It is noted that the inertia force F increases in proportion to the square of the angular velocity, resulting in a large value for a high-speed engine.

Inertia Force of Connecting Rod

In order to obtain the inertia force of the connecting rod, an equivalent dynamical system having the same inertia force and couple as the connecting rod is considered. The upper part of the connecting rod is considered to have a reciprocating motion and the lower part of the rod is assumed to have a rotating motion. The connecting rod shown in figure 5-19(a) is considered to be equivalent to the system with two concentrated masses, m_{c1} and m_{c2}, located at distances a and b apart from the center of gravity G as shown in figure 5-19(b). For the system (b) to be dynamically equivalent to the system (a) in figure 5-19, the following conditions must be satisfied:

1. Total masses for the two systems are equal.
2. Locations of the centers of gravity are the same.
3. Moments of inertia for both systems about the center of gravity are equal.

From the conditions 1 and 2 the following expressions are derived, noting that $a + b = l$.

$$m_c = m_{c1} + m_{c2}$$

$$m_c b = m_{c1} l \quad \text{or} \quad m_{c1} = \frac{b}{l} m_c$$

$$m_c a = m_{c2} l \quad \text{or} \quad m_{c2} = \frac{a}{l} m_c$$

Thus the mass m_c of the connecting rod can be divided into two masses, m_{c1} and m_{c2}, located at distances a and b, respectively, from the center of gravity. For condition 3 the moment of inertia I for the equivalent system is expressed in the form

$$I_b = m_{c1} a^2 + m_{c2} b^2$$

If I_a is the moment of inertia of the original system [fig. 5-19(a)], the inequality $I_a < I_b$ holds in general. To compensate, a ring of radius R and mass Δm will be placed about the center of gravity as shown in figure 5-19(c). Then the moment of inertia ΔI of the ring about G is expressed as

$$\Delta I = \Delta m R^2$$

Hence ΔI can be determined to satisfy

$$I_a = I_b - \Delta I$$

Addition of ΔI increases the total mass of the system by Δm and results in condition 1 not being satisfied. However, this problem can be practically solved by taking the ring radius R to be large enough so that Δm can be made small enough to be negligible. Then, considering the inertia force of the connecting rod, only the inertia forces of the reciprocating mass m_{c1} and the rotating mass m_{c2} are taken into account, neglecting Δm. For the inertia couples about the center of gravity, the inertia couple due to ΔI must be considered in addition to the couples due to m_{c1} and m_{c2}.

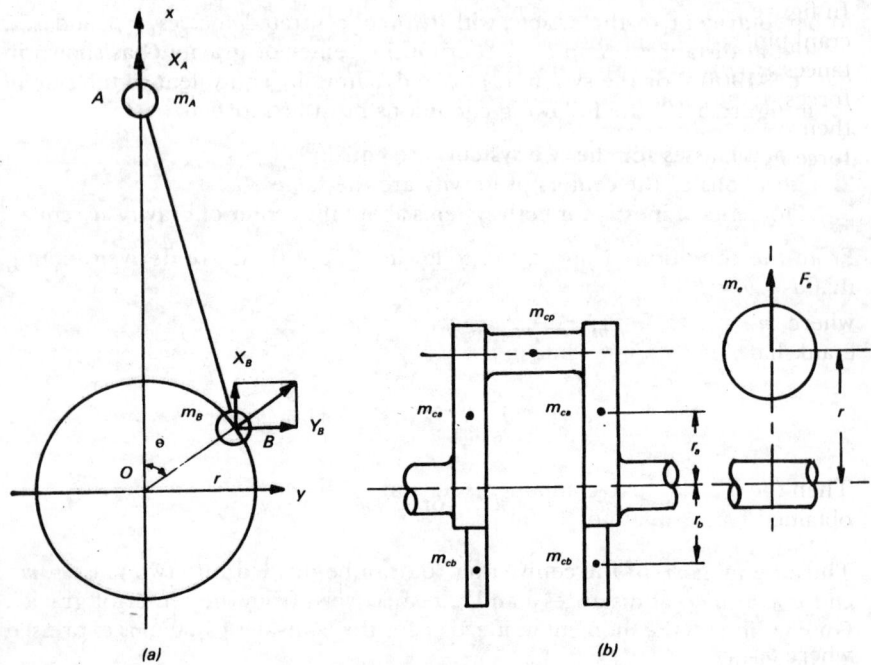

FIGURE 5-20   Inertia forces acting on engine parts.

Inertia Force of Single-Cylinder Engines

Taking the origin O at the center of the crankshaft and x and y coordinates as shown in figure 5-20(a), the mass m_A, consisting of the masses m_p and m_{c1} of the piston and the upper part of the connecting rod, respectively, has a reciprocating motion producing the inertia force X_A in the direction of Ox. Thus

$$X_A = m_A \omega^2 r(\cos \theta + (r/l) \cos 2\theta) \qquad (10)$$

where $m_A = m_p + m_{c1}$

The inertia force due to the mass m_{c2} of the lower part of the connecting rod can be expressed by the components X_r and Y_r of the centrifugal force as

$$\left. \begin{array}{l} X_r = m_{c2}\omega^2 r \cos \theta \\ Y_r = m_{c2}\omega^2 r \sin \theta \end{array} \right\} \qquad (11)$$

In figure 5-20(b) the centers of gravity of the masses m_{cp}, m_{ca}, and m_{cb} of the crankpin, crankarm, and balancing weight respectively are located at the distances r, r_a, and r_b from the center of the crankshaft. Denoting the inertia forces due to crankpin, crankarm, and balancing weight by F_{cp}, F_{ca}, and F_{cb}, then $F_{cp} = m_{cp}r\omega^2$, $F_{ca} = 2m_{ca}r_a\omega^2$, and $F_{cb} = 2m_{cb}r_b\omega^2$. Thus the total inertia force F_e of the crankshaft can be expressed in the form

$$F_e = F_{cp} + F_{ca} - F_{cb}$$

$$= \omega^2(m_{cp}r + 2m_{ca}r_a - 2m_{cb}r_b) = m_e\omega^2r$$

where $m_e = (m_{cp}r + 2m_{ca}r_a - 2m_{cb}r_b)/r$ is called the equivalent mass of the crankshaft. The components in the x and y directions are

$$\left. \begin{array}{l} X_e = m_e\,\omega^2r\cos\theta \\ Y_e = m_e\,\omega^2r\sin\theta \end{array} \right\} \tag{12}$$

Then the inertia force components X_B and Y_B due to the rotating part can be obtained from equations 11 and 12 as

$$\left. \begin{array}{l} X_B = X_e + X_r = m_B\omega^2r\cos\theta \\ Y_B = Y_e + Y_r = m_B\omega^2r\sin\theta \end{array} \right\} \tag{13}$$

where $m_B = m_e + m_{c2}$.

These inertia forces vary with crankshaft rotation and thus become one source of engine vibration.

Crank Effort

The gas pressure and the inertia forces may be combined to determine the net force P acting along the cylinder axis (fig. 5-18). The angularity of the connecting rod causes the net force to be divided into components, one producing piston thrust against the cylinder wall, the other acting along the axis of the connecting rod. As shown in figure 5-18, the tangential force Q_t acting at the crankpin is obtained by resolving the force Q acting along the rod into two components, one acting tangentially on the crank circle at the crankpin, the other acting radially at the crankpin.

Since $Q = P/\cos\phi$ and $Q_t = Q\sin(\theta + \phi)$, the torque T acting on the crankshaft is expressed in the form

$$T = Q_tr = Pr\sin(\theta + \phi)/\cos\phi \tag{14}$$

Also, $\cos\phi = \sqrt{1 - (r/l)^2\sin^2\theta}$ and $l\sin\phi = r\sin\theta$. Thus

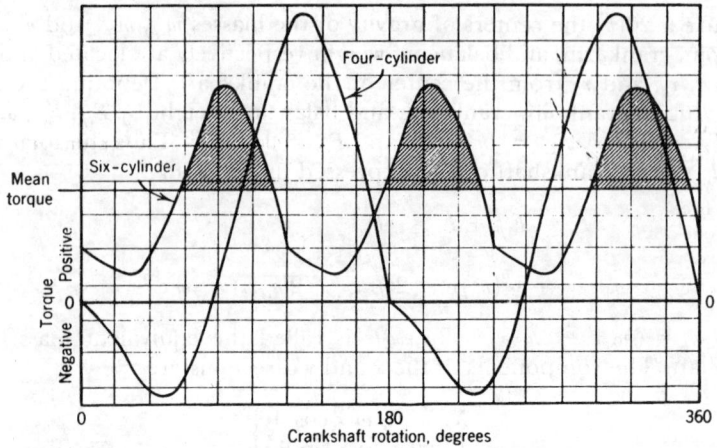

FIGURE 5-21 Crank-moment diagrams of four- and six-cylinder engines.

$$\sin (\theta + \phi) = \sin \theta \cos \phi + \cos \theta \sin \phi$$
$$= \sin \theta \sqrt{1 - (r/l)^2 \sin^2 \theta} + (r/l) \sin \theta \cos \theta$$

Substituting these values into equation 14 results in

$$T = \left(\frac{Pr}{\sqrt{1 - (r/l)^2 \sin^2 \theta}} \right) (\sin \theta \sqrt{1 - (r/l)^2 \sin^2 \theta} + (r/l) \sin \theta \cos \theta)$$

$$= Pr \sin \theta \left(1 + \frac{r \cos \theta}{\sqrt{l^2 - r^2 \sin^2 \theta}} \right)$$

$$= Pr \sin \theta \left(1 + \frac{\cos \theta}{\sqrt{(l^2/r^2) - \sin^2 \theta}} \right) \tag{15}$$

The torque-crank angle relation for a multicylinder engine is determined by combining the relations for the individual cylinders, which are assumed to be identical. The greater the number of cylinders, the smoother will be the torque, as shown in figure 5-21.

Flywheels

Speed variations are inherent in a piston-type engine because of the variation of input torque. The primary purpose of a flywheel is to reduce the speed variation to some acceptable value determined by its use. A minor purpose

TABLE 5-2   Approximate Values of Flywheel Constants

Engine Type	ΔE/(Indicated Work per Revolution)
One-cylinder, four-stroke-cycle, gasoline	2.40
Two-cylinder, four-stroke-cycle, cranks at 180°, gasoline	1.50
Three-cylinder, four-stroke-cycle, cranks at 120°	0.70
Four-cylinder, four-stroke-cycle, gasoline	0.30
Six-cylinder, four-stroke-cycle, gasoline	0.10

of a flywheel is to provide for a source of energy when the clutch is engaged under load.

For several reasons it is not desirable to design the flywheel excessively large because such a flywheel would cause the engine to be slow in accelerating; it would be needlessly heavy and expensive; it would increase the gyroscopic moment during turning; and, for machines not protected by slip clutches and shear pins, the excessive kinetic energy in the engine flywheel could be a hazard to the power transmission system. Obviously, then, an optimum flywheel design should exist for a particular engine to be used under certain conditions.

The inertia of the flywheel clutch or clutches must also be included with the flywheel. For a six-cylinder tractor engine having both a transmission clutch and a power takeoff clutch, the two clutches may account for 40 percent of the mass moment of inertia of the flywheel and clutch assembly.

The design of a flywheel that considers the variation of torque input depends on the permissible fluctuation of engine speed during the time from one power impulse to the next and on the speed at which the engine is required to idle. With reference to figure 5-21, the area above the mean torque line represents the excess energy periodically absorbed and given up by the flywheel and is generally denoted by the symbol ΔE. Experience has shown that the ratio between the ΔE and the indicated work of one revolution is approximately constant for a given type of engine. Values of this ratio for the most common types of tractor engines are given in table 5-2 and represent values as averaged from several sources.

Values of ΔE/(indicated work per revolution) for diesel engines are higher than those shown in table 5-2.

An increase in angular velocity from ω_1 to ω_2 changes the kinetic energy of the flywheel and other rotating parts according to the relation

$$\Delta E = \frac{I}{2}\omega_2^2 - \frac{I}{2}\omega_1^2 = \frac{I}{2}(\omega_2^2 - \omega_1^2) \tag{16}$$

where I is the mass moment of inertia of the flywheel. From the relation it is clear that speed variation is reduced by increasing the mass moment of inertia of the flywheel. Equation 16 becomes

$$\Delta E = \frac{I\pi^2}{1800}[N_2^2 - N_1^2]$$

where N is engine speed in rpm and subscripts 1 and 2 are lower and upper limits of speed respectively. Since $I = r^2 M$

$$\Delta E = \frac{\pi^2 r^2 M}{1800}[N_2^2 - N_1^2] \qquad (17)$$

where r is the radius of gyration and M is the flywheel mass.

Also

$$N_2^2 - N_1^2 = (N_2 - N_1)(N_2 + N_1) \qquad (18)$$

and

$$N_2 + N_1 = 2N \qquad (19)$$

where N is the average speed.

The ratio of the average speed to the variation in speed is defined as the coefficient of speed variation, k. Thus

$$k = N/(N_2 - N_1) \qquad (20)$$

Substituting equations 18, 19, and 20 into equation 17 results in

$$\Delta E = \pi^2 r^2 M N^2 / (900k)$$

or

$$Mr^2 = \Delta E\, 900\, k/(\pi^2 N^2) \qquad (21)$$

Balance of Single-Cylinder Engines

The vibration caused by the unbalance of the reciprocating engine is attributed to the following:

1. Unbalanced centrifugal force of rotating parts
2. Unbalanced inertia force of reciprocating parts
3. Piston side thrust
4. Flexural and torsional vibration of the crankshaft

Engine balancing is attained mainly by removing or minimizing the unbalanced forces described in items 1 and 2 using balancing weights or through the design of the crankshaft. The problem of piston side thrust can be practically solved from design and structural considerations assuming the cylinder is a rigid body. The flexural and torsional vibrations take a complex form, combining natural and forced vibrations. However, the crankshaft of the single-cylinder engine is short, which provides adequate rigidity, and thus the vibration is usually small and negligible. Thus the analysis is made here on the unbalance caused by the centrifugal force and the inertia force.

The centrifugal force caused by the rotating mass m_B (fig. 5-20) consisting of the equivalent mass m_e of the crankshaft and the mass m_{c2} of the lower part of the connecting rod can be balanced by placing a balance weight having mass m_{cb} at the opposite side of the crankpin. Denoting r_b as the distance between the center of the crankshaft and the center of gravity, G_b of the balance weight, the mass m_{cb}, and the distance r_b can be determined to satisfy the following relation

$$m_{cb}r_b\omega^2 = m_B r\omega^2 \tag{22}$$

The inertia force caused by the reciprocating part is expressed from equation 10 as

$$X_A = m_A r\omega^2(\cos\theta + (r/l)\cos 2\theta)$$
$$= m_A r\omega^2 \cos\theta + (m_A r)/(4l)\, r\, (2\omega)^2 \cos 2\theta \tag{23}$$

It is noted from equation 23 that the first harmonic of the inertial force is equal to the component in the x direction of the centrifugal force caused by an imaginary mass, m_A, rotating on the crank circle of radius r at an angular velocity of ω. Also, the second harmonic of the inertial force in the equation is equal to a component in the x direction of the centrifugal force caused by an imaginary mass of $m_A r/(4l)$ rotating on the crank circle at an angular velocity of 2ω. To balance the first harmonic of the inertial force, a counterweight having mass m_A is attached at the opposite side of the crankpin. The component of the centrifugal force in the x direction, which is caused by the balancing mass m_A, then counterbalances the first harmonic of the inertial force. However, a new unbalanced force, $m_A r\omega^2 \sin\theta$, is produced in the y direction. Thus the counterweight does not remove the unbalanced force completely but does minimize it.

Denoting αm_A ($\alpha < 1$) as the mass of the counterweight, the components F_x and F_y of the inertial force in the x and y directions become

$$\left. \begin{array}{l} F_x = (1 - \alpha)m_A r\omega^2 \cos\theta \\ F_y = -\alpha m_A r\omega^2 \sin\theta \end{array} \right\} \tag{24}$$

For $\alpha = 0.5$, the inertia forces are equal in magnitude. For $\alpha > 0.5$, the inertia force in the y direction becomes larger, whereas for $\alpha < 0.5$ the inertia force becomes larger in the x direction. For small single-cylinder engines, α is often taken as greater than 0.5 and is called overbalance. Because humans are more sensitive to vertical vibrations than to horizontal vibrations, the overbalance condition is considered an effective compromise.

The frequency of the second harmonic is twice the frequency of the first harmonic, but the amplitude is considerably smaller. However, this inertia force cannot be easily counterbalanced.

Balance of Multicylinder Engines

Common crankshaft and firing arrangements on tractor engines are shown in figure 5-22. The two-throw crankshaft is normally in static balance, since the center of gravity of the rotating parts is on the axis of rotation. However, the rotating masses M_1 and M_2 create a moment (periodic) in a vertical plane; therefore, they are resisted by an equal moment at the bearings. The two rotating masses may be balanced by the introduction of two additional masses so that the centrifugal forces set up are equal and opposite to the centrifugal forces resulting from the unbalanced masses.

Rotating masses may be balanced by other rotating masses, but the problem of balancing reciprocating masses, such as the piston or the swinging motion of the connecting rod, is much more difficult and cannot always be accomplished completely.

The method of calculating the unbalance in a multicylinder engine is discussed by Mabie and Ocvirk (1975).

It is possible to remove all of the unbalance on a six-cylinder engine by crankshaft counterweights. Such is not the case with one-, two-, three-, and four-cylinder engines. The secondary shaking force in a four-cylinder engine

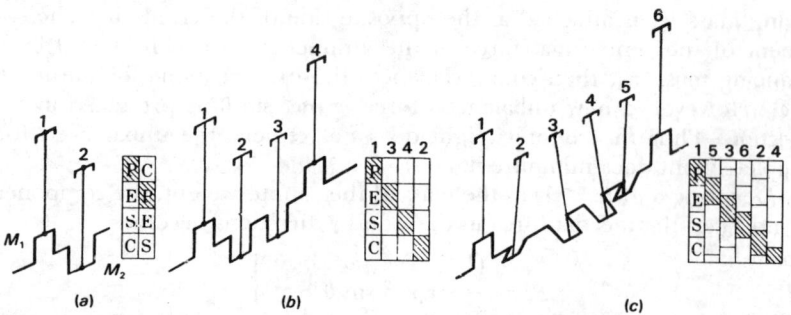

(a)   (b)   (c)

FIGURE 5-22 Common crankshaft and firing arrangements.

(see table 5-3) is a special problem, since so many tractors have four-cylinder engines.

Reciprocating forces are of two main types: primary and secondary. The forces resulting from piston assembly movement are given by the formula

$$F = M \, \omega^2 r \left(\cos \theta + \frac{r}{l} \cos 2\theta \ldots \right) \qquad (25)$$

where M = reciprocating mass
ω = angular velocity of crankshaft
r = radius (crank throw)
l = connecting rod length
θ = angle between connecting rod axis and cylinder axis

Equation 25 can be written as two equations,

$$M \, \omega^2 r \cos \theta = \text{primary force}$$

$$M \, \omega^2 \frac{r^2}{l} \cos 2\theta = \text{secondary force}$$

which are also known as the first and second harmonics, respectively. (Note that there are also higher harmonics present, as the formula is a series not given in full, but these are negligible in practice, being almost damped out.)

The primary force occurs once per crankshaft revolution, but the secondary, a function of cos 2θ, occurs twice every revolution.

In many cases, it is thought desirable to eliminate the secondary forces as much as possible for mechanical reasons or usually for operator comfort.

Since the counterbalance weights have to operate at twice engine speed, they are gear driven by the crankshaft. The weights are arranged so that one side is heavier than the remainder of the weight. At least two weights, or multiples of two, are required, and these are connected by the gears to rotate in opposite directions. The reason for the opposite rotation is that the forces resulting from the balance weights are only required to act in a vertical direction, and when so arranged, all horizontal forces from these weights cancel each other out until the weights are disposed vertically. The weights are synchronized so that the vertical forces coincide exactly opposite to the reciprocating out-of-balance force.

Ideally, the center of the weights, which are usually positioned longitudinally, along the crankshaft axis, underneath, should be located midway between cylinders 2 and 3, achieved by driving from a shaft arrangement. It is quite satisfactory to balance a major part of the force only, using slightly smaller balance weights.

TABLE 5-3   Inherent Balance of Single-acting Piston Engines

Type	Crank Arrangement	Inertia Forces, Max Value, N Primary	Secondary	Couples, Max Value, Nm Primary	Secondary
1-cyl	(crank diagram 1)	kM[a]	$\dfrac{r}{l}kM(v)$[b]	None	None
2-cyl vertical	(crank diagram 1-2)	Balanced	$2\dfrac{r}{l}kM(v)$[b]	$kMa(v)$[b]	None
3-cyl vertical	(crank diagram 1-2-3)	Balanced	Balanced	$1.732kMa(v)$[b]	$1.732\dfrac{r}{l}kMa(v)$[b]
4-cyl vertical	(crank diagram 1-2,3-4)	Balanced	$\dfrac{4r}{l}kM(v)$[b]	Balanced	Balanced
6-cyl vertical	(crank diagram 1-2-3,4-5-6)	Balanced	Balanced	Balanced	Balanced

SOURCE: Condensed from L. C. Lichty, *Combustion Engine Processes*, 7th ed. McGraw-Hill Book Co., New York, 1967.
[a]Half of this force can be eliminated by counterbalancing.
[b](v) indicates that the force or couple is in a vertical plane.
Key:
M = reciprocating mass, kg (piston plus part of rod)
r = radius of crank, m
l = length of connecting rod, m
a = distance between cylinders, m
$k = \dfrac{r}{2}\left(\dfrac{\pi n}{30}\right)^2$, where n = rpm

The secondary balancer (fig. 5-23) in its basic form was invented by F. W. Lanchester and is often called by that name. It has been used on most four-cylinder diesel and some gasoline engines for many years.

Rocking vibrations in the plane formed by the cylinders of an in-line engine can exist in two- and three-cylinder engines. This type of unbalance is not a problem in one-, four-, and six-cylinder engines of the types shown in table 5-3.

Inertia forces perpendicular to the crankshaft cause the engine to shake. From table 5-3 it can be seen that primary inertia forces do not exist in the usual three-, and four-, and six-cylinder engines. However, secondary inertia forces exist in a vertical direction in the usual four-cylinder vertical engine, which was discussed previously.

Torsional vibration of small farm tractor engines is not generally a problem because the crankshaft is relatively short. With a short crankshaft the

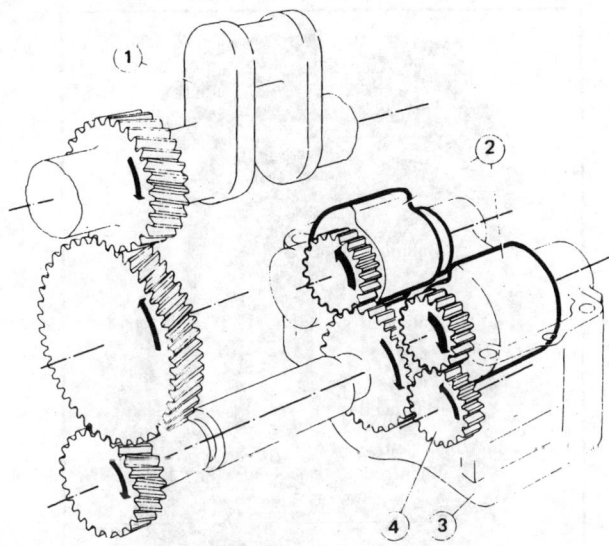

FIGURE 5-23 Schematic diagram of a Lanchester balancer used on four-cylinder diesel engines to counter-balance the secondary shaking forces. (Courtesy Fiat Trattori.)
Key:
1. Crankshaft
2. Two counter-rotating masses
3. Frame holding balancer is bolted to bottom of crankcase. Masses are located midway between the second and third cylinders.
4. Reversing gear so that the masses rotate opposite to each other.

natural frequency will be so high that it will not be excited by the normal pulse from the ignition of each charge in the cylinder. Longer crankshafts (large engines) require torsional dampers that fit on the front of the crankshaft and keep the amplitude from becoming excessively large (fig. 5-24). It is convenient to combine the damper with the V-belt sheave attached to the front of the crankshaft. Torsional vibrations can never be eliminated completely from a piston-type engine, but proper design can reduce the magnitude to an acceptable level.

Vibration analysis, isolation, and damping are very specialized and complex subjects. For the student having the necessary background, or having further interest in this subject, the suggested readings at the end of the chapter will be valuable.

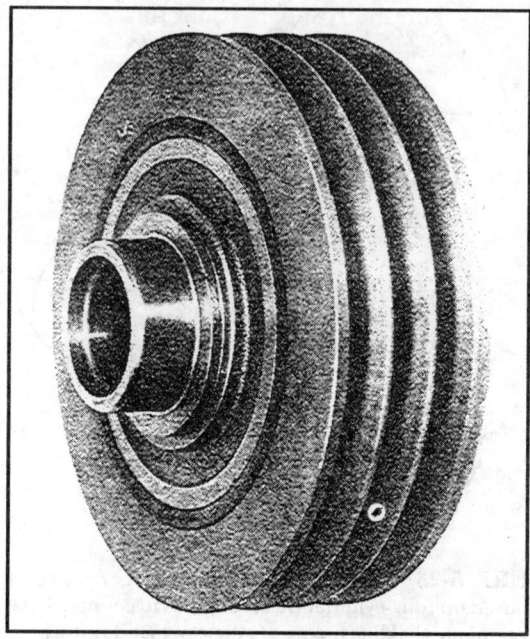

FIGURE 5-24   Torsional vibration damper that attaches to front of crankshaft. (Courtesy Perkins Engine.)

Principles of CI* Engine Operation

A modern, high-speed, automotive-type diesel engine differs little in its outward physical appearance from a spark ignition engine. There are, however, three important differences in the principles of operation of the compression ignition (diesel) engine as compared to the spark ignition engine:

1.   The ignition of the charge of fuel and air results from the high temperature of the compressed air and not from an electrical spark.
2.   Fuel is injected directly into the cylinder instead of being metered through a carburetor.

*The present diesel engine differs greatly from the original engine by Dr. Rudolf Diesel. The words *diesel, CI engine,* and *compression ignition* engine are used interchangeably.

3. The charge of air into the cylinders is not throttled as in the case of a spark ignition engine.

A diesel engine is like a spark ignition engine in that the cycle of events can be carried out in either two or four strokes, the latter being much more common for both diesel and spark ignition engines. Figures 5-25 and 5-26 clearly show the operating details of a two-stroke and a four-stroke-cycle diesel engine. The events that occur in a diesel engine are the same as those in a spark ignition engine. The events are:

1. *Intake* of air, and also of fuel in the case of a carbureted engine.
2. *Compression* of the air and the addition of enough heat (by means of the compression process) to ignite the fuel in the case of the diesel engine.
3. *Ignition* of the charge of fuel and air.
4. *Power* stroke. In this step the heat energy in the fuel is converted into mechanical energy.
5. *Exhaust* stroke. In this step the burned gases are pushed out by one stroke of the four-stroke engine. In the case of the two-stroke engine, a Roots blower, or some other positive displacement air pump, scavenges the burned gases out of the cylinder by forcing clean air into the cylinder after the gases expand far enough to expose the air intake ports in the side of the cylinder (see fig. 5-25).

Construction of Diesel Engines

The major difference between a spark ignition engine and a compression ignition engine is the compression ratio. Theoretically, there is a wide range in the type or kind of fuel that can be used in a diesel engine. However, in 1986 the fuels most commonly used in diesel engines were the heavy hydrocarbon (petroleum) fuels called distillates, which, when used in diesel engines, are called diesel fuels. Because diesel fuels require a temperature above their autoignition temperature in order to ignite, it has been necessary to build diesel engines with a compression ratio of 14:1 or greater. The higher compression of the diesel also, of course, accounts primarily for its higher thermal efficiency as compared to spark ignition engines, which normally have a lower compression ratio.

As a result of the high compression ratio, and also partly because the burning process cannot be carried out with a constant-pressure process, as was originally proposed by Dr. Rudolf Diesel, the peak pressures in the cylinder of a diesel engine are much higher than in a spark ignition engine. The strength of the parts of the engine that are exposed to the additional cylinder pressure must necessarily be increased.

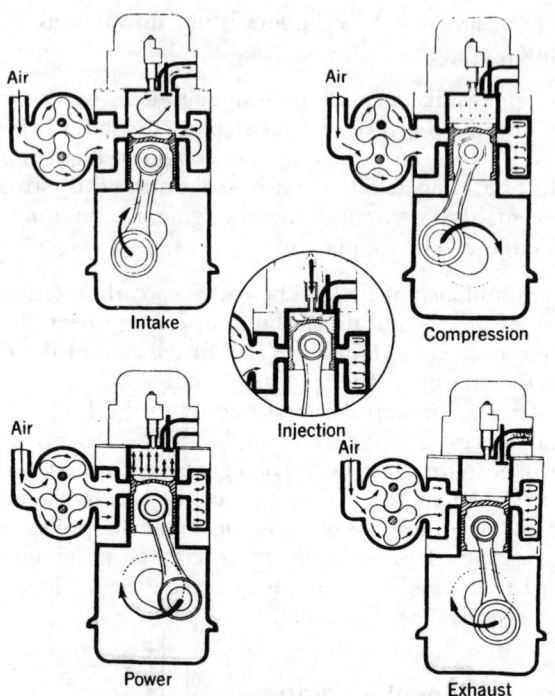

FIGURE 5-25   Diagram of a two-stroke-cycle diesel engine showing cycle of events. (Courtesy Detroit Diesel Division, General Motors Corp.)

Combustion Chamber Design

There are as many variations in the design of the combustion chamber of diesel engines as there are manufacturers making engines. Some variations are due to the different methods of starting the combustion process in the cylinder. Because the fuel-to-air ratio is very low as the fuel is first injected into the cylinder, ignition will not begin until the ratio is in the combustible range. Thus, if the fuel were injected directly into the combustion chamber, there would be an unnecessarily long delay before ignition begins, resulting in a very high initial pressure.

However, direct injection (fig. 5-27) of the diesel fuel into the combustion chamber is the most common method used. High-speed diesel engines employ several methods to reduce the ignition lag and improve the combustion. Four

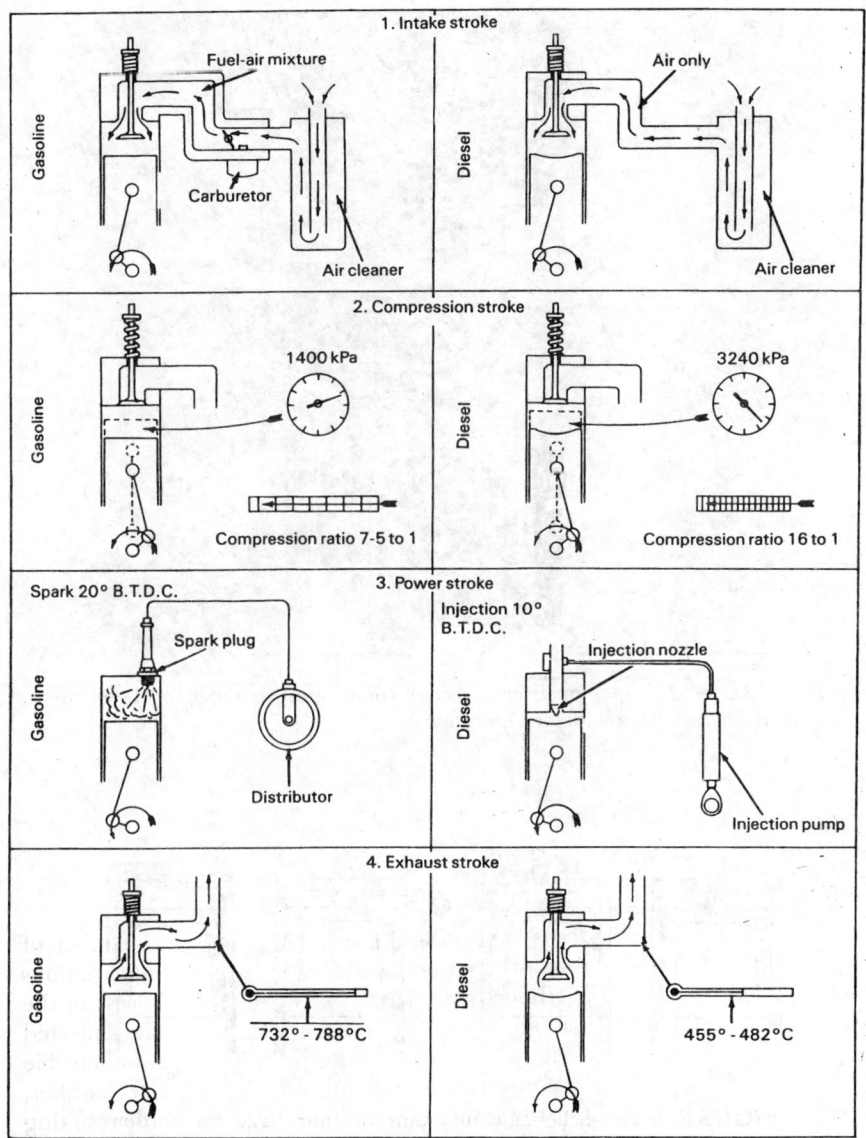

FIGURE 5-26 Comparison of operating principles of four-stroke-cycle gasoline and diesel engines. B.T.D.C. = before top dead center. (Courtesy Deere & Company.)

FIGURE 5-27 Open or direct injection combustion chamber for diesel engine. (Courtesy Ford Tractor Operations.)

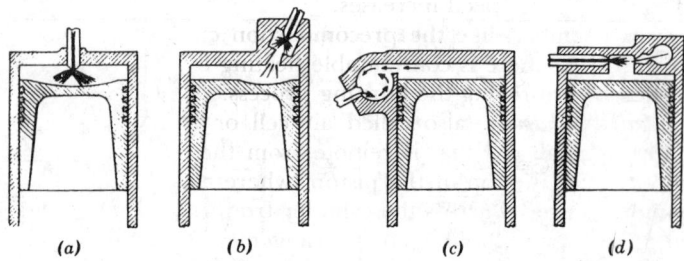

(a)   (b)   (c)   (d)

FIGURE 5-28 Schematic diagrams of four basic combustion chambers used in diesel engines. (a) Direct injection. (b) Precombustion chamber. (c) Turbulence chamber. (d) Auxiliary chamber. (*Standard Oil Co. Engr. Bull.*, TB-218.)

types of combustion chambers used in diesel engines are shown in figure 5-28.

The *open-chamber*, or *direct-injection*, combustion chamber generally employs a concave piston head. Mixing of the fuel and air is often aided by an induction-produced air swirl or by a movement of air from the outer rim of the piston toward the center of the piston commonly known as "squish."

The *precombustion chamber* is sometimes a part of the injection nozzle, or it may be part of the cylinder head. The entire fuel charge is injected into the precombustion chamber, which contains 25 to 40 percent of the clearance volume. Compared to the open combustion chamber, it is claimed the advantages of the precombustion chamber are (1) lower fuel-injection pressure required and (2) the ability to use a wider range of fuels. The disadvantage is a higher specific fuel consumption (sfc).

The *swirl combustion chamber*, which is sometimes called a *turbulence chamber*, is designed so that the burning fuel-air mixture is caused to swirl, thereby improving the mixing and subsequent combustion in the main combustion chamber. The swirl chamber contains from 50 to 90 percent of the compressed volume when the piston is on top dead-center.

The swirl chamber is sometimes referred to as the Ricardo turbulent chamber since Ricardo first used that design extensively. Swirl chambers are also often called precombustion chambers, and vice versa, because they are similar and each is usually close to the injection nozzle.

On the compression stroke, air enters the swirl chamber in a circular motion. When combustion begins, a reversal of flow occurs and the burning gases stream out of the chamber into the cylinder. Turbulence in the chamber is directly related to the speed of the engine, and therefore, ignition delay decreases as the engine speed increases.

The swirl chamber, like the precombustion chamber, is generally not water-cooled so that there is considerable heating of the air as it enters the chamber, thereby improving the burning process.

The *auxiliary chamber* (also called air cell or energy cell) is an open chamber with a small cell that is remote from the injection nozzle. It can be located either in the top of the piston (where it is called an air cell) or, more commonly, directly across the cylinder from the injection nozzle (then it is called an energy cell). The latter arrangement allows for easy removal for servicing and cleaning. In operation, approximately 60 percent of the fuel from the injection nozzle is directed into the auxiliary chamber, which contains about 10 percent of the clearance space. The blast from the fuel igniting in the auxiliary chamber will then be directed against the remainder of the fuel being sprayed from the nozzle. Thus, a fuel-air mixture is swept around the cylinder. The turbulence is sometimes aided by a swirl chamber in the head or the top of the piston located between the auxiliary chamber and the injection nozzle.

The starting characteristics of a diesel engine using an auxiliary chamber are good since the fuel is first injected through the main chamber where the air is hottest. The precision and care of the injection nozzle must be better than for the other types of combustion chambers, because the fuel must be directed accurately into a small opening that is several centimeters away from the nozzle.

The duration of combustion is longer for this type of chamber, resulting in lower peak pressures at the expense of slightly higher fuel consumption.

The maximum pressure occurring in the combustion chamber of the four designs previously mentioned is graphically illustrated in figure 5-29. The peak pressure is the highest for the open-chamber design.

Figure 5-30 illustrates some of the care required in the design of diesel combustion chambers. Note that a 20 percent variation in specific fuel consumption results from changing the direction of the injection charge into the swirl chamber. It is obvious that the performance of a diesel engine would be significantly affected by the adjustment and condition of the injection nozzle.

Fuel-Injection Systems

Although fuel injection is sometimes employed on spark ignition engines, its primary use is, and has been, on diesel engines. Its application to spark ignition engines has been limited because of the comparative simplicity, dependability, and low cost of the carburetor. The following discussion is limited to injection systems as applied to compression ignition engines used on tractors.

All fuel pumps that produce the high pressures necessary for fuel injection are of the piston type. There are many variations of fuel-injection systems used on diesel engines. Only those that are used on high-speed tractor engines will be discussed. The three systems most commonly used on tractors are:

1. Individual or *in-line* injection pumps of the timed, metered type (figs. 5-31 and 5-32)
2. The *distributor* system in which one injection pump serves all the injection nozzles by delivering a metered fuel charge at the correct instant through a distributor (fig. 5-33)
3. The *unit injector* system in which the fuel-injection pump is combined with the injection nozzle in one assembly on the cylinder head (fig. 5-36)

The pumping principle of the in-line type is shown in figure 5-32. This pump is of the constant-stroke, lapped-plunger type and is operated by a

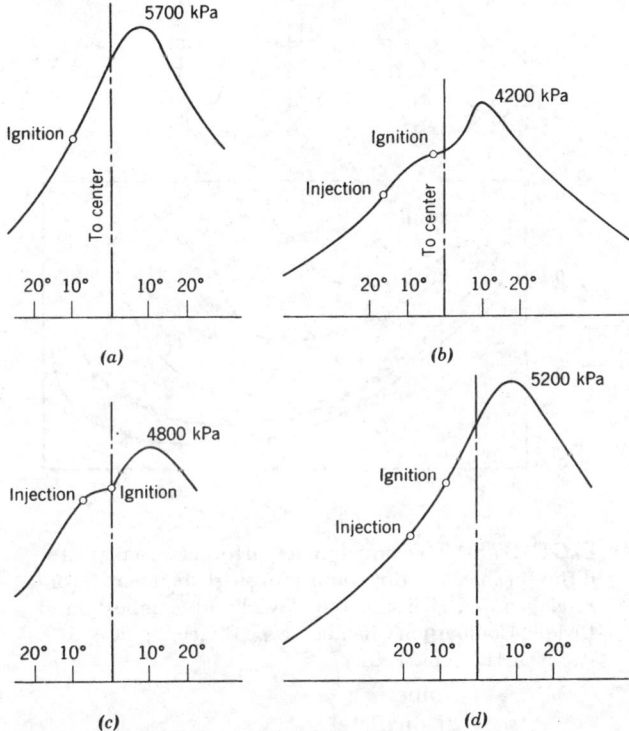

FIGURE 5-29 Pressure-time diagrams for diesel engines having four types of combusion chambers. (*a*) Open chamber. (*b*) Precombustion chamber. (*c*) Swirl chamber. (*d*) Air cell. (From P. H. Schweitzer, "Interpretation of Diesel Indicator Cards," DEMA Conference, Pennsylvania State University, June 1946.)

cam. Fuel enters the pump from the supply system through the inlet connection and fills the fuel sump surrounding the barrel. With the plunger at the bottom of the stroke, the fuel flows through the barrel ports, filling the space above the plunger and also the vertical slot cut in the plunger and the cutaway area below the plunger helix. As the plunger moves upward, the barrel ports are closed. As it continues to move upward, the fuel is discharged through the delivery valve into the high-pressure line to the injector. Delivery of fuel ceases when the plunger helix opens onto the bypass and the spring-loaded valve (top) leading to the nozzle again closes.

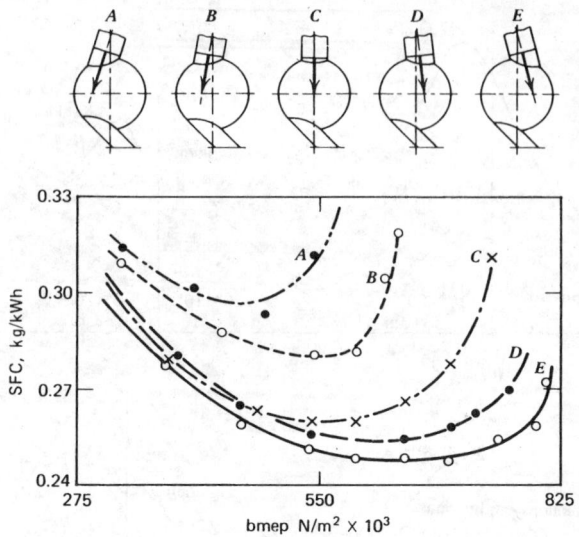

FIGURE 5-30 Comparative performance curves for different injection directions into swirl chamber. (From F. Nagao and H. Kakimoto, "Swirl and Combustion in Divided Combustion Chamber Type Diesel Engines," SAE Paper 401B, Sept. 1961.)

The quantity of fuel delivered by each stroke of the pump is controlled by rotation of the piston. Rotation of the piston is controlled by a rack, which in turn is attached directly or indirectly to the governor. Spring-loaded delivery valves are incorporated in the nozzles to prevent dribbling of excess fuel into the cylinder after the injection stroke of the piston has been completed.

Figure 5-33 illustrates a *distributor system* that incorporates pump, metering device, distributor, and governor in one unit. The timing principle is shown in the phantom views in figure 5-34.

In function the distributor system and the *in-line* or *individual pump* system are the same. The difference is that the distributor system replaces the individual pumps with a single pump plus a distributor rotor. Also, the distributor system incorporates a governor in the housing.

Because of the compact nature of a distributor-type pump, its principle of operation is more difficult to understand. The distributor pump, however, is like the individual pump system in that it performs the following function:

1. Meters a predetermined quantity of fuel as determined by the governor.

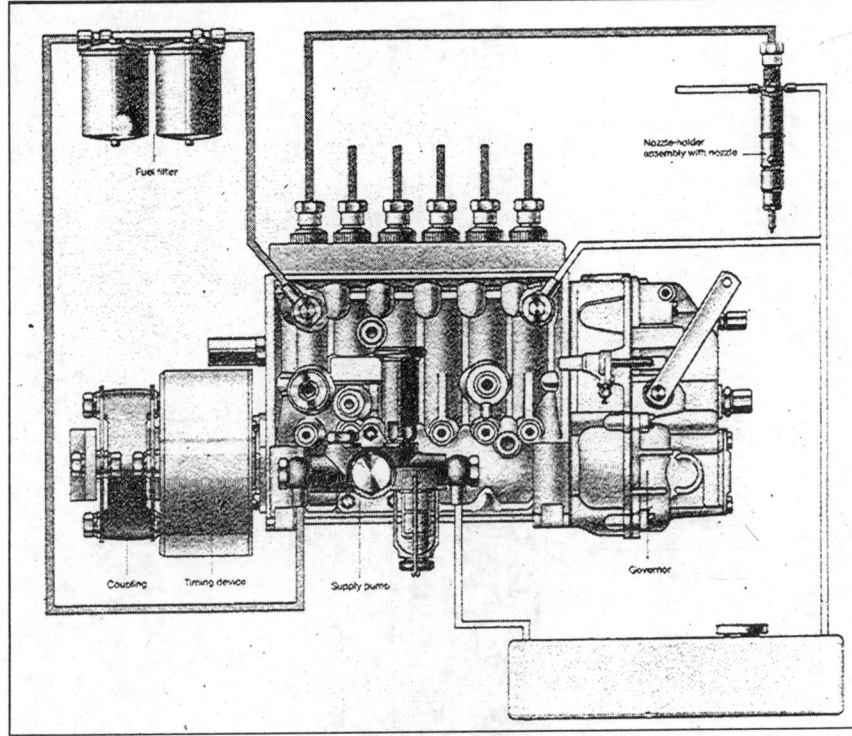

FIGURE 5-31   In-line or individual pump system. (Courtesy Robert Bosch GmbH.)

2. Delivers the fuel at high pressure to the nozzle, usually in the range from 10,000 to 20,000 kPa.
3. Times the fuel injection. Like a spark ignition engine, the timing of the fuel injection is a function of the engine speed.

Fuel Injectors

The type of fuel injector used depends on the type of pump and combustion chamber employed. A cross-sectional view of one fuel injector is shown in figure 5-35. This type is used in connection with the open-chamber or direct-injection combustion chamber.

A unit injector combines the functions of the high-pressure pump and injector into one unit (fig. 5-36). One advantage of such a system is that there is no high-pressure line to complicate the timing and metering of the fuel. The unit injector is commonly driven by a separate cam shaft.

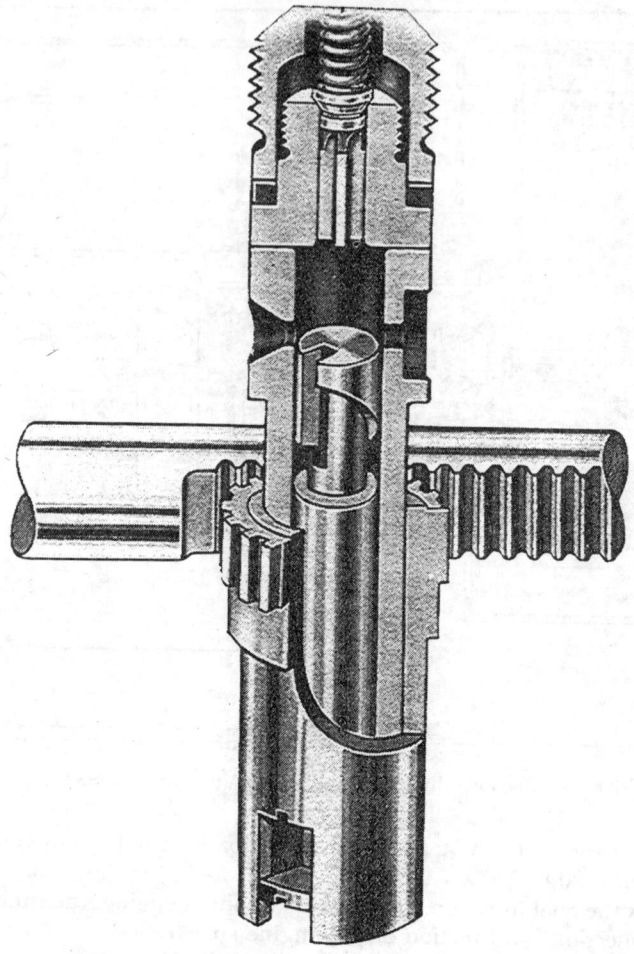

FIGURE 5-32 Method of metering fuel to injectors with in-line system. (Courtesy Robert Bosch GmbH.)

Turbochargers*

The most efficient method of increasing the power of an engine is the addition of a turbocharger. A turbocharger, by definition, is an engine supercharger

---

*There are many methods of supercharging or increasing the flow of air into an engine. The use of a turbocharger is the most cost-effective method of supercharging.

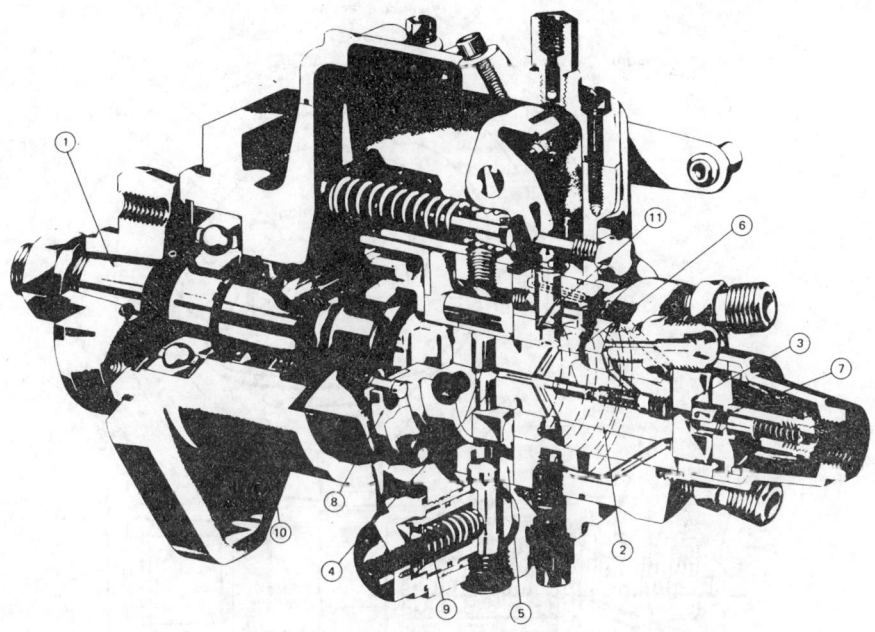

FIGURE 5-33   Sectional view of a distributor-type fuel injection pump. (Courtesy Roosa Master.)

Key:

1. Drive shaft
2. Distributor rotor
3. Transfer pump
4. Pumping plungers
5. Internal cam ring
6. Hydraulic head

7. Pressure regulator
8. Governor
9. Governor weights
10. Metering valve
11. Discharge fitting

driven by an exhaust gas turbine. Figure 5-37 shows a typical turbocharger. Exhaust gases from the engine enter the turbine housing radially and drive the turbine wheel, which drives the compressor wheel, both being mounted on the same shaft. Turbocharging increases the density of air delivered to engine cylinders above that available in natural aspiration and thereby allows an engine to burn more fuel and, in turn, develop more power.

A turbocharged engine operates with lower cylinder temperatures and reduces fuel consumption for the power produced. The turbocharger is sometimes equipped with a waste gate mechanism comprising a valve and its actuator. In high-speed operation, this mechanism acts to allow exhaust gases to bypass the turbine wheel in order to prevent overrunning of the turbocharger and abnormal pressure rise in the exhaust manifold.

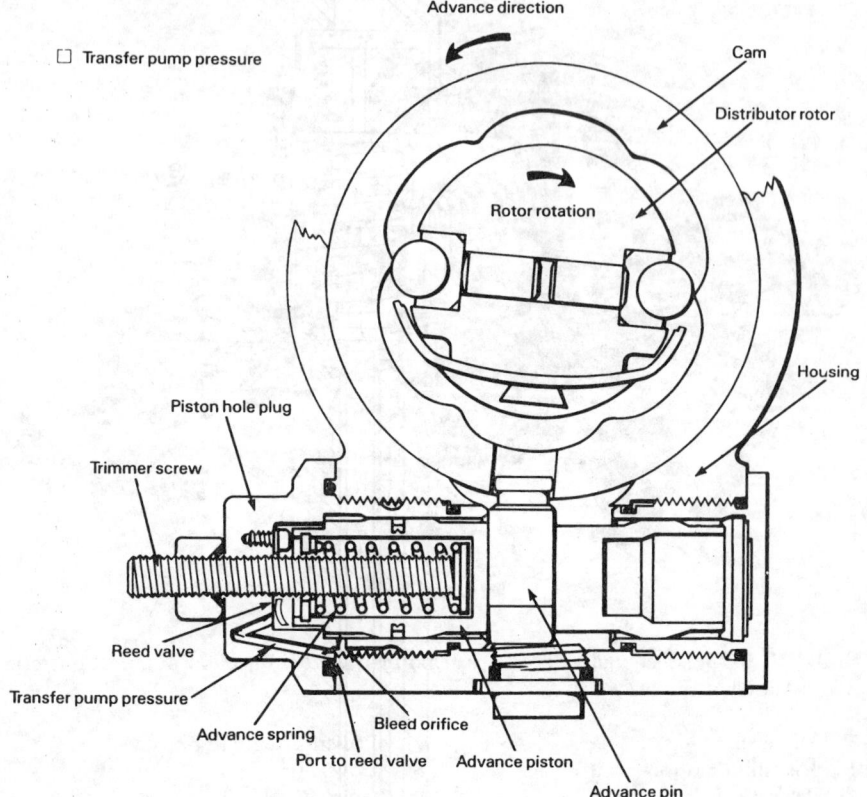

Advance direction

☐ Transfer pump pressure

Cam

Distributor rotor

Rotor rotation

Housing

Piston hole plug

Trimmer screw

Reed valve

Transfer pump pressure

Advance spring

Port to reed valve

Bleed orifice

Advance piston

Advance pin

FIGURE 5-34 Automatic advance mechanism for distributor-type fuel-injection pump. (Courtesy Roosa Master.)

The bearings are lubricated with engine oil supplied under pressure from the engine. This oil leaves the turbocharger through the outlet port in the bottom of the bearing housing and returns to the engine oil pan.

A typical turbocharger performance map is shown in figure 5-38.

Matching of Turbocharger to Engine*

The matching of a turbocharger to an engine has progressed from a long, painstaking procedure of trial and error in the laboratory to a simple process

*This section is condensed from a paper by W. E. Woolenweber entitled "The Turbocharger—A Vital Part of the Engine Intake and Exhaust System," SP-359, "Engineering Know-how in Engine Design—Part 18," Society of Automotive Engineers, Warrendale, PA, 1973.

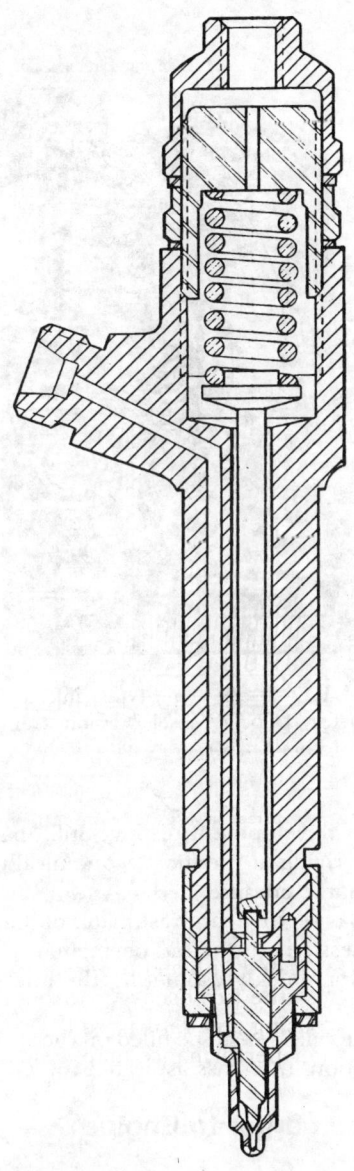

FIGURE 5-35 Cross section of fuel injector. (Courtesy Robert Bosch GmbH.)

FIGURE 5-36 Unit-type injector. (Courtesy Detroit Diesel Division, General Motors Corp.)

requiring a minimum of development testing. Still, many important factors must be considered in arriving at the most economical package with the best possible performance characteristics.

A new application may require some estimates of the airflow requirements of the engine to obtain desired output and performance. A relatively accurate and simple expression for quickly estimating the inlet airflow requirement may be derived as follows:

Assuming the engine cylinders to be filled on the suction stroke with fresh air available for combustion, the mass aspirated is

$$M_a = D_c V_c N_s E_s \qquad (26)$$

where
M_a = total mass of aspirated air
D_c = air density in the cylinders
V_c = displacement volume of cylinders
N_s = suction strokes per minute
E_s = supercharged volumetric efficiency

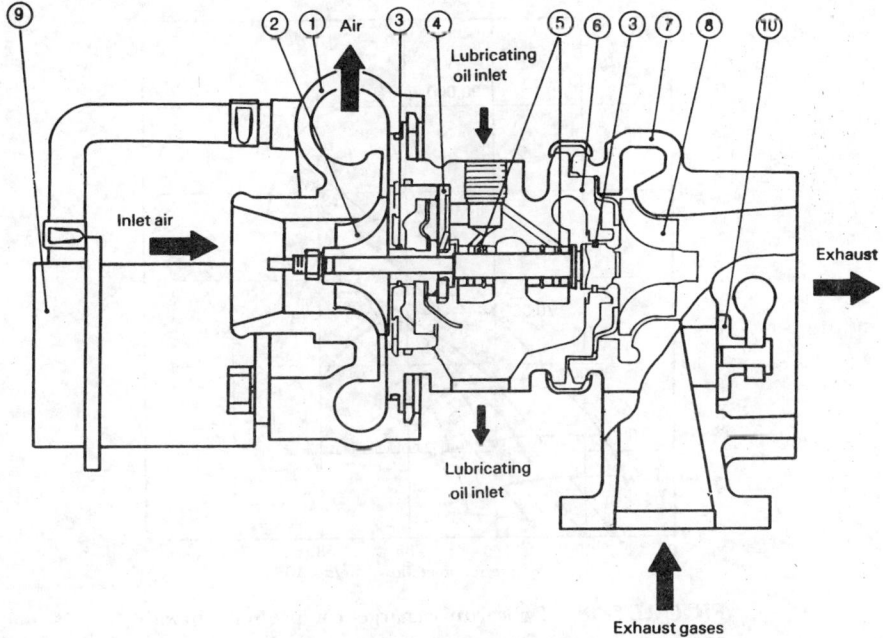

Air

Lubricating
oil inlet

Inlet air

Exhaust

Lubricating
oil inlet

Exhaust gases

FIGURE 5-37   Cross-sectional view of a turbocharger. (Courtesy Mitsubishi Heavy Industries, Ltd.)
Key:

1. Compressor cover	5. Bearings	9. Waste gate actuator
2. Compressor wheel	6. Bearing housing	10. Waste gate valve
3. Piston rings	7. Turbine housing	
4. Thrust bearing	8. Shaft and turbine wheel	

If volumetric efficiency values for a particular supercharged engine are unknown, then it may be assumed that losses through the manifold, ports, and valves, as well as temperature effects, are offset by the inertia effect of the supercharging airflow. It then follows that the air density in the intake manifold may be substituted for the air density in the cylinder. Usual values of volumetric efficiency for supercharged engines are close to 1.0.

Then equation 26 becomes

$$M_a = \frac{P_m}{RT_m} V_c N_s \tag{27}$$

where P_m = absolute pressure in the intake manifold
T_m = absolute temperature in the intake manifold

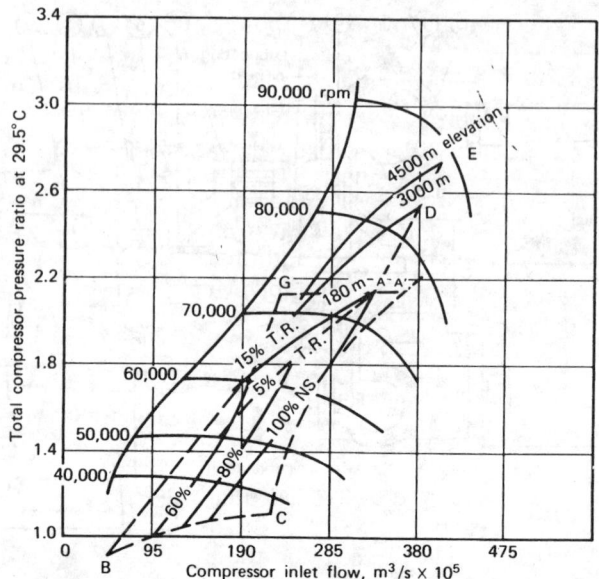

FIGURE 5-38 Typical turbocharger compressor map with engine airflow requirements superimposed. (From Wollenweber 1973.)

For a steady flow process,

$$M_a = Q_m D_m = Q_c D_o \tag{28}$$

Then

$$Q_o D_o = \frac{P_m}{R T_m} V_c N_s \tag{29}$$

$$Q_o = \frac{P_m}{P_o} \frac{T_o}{T_m} V_c N_s \tag{30}$$

Thus the inlet volume flow, Q_o, of the turbocharged engine may be calculated from the static pressure ratio, P_r; the inverse of the temperature ratio, T_r; displacement volume, V_c; and speed, N_s. To obtain inlet air volume flow in cubic meters per second, equation 30 reduces to

$$Q_o = \frac{P_r T_r V_c N_s}{60,000} \text{ for four-stroke-cycle engine} \tag{31}$$

where V_c = displacement volume in liters

   N_s = engine speed in rpm

Pressure ratios required to obtain desired power output must be assumed or known for a given type of engine, and temperature rise across the compressor may be calculated from assumed values of compressor efficiency.

Once approximate airflow values are obtained, the next step in the matching process is to select a compressor characteristic on the basis of a number of important considerations. One must have a good idea of what maximum pressure ratio can be used at sea level at the full-load, full-speed rating point.

The complete field of required airflow over the entire speed and load range of the engine may be established and then superimposed on the compressor performance characteristic. The following factors have the greatest influence:

1. Desired torque rise, if any
2. Speed range of the engine
3. Altitude operating requirements
4. Aftercooling the compressed air

Selection of a compressor characteristic that will give the maximum efficiency over the engine operating range is the primary consideration. However, some compromises may be made on the basis of size and cost of the superchargers that might be applicable.

Referring to figure 5-38, a vehicle engine application will usually have a rather broad speed range over which it will be required to operate at various loads. This load variation requires the turbocharger to operate at various speeds, producing the operating line referred to as 100 percent, N_s. Lower values of engine speed move the air requirement line to the left, producing the other speed lines shown as 80 percent and 60 percent. Reducing the engine speed at full load will follow the dotted line from A in a manner determined by the amount of torque rise to be produced by the engine. Higher torque rise requirements result in broadening the engine air requirement field shown by the 5 percent and 15 percent torque rise lines. The low-idle point is shown at B, and high idle or governed speed at no-load is represented by point C.

As the turbocharged engine is taken to higher altitudes without derating, the rotational speed of the supercharger increases, as indicated by points D and E. Pressure ratio and airflow delivered to the engine both increase. Since there is no mechanical connection between the tubocharger rotor and the engine, it can "float" at various speed values depending on the altitude of operation and the energy in the engine exhaust gas. Thus the turbocharger can partially compensate for losses in ambient air density as the engine is taken to high altitude.

Another important consideration that affects the compressor match is the use of aftercooling of the compressed air. Aftercooling presents a means of reaching higher brake mean effective pressure values with a given turbocharged engine while still maintaining reasonable exhaust temperature to ensure good engine life. Again referring to figure 5-38, the full load and speed point, as well as the slope of the engine speed lines, are altered by aftercooling, moving point A to point A'. The location of point A' primarily depends upon whether water or air-to-air aftercooling is employed.

Aftercooling affects the compressor match, since the compressor must be capable of a broader flow range to ensure minimum loss in compressor efficiency at point A'. Although the aftercooler is capable of removing added heat due to lower compressor efficiency, the higher compressor power required results in added back pressure in the engine exhaust system, thereby affecting engine performance.

Pressure drop through the aftercooler core must be as low as possible commensurate with the size limitations of the installation. Aftercooler pressure losses must be compensated for by increasing the turbocharger compressor output to maintain a given charging pressure level to the engine cylinders. Any increase in charging pressure must be accompanied by an increase in exhaust pressure, provided the exhaust temperature remains essentially constant and, again, causes a loss in engine performance.

If a pressure ratio control system is employed using an exhaust gas bypass valve, the air requirement will be broadened considerably, as shown by the extension of the 60 percent engine speed line to G. Point G is determined by the amount of air needed to produce the desired low engine speed performance. At higher engine speeds and full load, the pressure ratio control functions to hold the pressure ratio constant. The condition is represented by the dotted line G-A.

Thus, by considering the engine speed range, torque rise, operating altitude, the use of aftercooling, the use of an exhaust gas bypass system, and the low and high idle points, the entire field of the engine air requirement can be represented on the turbocharger compressor map.

Knowledge of the engine's complete air requirement field has enabled small centrifugal compressors to be developed to cover this field at the highest possible compressor efficiency. Compressor surge lines have been tailored to follow the shape of the field along the torque curve. Maximum speeds have been extended so that high-altitude requirements could be easily met at high compressor efficiency. Maximum flow capacity has been extended while still maintaining an adequate surge line so broad engine operating speed ranges, with aftercooling and wastegating, can be covered. Adequate operating margins away from the surge line and choke area must be maintained to allow for variations in production hardware and any differences between laboratory data and actual field operating conditions.

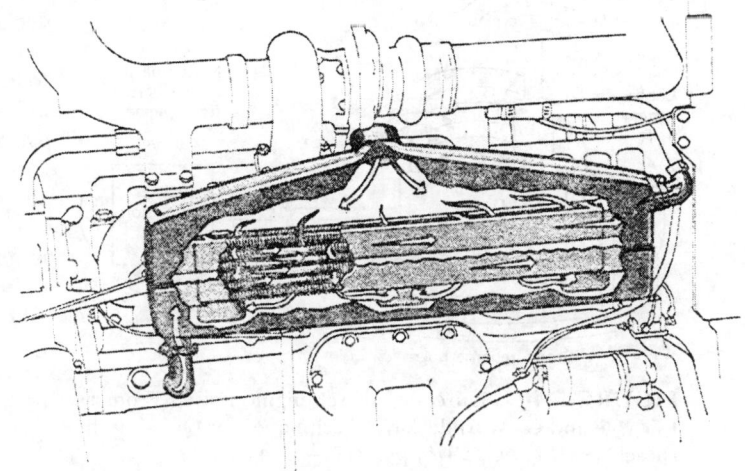

FIGURE 5-39   Aftercooler (or intercooler) used on turbocharged tractor engine. (Courtesy Deere & Co.)

Aftercooling

The turbocharger unfortunately heats the air going into the engine. The heat reduces the density of the air and also increases the temperature of the combustion chamber. Aftercooling (or intercooling) reduces the temperature of air entering the combustion chamber, thereby increasing the air density and in turn increasing the power output.

Figure 5-39 shows clearly a typical tractor engine installation of an aftercooler. In most mobile diesel engine installations the aftercooler also uses the engine water-jacket coolant to cool the air. Such installations limit the aftercooling to about 25°C above the water-jacket temperature.

Using air to cool the air from the turbocharger is more effective in lowering the temperature; however, the size of an air-cooled aftercooler may prohibit its use on some mobile equipment.

Engine Noise

Most of the noise coming from a tractor originates in the engine. The solution of different noise problems is beyond the scope of this text. Several excellent

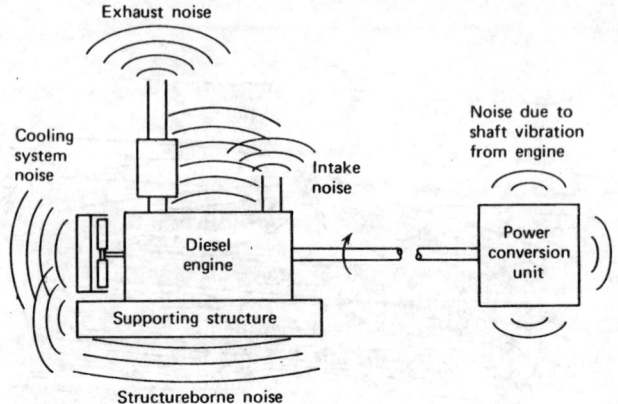

FIGURE 5-40 Sources of diesel engine noise. (From D.
F. Kable and G. A. Anderkay, "Techniques for Quieting the
Diesel," *SAE Trans.*, 1975, pp. 2176–2184.)

references on noise control are listed at the end of the chapter. Students who
have had a course in vibrations are encouraged to study and work problems
on the subject of engine noise control.

Reports by Kable and Anderkay (1975) and Thein and Fachbach (1975)
point out clearly the sources of noise on diesel engines. Figure 5-40 shows
the noise source on a typical diesel engine. Kable states,

> Generally speaking, reduction at the source is the most cost effective although
> it often can be the most complicated. Typical source treatments and the resulting
> noise reduction include the following:
>
> 1. Smoothing combustion, 3 dB (*A*)
> 2. Reducing piston slap, 1–4 dB (*A*)
> 3. Block stiffening, damping, cover isolation and shielding
>   (a) Radical redesign, 10–15 dB (*A*)
>   (b) Modify existing structure, 5–8 dB (*A*)
> 4. Total enclosure, 15–20 dB (*A*)

From figure 5-40 and the list of source treatments is should be clear that
the sources of sound in most cases are radiating surfaces. The techniques
used serve to:

1. Modify stiffness and mass distribution
2. Add shielding
3. Add damping
4. Provide isolation

PROBLEMS

1. A three-cylinder tractor engine has a 90-mm bore, a 90-mm stroke, and a governed speed of 2400 rpm. It develops 24 kW on the pto at maximum load. The intake valve face angle is 30°, and the exhaust valve face angle is 45°. The specified valve seat width is 1.6 mm. The maximum valve head diameter G (fig. 5-15) is 35.3 mm for the intake and 30.6 mm for the exhaust. Assume that the dimension 1–2 (fig. 5-15) is 6.4 mm and that the valve seat rests against the middle of surface 1–2 when the valve is closed. The cam lift for both intake and exhaust valves is 7.11 mm. The valve stem is 9.5 mm for both intake and exhaust. Assume the valve port diameter to be the distance measured across the lower edges of the valve seat. (a) Compute the valve-opening area for the intake valves. Note the possibility that the line 1–5 (fig. 5-15) may or may not fall outside the point 4 and that the line 1–4 may become the minimum slant height for computing the valve-opening area. (b) Compute the valve-opening area for the exhaust valves. (c) What would be the effect on the valve-opening area for the exhaust valves if the valve seats were ground so that the seats were 4.7 mm wide? Assume the valve port diameter remains the same. (d) Compute the maximum instantaneous velocity through the intake valve in meters per second. The maximum piston speed in meters per second for this engine is approximately 1.04 times the velocity of the crankpin. You may assume (approximately true) that the valve is open the maximum distance at this point.

2. Construct a valve-timing diagram similar to that shown in figure 5-4 for a tractor engine with the following specifications: intake valve opens 5° after hdc; intake valve closes 40° after cdc; exhaust valve opens 45° before cdc; exhaust valve closes 10° after hdc.

3. Construct (to three times actual size) a cam similar to that shown in figure 5-16 to give the inlet valve timing shown in problem 2. Let the radius of the base circle be 20.3 mm. Let the flanks be straight lines, and determine the nose radius necessary to give a maximum lift of 5.1 mm. Neglect the tappet clearance and ramp on the cam. Determine graphically the lift for each 5° of cam rotation, and plot the results on cross-section paper.

4. For an assigned tractor engine, determine the average and maximum gas velocity through the intake and exhaust valve ports. Data needed are stroke, bore, piston rod length, valve lift and timing, and the diameters of the valve stem and valve face. At what crank angle is the piston velocity a maximum? If all the necessary data cannot be readily obtained, use assumed data and so note.

5. A diesel engine is to start cold at $-30°C$. The fuel has an autoignition temperature of 350°C. What minimum compression ratio must the engine have to facilitate starting? It is found by test that this engine consumes 0.24 kg/kWh. What is the thermal efficiency? How many kilowatt hours per liter are needed?

6. A 105-kW, six-cylinder gasoline engine is to run at 2400 rpm. Design a flywheel (i.e., find I) for the engine if the speed is to vary no more than ±5 percent from the average speed. Hint: You may calculate the indicated work per revolution from the following equation.

$$\text{Indicated work per revolution} = \frac{60{,}000P}{NE_m} = \text{joules/rev.}$$

where P = brake power output, kW
       N = average engine speed, rpm
       E_m = decimal mechanical efficiency
       Assume E_m = 0.8
  Note $I = Mr^2$

7. If a four-cylinder four-stroke cycle diesel operates at 2500 rpm and uses 18.5 liters of fuel per hour, what is the average volume in cubic millimeters of the individual injections? This volume would make a sphere of what diameter? If the compression ratio is 16.8:1 and the engine displacement is 3920 ml, what is the clearance volume of each cylinder (i.e., the volume with the piston at head dead-center)? What is the ratio of the clearance volume to the volume of an injection?

REFERENCES

Kable, D. F., and G. A. Anderkay. "Techniques for Quieting the Diesel." *SAE Trans.*, 1975, pp. 2176–2184.

Mabie, N. H., and F. W. Ocvirk. *Mechanisms and Dynamics of Machinery*, 3d ed. John Wiley & Sons, New York, 1975.

Thein, G. E., and H. A. Fachbach. "Design Concepts of Diesel Engines with Low Noise Emissions." *SAE Transactions*, 1975, pp. 2160–2175

SUGGESTED READINGS

Beranek, L. L., et al. *Noise and Vibration Control*. McGraw-Hill Book Co., New York, 1971.

Buchi, A. J. "Exhaust Turbocharging of Internal Combustion Engines." *Jour. of the Franklin Institute*, Philadelphia, July 1953.

"Computer in Internal Combustion Engine Design." Symposium of the Institute of Mechanical Engineers (England), 1968.

Crocker, Malcolm J. *Noise and Noise Control*. John Price, Cleveland, 1975.

den Hartog, J. P. *Mechanical Vibrations*, 4th ed. McGraw-Hill Book Co., New York, 1956.

Engineering Know-how in Engine Design—Part 18, SP-359. Society of Automotive Engineers. Warrendale, PA, 1970.

Ferguson, Colin. "Internal Combustion Engines." John Wiley & Sons, New York, 1986.

Givens, Larry. "The Diesel Engine: Today and Tomorrow." *Auto. Eng'g.* June 1976.

Hare, C. T., K. J. Springer, and T. A. Huls. "Exhaust Emissions from Farm, Construction, and Industrial Engines and Their Impact." SAE Paper No. 750788, 1975.

Holler, H. G. "Tomorrow's Diesel. What Will It Offer?" Published in SP-270, "Powerplants for Industrial and Commercial Vehicles—A Look at Tomorrow." SAE Paper No. 650479, 1965.

Holt, R. C., R. R. Yoerger, and J. A. Weber. "Why Early Tractor Intake Valve Failure?" Paper No. 60-140 presented at the ASAE meeting, Columbus, OH, June 1960.

Holzhausen, G. "Turbocharging Today and Tomorrow." Paper 660172 presented at the SAE Mid-Year Meeting, Detroit, June 1966.

Lichty, L. C. Combustion Engine Processes, 7th ed. McGraw-Hill Book Co., New York, 1967.

Mitchell, J. E. "An Evaluation of Aftercooling in Turbocharged Diesel Engine Performance." SAE Trans., vol. 67, 1958.

Mitchell, J. E. "Power Producing Characteristics of Diesel Engines." Published in SP-243, "Engineering Know-how in Engine Design—Part 11." Society of Automotive Engineers, New York, 1963.

Nancarrow, J. H. "Influence of Turbocharger Characteristics on Supply of Air for High Speed Diesel Engines." SAE Trans., vol. 75, 1966.

Obert, E. F. Internal Combustion Engines, 3d ed. International Textbook Co., Scranton, PA, 1973.

Peterson, Arnold P. Handbook of Noise Measurement. GenRad, Concord, MA, 1974.

Rogowski, A. R. Elements of Internal Combustion Engines. McGraw-Hill Book Co., New York, 1953.

Schweitzer, P. H. "Must Diesel Engines Smoke?" SAE Trans., vol. 1, July 1947.

Sprick, W. L., and T. H. Becker. "The Application and Installation of Diesel Engines in Agricultural Equipment." ASAE Distinguished Lecture Series, Tractor Design, no. 11. American Society of Agricultural Engineers, Dec. 1985.

"Symposium on Diesel Engines—Breathing and Combustion." Institute of Mechanical Engineers, England, 1966.

Taylor, C. F. "The Internal Combustion Engine in Theory and Practice," vols. I and II. The MIT Press, Boston, MA 1985.

"Turbocharged and Intercooled Diesel for Farm Machinery." Diesel and Gas Turbine Progress, September 1970.

Woolenweber, W. E. "The Turbocharger—a Vital Part of the Engine Intake and Exhaust System." Published in SP-359, "Engineering Know-how in Engine Design—Part 18," Society of Automotive Engineers, Warrendale, PA 1973.

6
ELECTRICAL SYSTEMS

The design of electrical systems is very specialized. Components are rarely designed and manufactured by the tractor manufacturers but instead are purchased from vendors.

Electrical systems are becoming much more complicated and promise to become even more so, paralleling the changes that are taking place in hydraulic systems. Figure 6-1 shows a basic electrical system on a tractor. Numerous components can be added to this system.

Battery

An energy storage device is always needed for starting a large tractor and also for furnishing energy when the engine is running at a very low speed. Electrical current in a *lead-acid battery* is produced by the chemical action illustrated by Figure 6-2. The lead peroxide (PbO_2) in the positive plate is a compound of lead (Pb) and oxygen (O_2). Sulfuric acid is the electrolyte between the positive and negative plates. During the discharge process, oxygen (O_2) combines with the hydrogen (H) in the sulfuric acid to form water (H_2O). This change is used to determine the degree or percentage of battery charge. A fully charged lead-acid battery has about 36 percent sulfuric acid and 64 percent water, and the solution has a specific gravity of 1.27. A completely discharged battery will still have some sulfuric acid and will have a specific gravity of about 1.1.

The chemical reaction during the charging cycle (see fig. 6-3) is essentially the reverse of the discharge. The lead sulfate is split into Pb and SO_4, and the water (H_2O) is split to form sulfuric acid (H_2SO_4). At the same time, the free oxygen (O_2) combines with the lead at the positive plate to form lead peroxide (PbO_2).

134

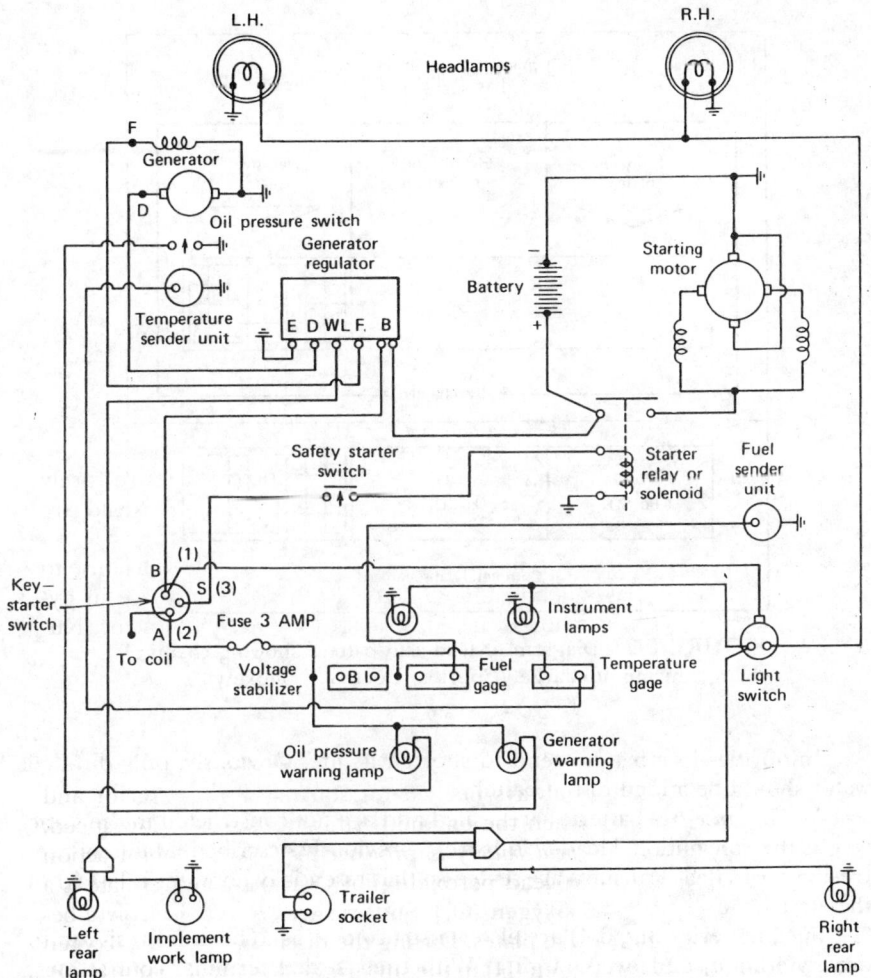

FIGURE 6-1 Basic electrical system on diesel tractor. (Courtesy Ford Tractor.)

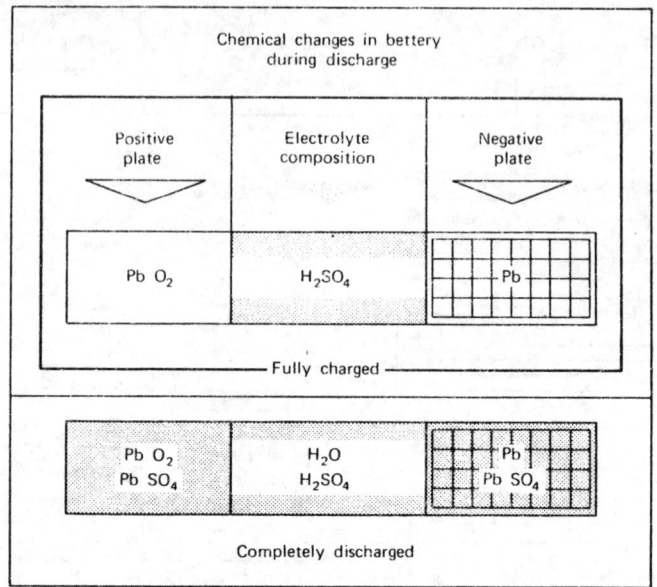

FIGURE 6-2   Diagram of lead-acid battery showing chemical change during discharge. (Courtesy Deere & Company.)

Impurities in a battery tend to shorten its life. Obviously, only distilled water should be added to the battery.

Maintenance-free batteries are sealed and are not constructed in the same way as the conventional lead-acid battery previously described. Maintenance-free batteries (lead-calcium) are sealed so that essentially no water is lost from the electrolyte.

Such batteries (fig. 6-4) are also constructed so as to be more resistant to heat, vibration, and overcharging. With their sealed terminal connections, such batteries are more adapted to the environment of dust and heat associated with agricultural tractors.

Capacity ratings of batteries have been standardized by the Society of Automotive Engineers. Of importance is the *cold power rating,* which gives the amperes the battery can discharge for 30 s without dropping below 1.2 V per cell (see SAE J537 [June 1982] for details of the ratings). The cold power rating is of course valuable when rating the battery for starting an engine on a cold day.

The *reserve capacity* of a battery is another battery rating. It rates the battery on the basis of the minutes that a new fully charged battery at 26.6°C will deliver 25 A while maintaining a voltage of 1.75 V per cell. Such a rating

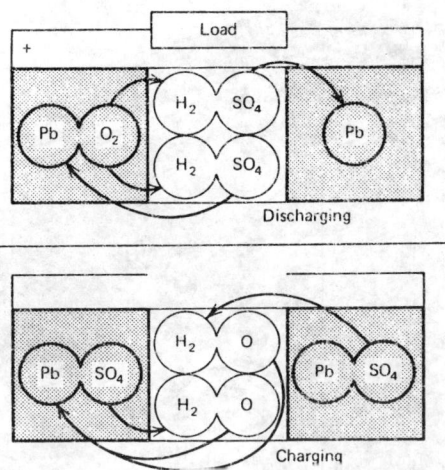

FIGURE 6-3 Chemical action in a battery during discharge and charge cycles. (Courtesy Deere & Company.)

is valuable in determining how long the tractor could run after the charging system fails.

The problem of starting an engine, especially a diesel, is affected by several factors, including the viscosity of the oil. One test of a 9.4-L engine (see fig. 6-5) shows the effect of oil viscosity and temperature on cranking speed. This chart shows that with SAE 30 oil in the engine the minimum cranking speed of 100 rpm could not be reached unless the temperature was at least −4°C. With SAE 10W oil in the engine, the minimum starting speed of 100 rpm could be reached when the temperature was −15.5°C.

The high compression of a diesel engine, plus the added friction in cold weather, places a large demand on the battery and starting motor.

One improvement in the diesel cranking problem can be made by using two 12-V batteries that are normally in parallel for 12-V lights, air conditioners, and so forth. For starting, however, switches place the two batteries in series to drive a 24-V starting motor. This series-parallel circuit is commonly used on tractors having more than 100 kW of power.

Starting Motor

A starting motor (see fig. 6-6) consists of a dc motor that is engaged to the ring gear on the engine by means of a sliding pinion gear. The pinion gear is meshed with a ring gear on the flywheel of the engine by a solenoid. When

FIGURE 6-4 Cross section of a maintenance-free battery. (Courtesy General Battery Corp.)

Key:

1. Manifold vent
2. Rib reinforced polypropylene case
3. Plate separator
4. Plate cells bonded with epoxy
5. Calcium alloy grids
6. Lugs
7. Stainless-steel terminals
8. Cell partition
9. Lifting slot

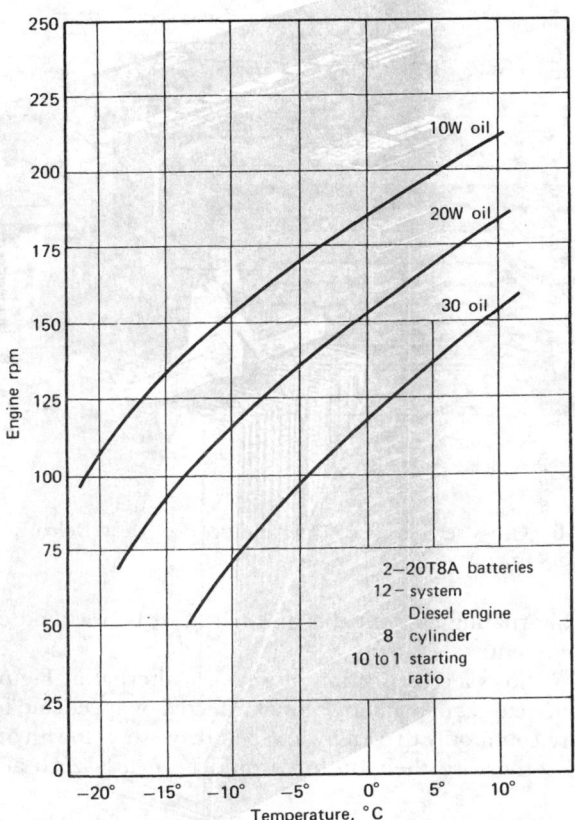

FIGURE 6-5   Effect of temperature and oil viscosity on starting motor cranking speed. (Courtesy Delco Remy Division, General Motors Corp.)

electrical power is switched to the dc motor, the solenoid is simultaneously activated, which in turn engages the pinion. The spring on the pinion disengages the pinion when power to the solenoid is shut off.

Electrical Charging

A battery must be regularly or continuously recharged. The most common type of recharging or generating system is often called an alternator, which

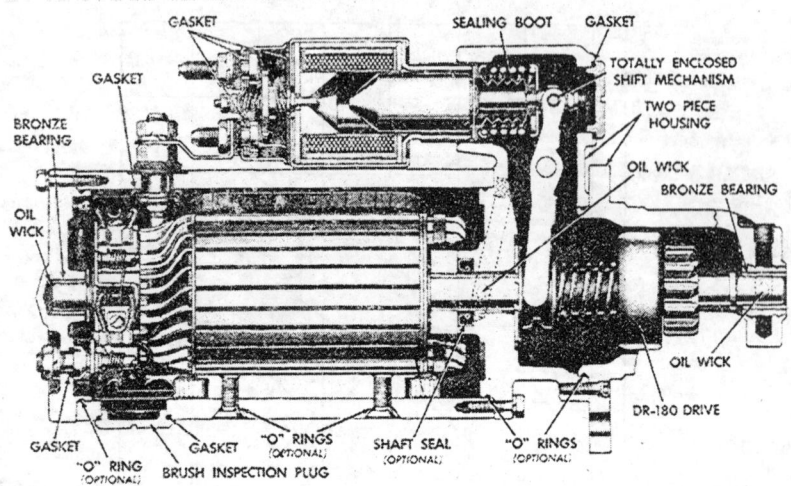

FIGURE 6-6   Cross section of starting motor. (Courtesy Delco Remy Division, General Motors Corp.)

produces an alternating current that in turn must be rectified or converted into a direct current.

Figure 6-7 shows a cross section of a typical alternator. Figure 6-8 shows the performance curves of an alternator as affected by speed and temperature. Alternators are commonly driven by a V-belt drive so as to run at about twice engine speed; otherwise their output would be nearly zero at engine idle speed.

A complete generating circuit is shown in figure 6-9. The primary parts of this circuit are the battery, the alternator with a built-in rectifier, and the voltage regulator.

Battery Ignition System with Breaker Points

A battery ignition system (fig. 6-10) has two electrical circuits. The primary circuit consists of the battery, ammeter, switch, primary coil winding, condenser, and breaker points. When the points are opened, the primary current charges the condenser instead of dissipating itself in an arc across the points. The condenser then discharges back through the primary coil. This discharge through the primary coil aids the rapid change and reverses the magnetic flux, which in turn induces a sufficiently high voltage in the secondary coil to result in a spark discharge at the spark-plug gap. Condensers for ignition systems range from 0.15 to 0.45 microfarads capacity.

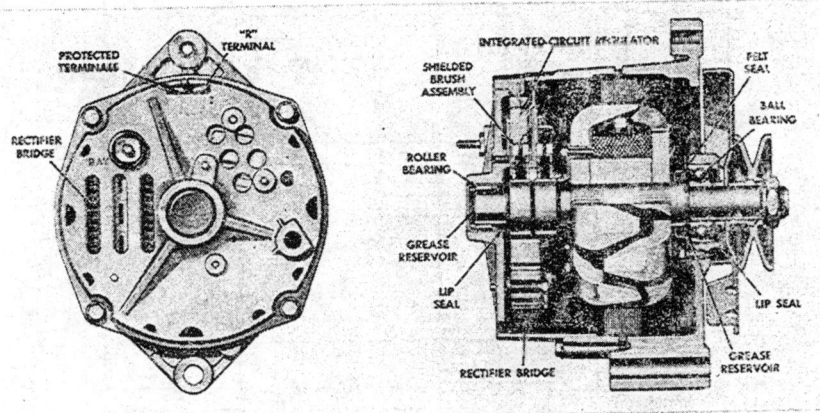

FIGURE 6-7 Cross section of alternator. (Courtesy Delco Remy Division, General Motors Corp.)

The secondary circuit consists of the secondary winding, the distributor, and the spark plugs, plus the necessary connecting wires. Whereas the primary winding consists of relatively few turns of heavy wire, the secondary winding has many turns of very fine wire. Although the ratio of turns varies for different applications, it is on the order of 100 to 1. The number of turns in the secondary coil may be on the order of 10,000 to 15,000, which induces secondary voltages of 10,000 to 20,000 V at the spark plug.

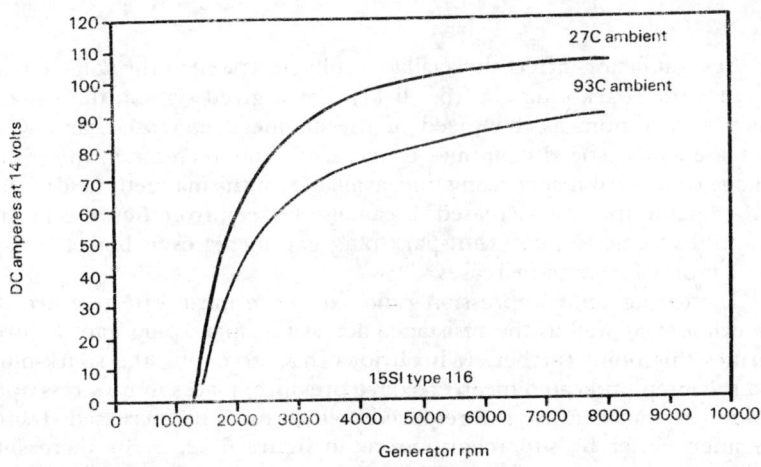

FIGURE 6-8 Charging system average performance. (Courtesy Delco Remy Division, General Motors Corp.)

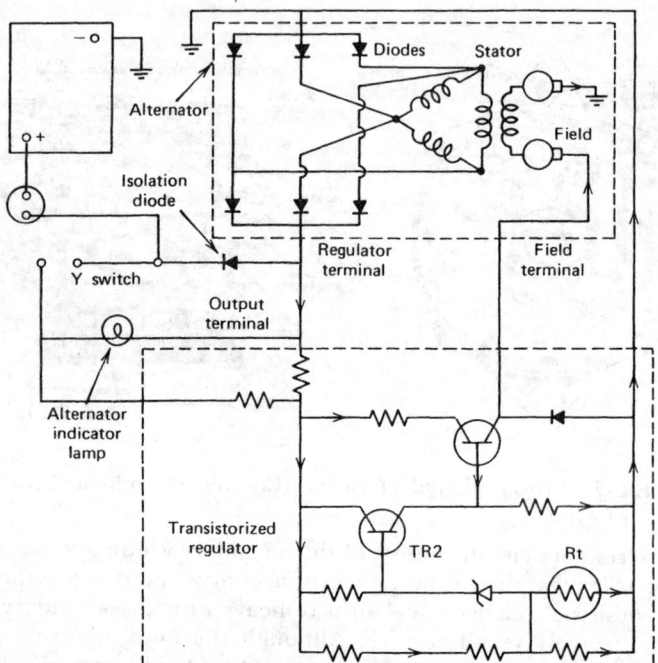

FIGURE 6-9 Type B circuit field for alternator. (Courtesy Deere & Company.)

Several factors affect the available voltage, as well as the voltage required to jump the spark-plug gap (fig. 6-11). For a given system, the voltage will reach a maximum as the speed of the engine is increased, and will then decrease as the speed continues to increase. The decrease in the secondary voltage results from decreasing time available for the magnetic field to develop as the engine speed is increased. It can also be seen from figure 6-11 that the required voltage to jump the spark-plug gap increases as both the gap and the compression ratio increases.

Increasing the compression ratio raises the mean effective pressure in the cylinder as well as the resistance across the spark-plug gap. Figure 6-11 clarifies this point further. It is obvious that increasing the spark-plug gap and the imep (indicated mean effective pressure) places more stress upon the ignition system because the required voltage must be increased. Increasing the imep either by supercharging, as in figure 6-12, or by increasing the compression ratio, as in figure 6-11, will increase the voltage required to jump a given spark-plug gap.

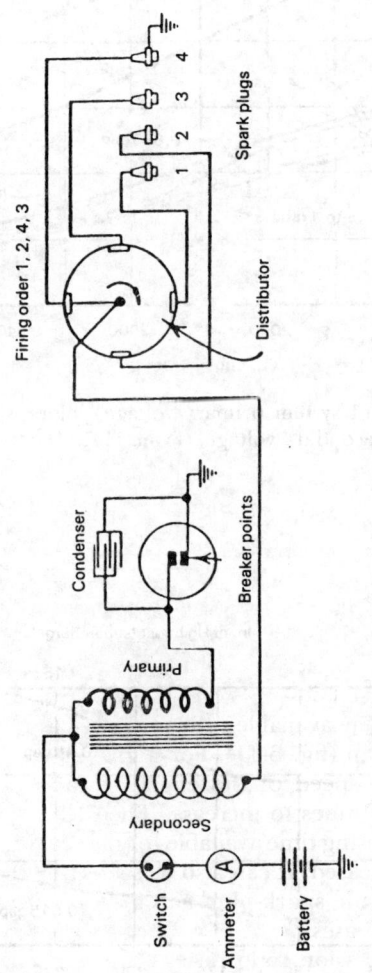

FIGURE 6-10 Schematic diagram of battery ignition system.

143

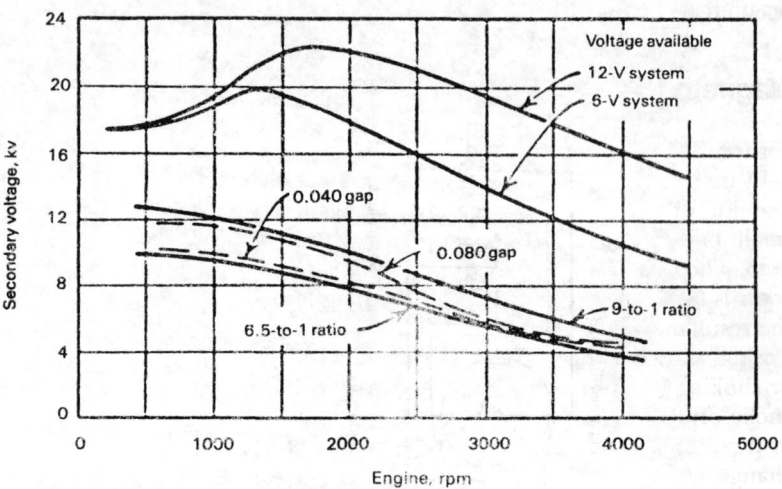

FIGURE 6-11 Effect of system primary voltage, engine speed, compression ratio, and spark gap on secondary voltage. (From H. L. Hartzell, *SAE Trans.*, vol. 53, p. 427.)

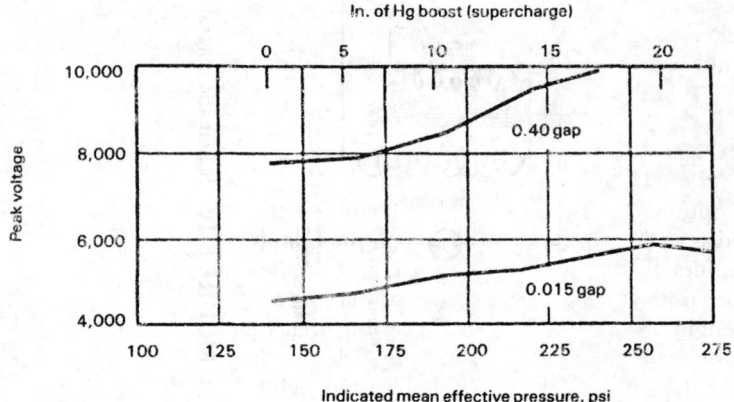

FIGURE 6-12 Effect of imep and spark gap on peak voltage required. (From T. Tognola and A. W. DeChard, *Automotive Inds.*, vol. 78, Jan. 1, 1938, p. 87.)

Magneto Ignition System

Figure 6-13 is a schematic diagram of a magneto with breaker points. Figure 6-14 shows graphically the operation of a magneto. The interval between the opening of the breaker points and the occurrence of the spark is extremely small, but oscillographic records show plainly the performance of the magneto. The light dotted line in figure 6-14 shows the static flux curve when the rotor is turned without any coil in the magneto. The heavy dotted line shows the resultant flux curve when the coil is in place with the points opening and closing, as the rotor is turned. The resultant flux arises because of retardation, or choking by the current in the primary circuit. When the points open, the choke effect of the coil is removed and there is an extremely rapid change in flux. The steepest part of the resultant flux curve indicates the greatest change in the magnetic circuit, and the maximum primary current (*heavy dashed line*) occurs as a result. The breaker points are then opened, whereupon the primary current falls to zero and the high-voltage current (*light dashed line*) is induced in the secondary. At the time of maximum secondary voltage, the rotor will be approximately as shown in figure 6.13(*b*). In figure 6-14 this point is shown as 9° past the neutral or 0° position and occurs each 180° for the two-pole magneto shown.

A magneto in general use on small farm engines, as well as on outboard motor boat engines, is the flywheel type (fig. 6-15). The rotor of the magneto serves as the flywheel of the engine and also as a fan in air-cooled engines. Breakerless ignition systems exist for both magneto ignition systems (fig. 6-15) and battery-powered ignition systems (not shown).

Ignition Timing

The optimum spark timing depends on the throttle position and the engine speed as well as on the plug location in the cylinder, compression ratio, mixture ratio, fuel distribution, valve timing, and octane number of the fuel used. The proper timing can be determined by making a full-load dynamometer test on the engine in question, and the setting that operates without objectionable ping or detonation can be determined for each speed.

Leaner mixes burn more slowly than rich mixes and require more spark advance. For example, in one test at wide-open throttle with a very lean fuel-to-air ratio of 0.052, the spark was advanced to 80° before head dead center to obtain maximum power (Youngren 1941). The knocking was about the same for a 30° advance and fuel-to-air ratio of 0.071. Actually, there is some variation in the ideal spark advance for different cylinders in engines.

The maximum power for any given engine fuel and engine speed occurs

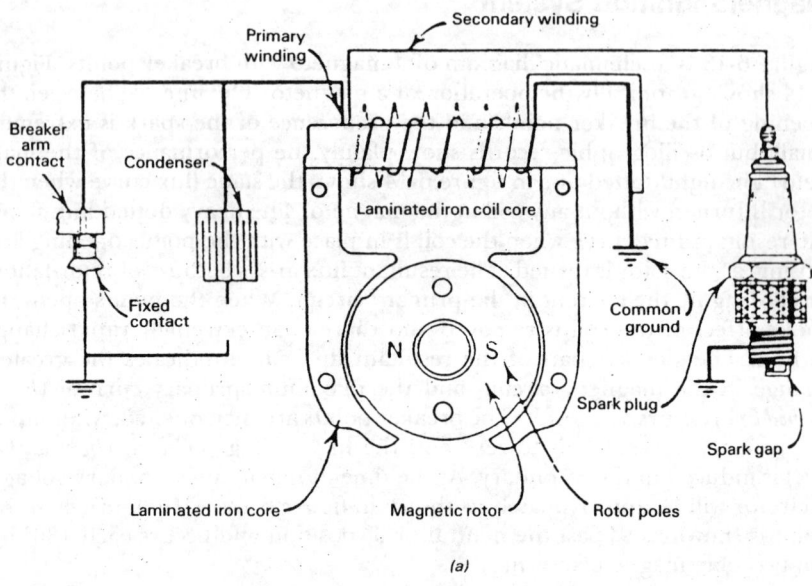

(a)

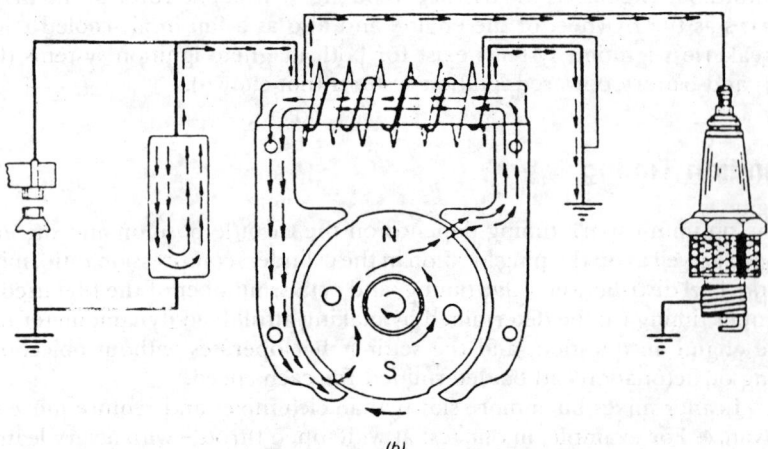

(b)

FIGURE 6-13 (*a*) Basic parts of magneto. (*b*) Magneto rotor at position of maximum magnetic flux change. (Courtesy WICO, Inc.)

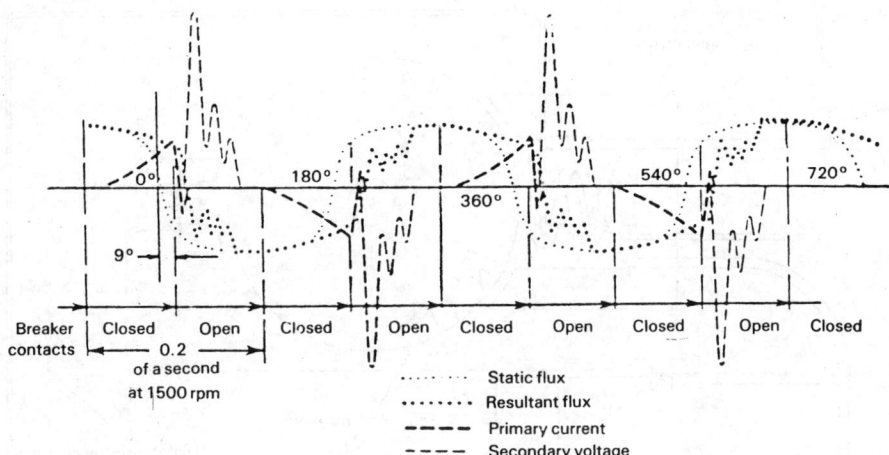

FIGURE 6-14   Oscillographic curves of operation of a magneto. (Courtesy WICO, Inc.)

at some specific ignition timing. It must be remembered that the combustion of the fuel is not instantaneous but requires a finite period, which is usually expressed in degrees of crankshaft rotation.

Curve A in figure 6-16 shows the relationship between engine speed and spark advance for maximum power in a typical engine. If the spark is retarded or advanced with respect to curve A, the engine will produce less power. Curve C shows a typical automatic spark advance for an engine. Curve B shows the spark advance, which should not be exceeded without producing audible knock for a given fuel. Thus it is evident that spark advance for any engine will depend on the type of service for which the engine is being used as well as on the fuel and the engine itself. Proper adjustment is made by rotating the distributor on the engine, the curve C moving up or down to approach curve B. Figure 6-17 shows the importance of correct spark timing on fuel consumption. As normally operated, this engine would run with an advance of about 30°. At this setting the fuel consumption was 16 percent less than at a 5° advance. When the spark was advanced past the normal operating position, a still lower fuel consumption resulted, but the engine operated roughly.

Spark advance with a battery ignition system can be obtained by means of a centrifugal weight arrangement. In automobiles that operate more often at lighter loads, a vacuum device may supply additional advance. In future spark ignition engines, the spark advance will likely be controlled by a microprocessor.

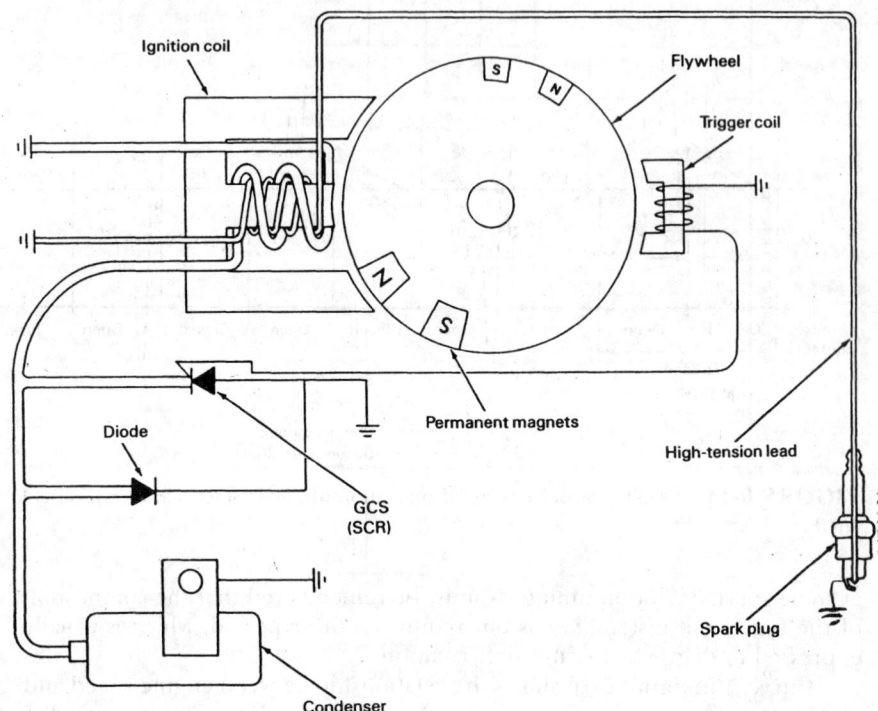

FIGURE 6-15 Breakerless ignition system components. (Courtesy Deere & Company.)

Spark Plugs

Spark plugs (fig. 6-18) must operate at the correct temperature. Operation at too low a temperature results in fouling by carbon; operation at too high a temperature results in preignition of the mixture and burning of the plugs. Temperatures of about 870°C to 925°C will likely cause preignition. Below 870°C, a range of about 360°C exists through which the carbon will be burned off the plugs and yet not preignite the mixture.

The temperature of the plug is, among other things, a function of the load under which it operates. Spark-plug gap affects the performance of an engine, particularly at light loads. Under conditions of part-throttle operation, when the mixtures may be lean and stratified, the wide gap provides a margin of safety. Oscillograph records show that the discharge across a wide gap lasts longer than the discharge across a narrow gap.

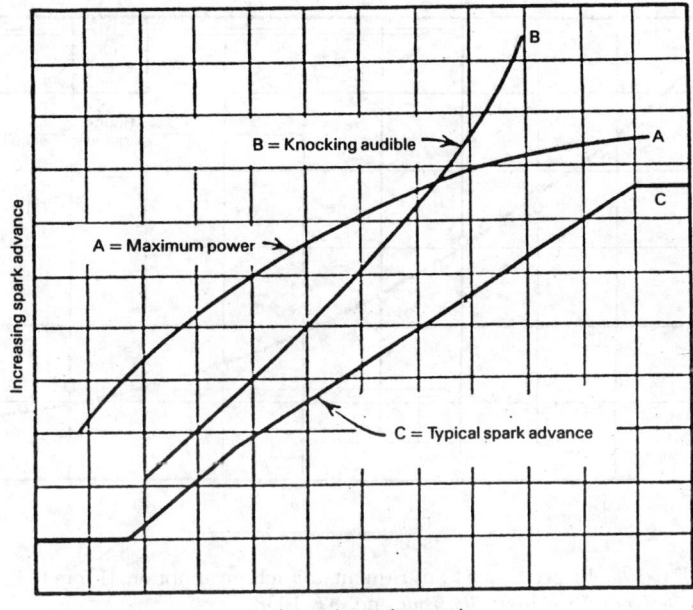

FIGURE 6-16 The relationship between engine speed and spark advance for maximum power. (From The Texas Co., *Lubrication*, Sept. 1942.)

Sensors

Sensors are devices that provide information to the driver or to automatic controls. If properly designed and tested for the environment in which they are to operate, sensors provide useful and reliable information that would otherwise be unavailable to the operator. If the tractor operator is alert and properly trained, the senses of sight, hearing, and touch provide much, but not all, of the information that the operator needs to control the tractor.

A. S. Farber (1982) of Deere & Co. points out that with larger, complex tractors the microprocessor can make many of the decisions and control many parts of the tractor for the operator. Of course, the tractor must have the necessary sensors for the microprocessor. Farber (1982) also points out that with microprocessors the vehicle (tractor) productivity can be increased. This can be done by selecting or automatically controlling the tractor transmission for the best combination of load and speed, thereby optimizing the tractor output or power.

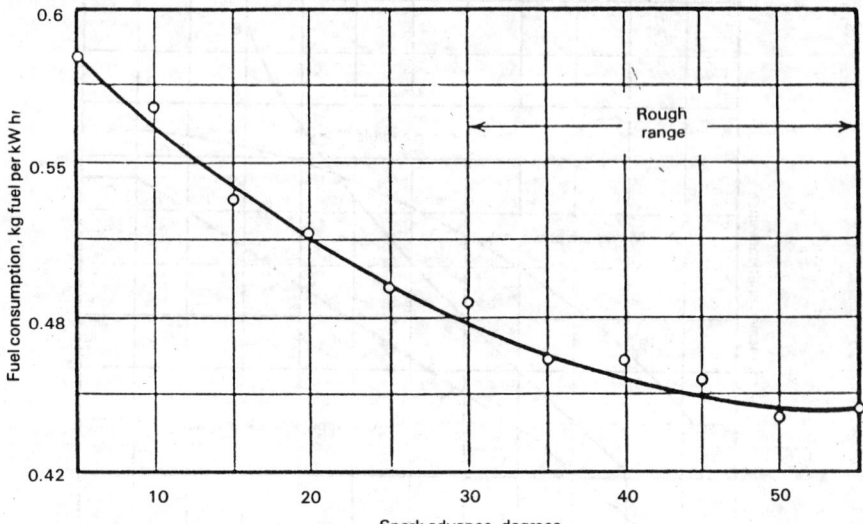

FIGURE 6-17   Effect of spark adjustment on fuel consumption. (From E. L. Barger, *Kansas State Univ. Engr. Exp. Sta. Bull.*, no. 37, 1938.)

     Farber (1982) also points out that the operator is not always a predictable control element because he or she has a variety of skills; may not always be alert; may be diverted by other functions; may be hampered by vision obstructions; and may be largely isolated from sound signals because of the tractor cab. Farber notes that variables that must be considered in making choices are difficult to process and combine except by electronics. Basic sensors are:

1. Pressure
2. Temperature
3. Torque (fig. 6-19)
4. Force
5. Flow
6. Position
7. Velocity
8. Voltage

Combinations that give additional information can supplement this list. For example, tire slip can be determined by making two velocity measurements (vehicle velocity and tire surface velocity), or $S = 1 - V_v/V_t$.

     In the following list, G. R. Mueller (1985) provides some of the operational

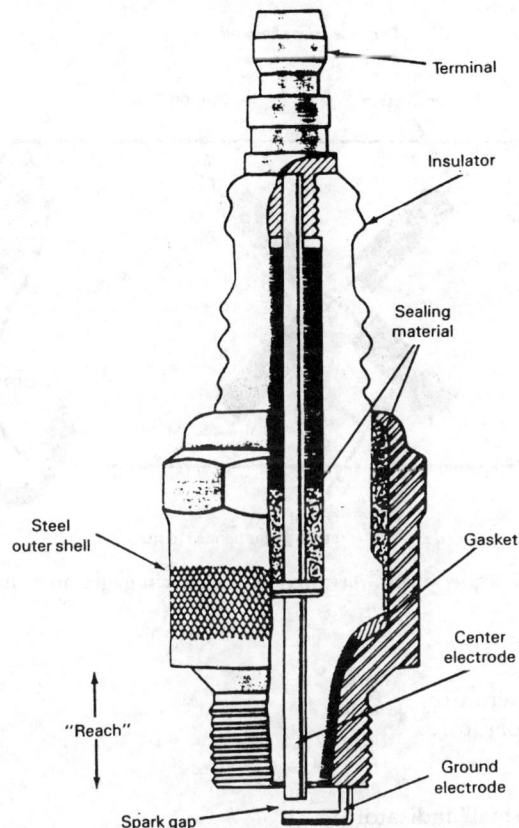

FIGURE 6-18 Spark-plug construction.

information that can be obtained from sensors. The sensor plus the read-out device is often called a monitor.

1. Low engine oil pressure
2. Air filter restriction
3. Coolant level
4. Alternator not charging
5. Park brake engaged
6. Transmission oil temperature
7. Transmission oil pressure
8. Transmission oil filter restriction

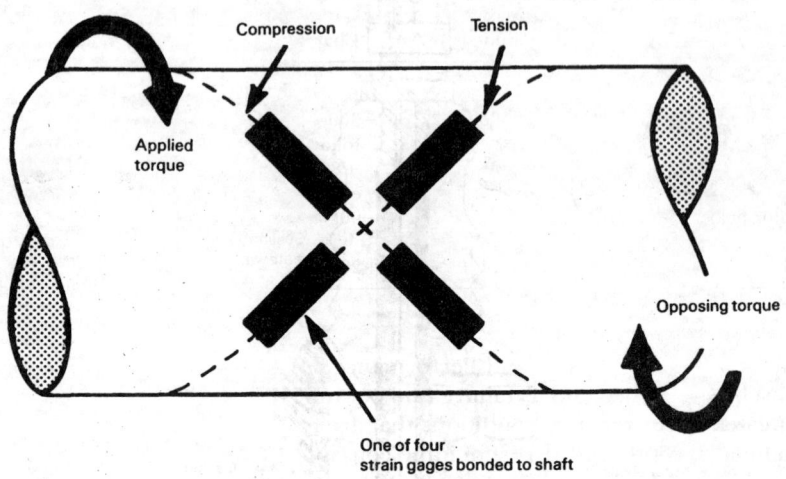

FIGURE 6-19 Torque sensor. Arrangement of strain gages on a shaft to measure torque.

9. Coolant temperature
10. Exhaust temperature
11. Fuel level
12. Voltage level
13. "Systems Normal" indicator

One common sensor for measuring torque is shown in figure 6-19. This sensor uses electrical resistance strain gages to measure the torque or moment on any shaft that needs to be monitored. For example, automatic control of a three-point hitch must measure the force on the lower links. This can be done by attaching the lower links of the three-point hitch to offsets on two shafts that are fixed on one end.

Environmental Problems

Electrical and electronic components on a tractor must be designed and tested to withstand a variety of environmental conditions. The reliability of each component must be determined through methods that are briefly discussed

TABLE 6-1   Vehicle Transient Voltage Characteristics

Type	Max Amplitude (V)	Characteristic	Remarks
Load dump	185	$106\varepsilon^{-t/.188} + 14$	Damage potential
Inductive load switching	-600	$-300\varepsilon^{-t/.001} + 14$ followed by $+80$ volt excursion	Logic errors
Alternator field decay	-90	$-90\varepsilon^{-t/.038}$	Occurs at shutdown only
Mutual	214	$+200\varepsilon^{-t/.001} + 14$	Logic errors

in chapter 14 "Tractor Tests and Performance." Each component's expected life must be established and its failure rate per unit of time measured in actual or simulated environmental conditions that are realistic. Following is a list of environmental conditions that should be considered.

1. Temperature
2. Humidity
3. Vibration and shock
4. Electrical transients (high-voltage surges)
5. Dirt
6. Corrosive fluids and salt
7. Immersion in water
8. Steam cleaning
9. Radio frequency and electromagnetic interference

Vibration is the most important environmental condition that affects the life of electrical components. In most cases, the sensors cannot be isolated from the vibration and therefore must be designed and tested to withstand the imposed vibrations. Components that are mounted on engines may experience vibrations with acceleration of 6 g's and a frequency that varies between 10 and 1000 Hz, according to Jones (1986).

Jones (1986) also points out that the electrical and electromagnetic environment in which the vehicle electronics must function is just as significant as the thermal and mechanical considerations. The voltage of the power source can vary considerably. Table 6-1 lists the expected voltage variation due to normal usage of a truck with a 12-V battery and starter.

Electronic systems on a farm tractor may also suffer serious failure modes if the effect of *electrical transients* is not adequately considered. Electrical tran-

TABLE 6.2   Truck/Tractor (12-V Starter)
Voltage Regulation Characteristics

Condition	Voltage
Normal operating vehicle	16 V maximum
	14.2 V nominal
	9 V minimum
Cold cranking at (−40°C)	3.5 to 6.0 V
Jump starts	+24 V
Reverse polarity	−12 V
Voltage regulator failure	9 to 18 V
Battery electrolyte boil-off	75 to 130 V

sients are high-voltage signals or surges of very short duration that occur on the supply circuit when high amperage loads are suddenly disconnected or when the battery is suddenly disconnected while the engine is running. Table 6-2 (from SAE J1455, draft) shows the danger to electronic equipment if it is not properly protected.

This chapter does not attempt to show how to design electrical circuits that will be protected from electrical transients. The design of such circuits should be left to electrical engineers. However, the engineer concerned with the design or servicing of tractors should be aware of the damage that voltage variations can do to inadequately protected electronic systems.

Electromagnetic interference or radio frequency interference (EMI/RFI) from power lines, radio, or radar can also disturb the performance of electronic equipment. Some techniques for keeping unwanted electromagnetic signals either in or out of electronic systems include a combination of shielding, grounding, and filtering.

REFERENCES

Jones, Travar O. "Commercial Vehicle Electronics." SAE Buckendale lecture (SP-647), 1986.
Youngren, H. T. "Engineering for Better Fuel Economy." *SAE Journal*, October 1941.

SUGGESTED READINGS

Batcheller, B. D. "Integrated Electronic Tractor Controls." *Convergence 84 Proceedings.* IEEE Catalog No. 84CHI988-5, Oct. 1984.
Beresa, Jonas. "Applications of Microcomputers in Automotive Electronics." *IEEE Transactions on Industrial Electronics*, vol. IE-30, no. 2, May 1983, p. 87.

Bischel, Brian J. "The 4994 Tractor Steering and Transmission Control System." *Convergence 84.* Proceeding of the International Congress on Transportation Electronics, Oct. 22–24, 1984.

Chancellor, William J., and Nelson E. Smith. Tractor Engine Torque Transducer Using Throttle Position and RPM. ASAE Paper 85-1557, 1985.

Deere & Co. "Electrical Systems." *Fundamentals of Service* 5th ed. John Deere Service Publications, Moline, IL, 1984.

Deere & Co. "Electrical Systems—Compact Equipment." *Fundamentals of Service*, Moline, IL, 1982.

Farber, A. S. "Electronic Transmission Controls for Off-Highway Applications." SAE Paper No. 820920, 1982.

Fleming, W. J. "Engine Sensors: State of the Art." SAE Paper 820904. *Convergence '82 Proceedings*, Oct. 1982.

Flick, J. F., and J. A. Salinger. "Vehicle Mounted Management Information Systems." SAE Paper 820908, Oct. 1982.

Holmes, R. C., et al "The Automation of Mechanical Transmissions." *Convergence '84 Proceedings*. IEEE Catalog No. 84CH1988-5, Oct. 1984.

Howes, P., et al. "The Electronic Governing of Diesel Engines for the Agricultural Industry." SAE Paper 860146, 1986.

Kainz, A., et al. "A New Concept for Electronic Diesel Engine Control." SAE Paper 860141, Feb. 1986.

Moncelle, M. E., and G. C. Fortune. "Caterpillar 3406 PEEC (Programmable Electronic Engine Control)." SAE paper 850173, Feb. 1985.

Mueller, G. R. "A Digital Instrumentation System for Agricultural Tractors." SAE Paper No. 85-1113, 1985.

Sokol, David G. "Radar II—A Microprocessor-Based Tire Ground Speed Sensor." Paper no. 85-1081. Presented at the 1985 Summer Meeting of ASAE, June 23–26, 1985.

Tooker, G. L. Semiconductor Technology—The Pervasiveness Continues." *Convergence '84 Proceedings*. IEEE Catalog No. 84CH1988-5, Oct. 1984.

Tschulena, G. R., and Selders, M. "Sensor Technology in the Microelectronic Age." Battelle Technical Inputs Report No. 40, 1984.

Tsuha, W. K., et al. "Radar True Ground Sensor for Agricultural and Off Road Equipment." SAE Paper 821059, Oct. 1982.

"Vehicle Condition Monitoring and Fault Diagnosis." *Institution of Mechanical Engineers Proceedings*, England, Mar. 1985.

7
ENGINE ACCESSORIES

When you can measure what you are speaking about, and express it in numbers, you know something about it.

LORD KELVIN, 1824–1907

An engine must have some accessories in order to function. The complexity of accessories on an engine is determined mostly by its intended use. Some of the common accessories are described in other chapters of this book. Devices to dynamically balance the engine are described in chapter 5, "Engine Design." Lubrication is used throughout the tractor and is covered in chapter 8. Electrical systems are also used throughout the tractor and are described separately in chapter 6. The accessories of importance to agricultural tractors are:

1. Governors or speed control devices
2. Engine cooling systems
3. Mufflers
4. Spark arresters
5. Air cleaners
6. Superchargers (chapter 5)
7. Fuel systems (chapter 5)
8. Electrical systems (chapter 6)
9. Lubrication (chapter 8)

Speed Control

The engine control device is usually called a governor. The governor can be a mechanical or an electronic device. In general, governors for internal combustion engines are of the centrifugal-force, spring-loaded type.

156

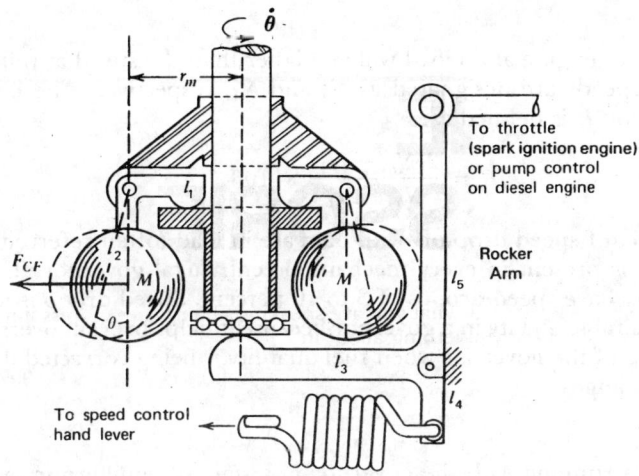

FIGURE 7-1   Diagram of centrifugal governor.

Principles of Centrifugal Governor Action

Although mechanical governors come in many shapes, they still operate on the same principle as the original governor designed by Watt for steam engines. The regulation of a centrifugal governor results from a change in centrifugal force when the speed of rotation changes. For equilibrium to be attained, the masses must assume a new position in response to any change in speed, and this movement controls the device that varies the flow of fuel to the engine, thus restoring its speed to the normal value.

Figure 7-1 is a schematic drawing of a simple spring-loaded, centrifugal-force governor. The centrifugal force F is exerted by a mass M rotating about the center line at a radius r with an angular velocity of $\dot{\theta}$ rad/s.

$$F = Mr\dot{\theta}^2 \tag{1}$$

The performance of a governor depends on the design, and certain definitions relating to design will now be discussed.

Stability

A governor is said to be stable when it occupies a definite position of equilibrium and does not oscillate for each speed within its working limits.

Regulation

The speed of an engine at no load will be higher than the speed at full-load. If these two speeds are designated as N_h and N_1, respectively, the percent speed regulation R is defined as:

$$R = \frac{2(N_h - N_1)}{N_h + N_1} \times 100 \qquad (2)$$

This effect of speed drop-off with increase in load (often referred to as speed droop) is present in every mechanical centrifugal governor. Tractor governors may have speed droops of 3 to 10 percent. Speed droop is not an entirely undesirable quality in a governor because it helps prevent overshooting or hunting of the governor when fuel quantity is being corrected during engine-load changes.

Sensitiveness

If an engine is running with the governor in a state of equilibrium, engine speed must change a certain rpm before the governor will act, because of friction and lost motion in the parts to be moved. Sensitiveness is defined as a percentage of the mean speed and may be calculated by equation 2. Excessive friction or lost motion will cause the governor to move large distances after it finally starts. It will then overshoot and cause large variations in the speed, which is known as hunting. Air-fuel mixture conditions also have an influence on the tendency of a governor to "hunt." If the throttle valve is permitted to close completely, the engine will start to die and the action of the governor tends to open the throttle, causing a hunting condition.

Governor Strength

In a simple mechanical governor, the flyballs (masses) must do all the work. The flyballs must not only move to bring about a speed change but also supply the force to move the connecting links to the fuel-metering shaft in the engine. This requirement emphasizes the necessity of minimizing friction in the governor and attached mechanisms.

Close regulation with a single spring and a given set of weights is possible over only a short range of speeds because the forces of the governor spring and the governor masses do not increase at the same rate. The force versus deflection or linear movement curve for a spring is nearly a straight line, whereas the centrifugal force versus speed curve varies as θ^2 as shown in equation 1. Governor equilibrium is reached when the engine attains the speed set by the throttle lever, and at this point the spring tension is balanced by the centrifugal force of the rotating weights. This point is shown on figure 7-2, and the governor will try to hold the engine at this speed.

Full-range governing from engine low-idle to high-idle speed is required

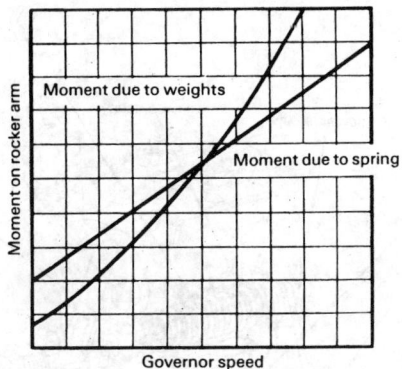

FIGURE 7-2 Governing occurs at intersection of spring and weight curves.

of tractor engines. The speed range is controlled by a speed-change lever that allows the operator to control the tension in the governor spring. Variations in design can produce weight curves such that the "weights-open" and "weights-closed" curves tend to parallel each other. Force curves that are not very divergent have relatively long sections that can be covered by the same spring rate.

The available work force at any given rpm—that is, the force available for moving governor and fuel metering mechanisms is the difference between the force generated by the weights revolving in the open position and the force generated in the closed position, *less* the force required to take up the deflection of the spring, plus friction. A set of weights revolving in the closed position (held closed by spring force) might exert a force of 210 N, whereas the same set at the same rpm would develop 334 N of force in the open position because of the increased radius of rotation. The available change in force for moving fuel-system linkages would then be 124 N, less the spring deflection force and friction.

The assumption that the flyweights on governors have their masses concentrated at their centers of gravity is an approximation for spherical weights. However, few governors are made in this manner because they are designed for ease of manufacture and compactness of the assembly. For a more nearly correct solution, the flyweights should be assumed as broken down into simple shapes and the moment should be found for each separately. These moments may then be summed to obtain the total moment for any position.

A phantom view of a distributor-type diesel fuel pump is shown in figure 7-3. The governor for the engine is an integral part of the distributor pump.

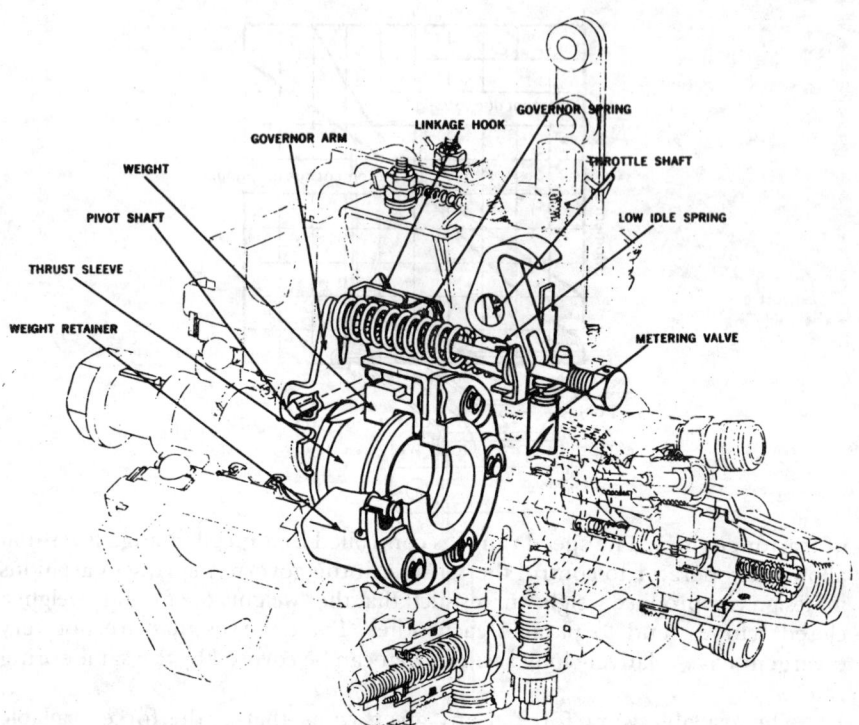

FIGURE 7-3   Mechanical governor built into a distributor-type diesel injection pump. (Courtesy Roosa Master.)

Spark Arresters

Fires may result from contact between dry vegetation and hot exhaust pipes or from the emission of hot carbon particles in the exhaust stream. Carbon particles have been found to ignite at a minimum temperature of about 480° to 540°C. Because temperatures within the cylinder may be from 1650° to 2200°C, loose carbon particles will be burning as they leave the exhaust system. Field tests indicate that fires may be started consistently in dry vegetation by glowing carbon particles.

Spark arresters are generally of two types, a centrifugal separator and a screen type. The chief criticisms of the screen type have been insufficient retention of smaller particles and clogging, which results in increased back pressure.

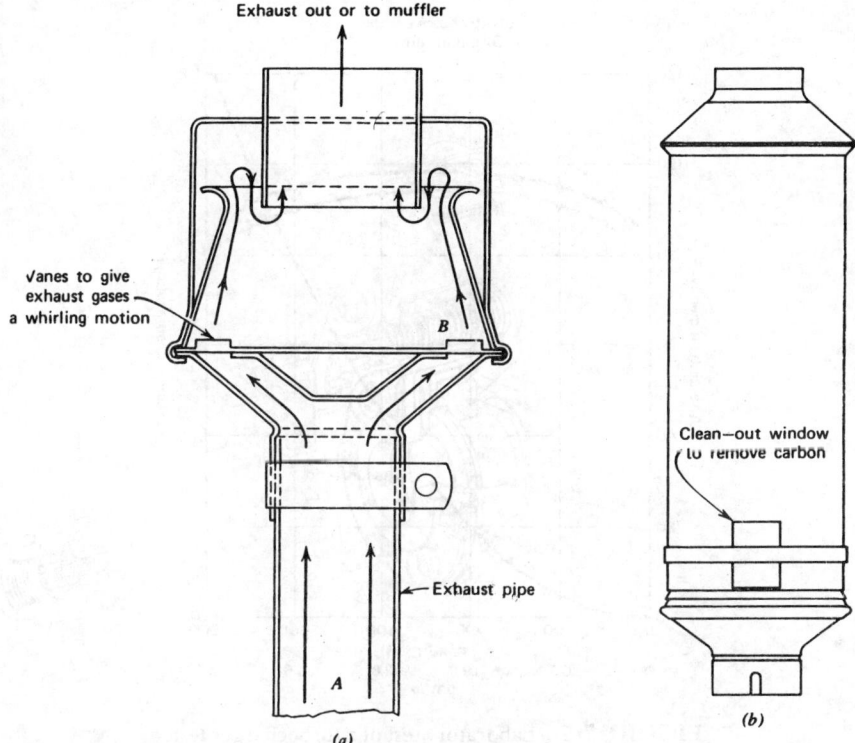

FIGURE 7-4   Centrifugal separator-type spark arrester. (a) Spark arrester only. (b) Combined with muffler. (Courtesy Hasco Manufacturing Co.)

Figure 7-4 shows a cross-sectional view of a typical centrifugal-type spark arrester. Exhaust gases entering at A are channeled to the outer walls by the baffles at B. Centrifugal force causes the carbon particles to go to the outside wall, where they are collected. The exhaust gases turn and go out the top of the exhaust pipe.

The U.S. Forest Service requires all vehicles to use approved spark arresters in specified forests. The Forest Service specifies the method of testing and the requirements that each spark arrester must meet.

The Society of Automotive Engineers* has issued guidelines in the form

*There are three practices (SAE J3356, J342 [Nov. 1980], and J350 [Jan. 1980]) to cover the range from small to large engines. A related standard (SAE J997 [Jan. 1980]) specifies the carbon to be used with the three SAE practices.

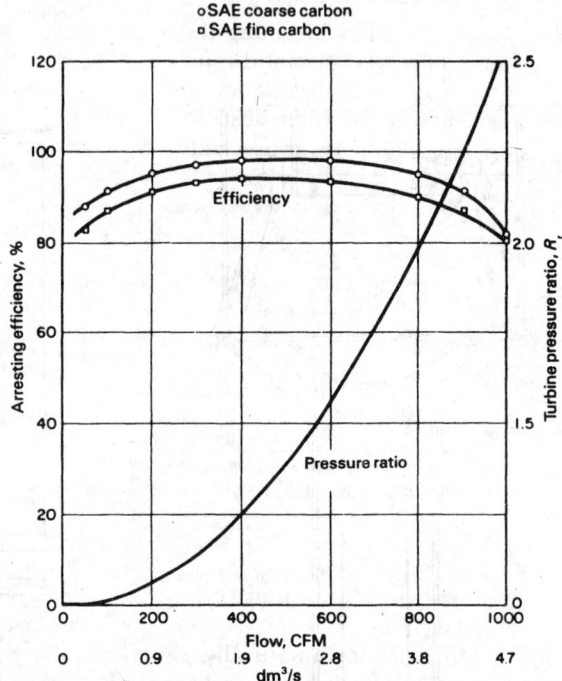

FIGURE 7-5 Laboratory test of a turbocharger tested
as a spark arrester. (From SAE J350 [Jan. 1980], "Spark
Arrester Test Procedures for Medium Size Engines.")

of "Recommended Practices" that provide a standard method of testing spark
arresters.

Turbochargers are generally very effective spark arresters. They break
up the larger carbon particles into fine carbon dust that burns up or cools
off much more quickly after leaving the exhaust pipe. Therefore, there are
no large red-hot carbon particles to land on combustible material. Figure
7-5 shows the results of a test of a turbocharger to determine its efficiency as
a spark arrester.

Mufflers

Exhaust mufflers or silencers are variations of low-pass acoustical filters
(Beranek 1971) consisting of a through tube to which closed cavities are

connected by small holes at intervals along the tube. A muffler of this type causes little increase in back pressure and much attenuation of sound waves with frequencies above a cut-off frequency determined by the size of the holes and cavities. Baffles or irregular obstructions producing devious flow paths produce muffling but also appreciably increase the exhaust back pressure.

A muffler and a spark arrester can be combined into one device.

Air Cleaners

Dust entering a tractor engine is often the principal cause of wear. The adverse effect of dust mixed with lubricating oil or grease is better realized when the mixture is likened to lapping compound.

The airflow requirement that must pass through a four-stroke-cycle engine is

$$Q(\text{m}^3/\text{s}) = \frac{\text{engine displacement (liters)} \times \text{rpm} \times \text{volumetric efficiency}}{1.2 \times 10^5}$$

This applies to engines of four or more cylinders because of the relatively uniform rate of intake airflow.

The volumetric efficiency for a naturally aspirated gasoline engine may be assumed to be about 75 to 80 percent, and for a naturally aspirated diesel engine, it is about 85 to 90 percent. In the case of turbocharged engines, the airflow should be as specified by the engine manufacturer.

The airflow requirement for two-cylinder engines is calculated as for four-cylinder engines, but because of the pulsating action that causes higher maximum flows, the computed flow should be multiplied by 2. The calculated value is also multiplied by 2 for two-cycle engines and by 4 for one-cylinder engines.

The ideal air cleaner should possess the following characteristics: high efficiency in dust removal from the air, small air restriction, small size, infrequent need for servicing, simplicity in design, ability to muffle air intake noises, durability, low cost, and ability to act as a backfire suppressor.

Air Inlet Location

The location of the air inlet affects the quantity of dust to be removed by the air cleaner. For example, field tests on a farm tractor showed a variation of 548 percent in dust concentration (Larson 1952) around the tractor in the field.

The dust concentration was highest near the engine and lowest in the region directly above the engine. For this reason, tractor designers usually select an air inlet above or near the top of the engine housing.

Precleaners

Under some conditions, a preliminary cleaning device is advisable to protect the air cleaner and to reduce the load on it. Under severe dust conditions and without any preliminary cleaning process, parts of the cleaner may become full of dust in a very short time. The precleaner will not increase the efficiency of the air cleaner, but it will prolong the required service interval by removing a large part of the dirt before it reaches the cleaner.

The dry-type air cleaner shown in figure 7-6 has a precleaner section. The air that enters the cleaner at (a) strikes the air vanes at (b), which causes the air and dirt to rotate clockwise. The rotating air mass is partly cleaned of the larger dirt particles, which are moved by centrifugal force to the outside of the steel container. Many of the larger particles are then carried into the end section through the exit at (c), where they fall to the bottom. The rubber valve (d) will open when the engine idles, expelling the larger dirt particles.

The air containing the finer particles of dirt is drawn through the dry-type air cleaner (e), and from there the cleaned air enters the intake manifold.

Some precleaners are equipped with an aspirator connected to a venturi in the exhaust pipe that continuously sucks the dirt out of the precleaner (fig. 7-7).

Dry-Type Air Cleaner

The trend toward higher-output engines and increased performance has also increased air-cleaning requirements. Research shows that the efficiency of dry-type air filters exceeds that of the oil-bath type (Siemens and Weber 1958). Because a tractor operates most of the time at less than full load, the relative cleaning efficiency at one-half load is significant. At less than rated airflow the efficiency of the oil-bath cleaner tends to decrease, whereas that of the dry-type filter remains high.

Although the maximum safe airflow rate through an oil-bath filter is determined by the point at which oil will be pulled out of the cleaner and into the airstream, the only limiting factor in a dry-type cleaner appears to be the allowable air restriction, which in turn is governed by the filter size and its dust-holding capacity. A pressure drop indicator between the outlet of the cleaner and the inlet is added to signal needed service.

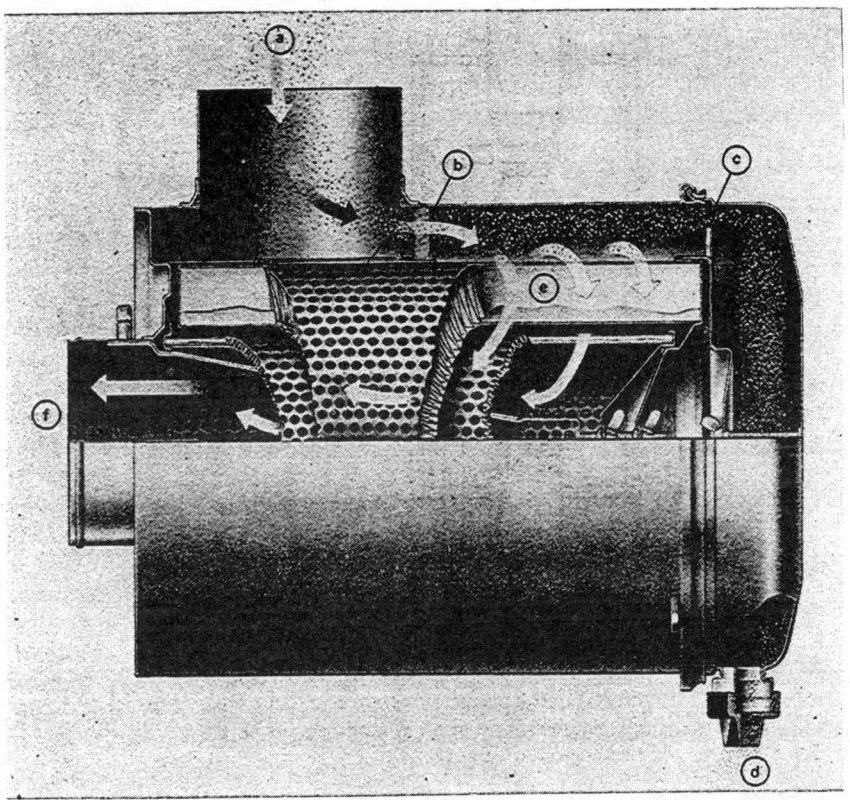

FIGURE 7-6   Dry-type air cleaner commonly used on small to medium tractors in North America. (Courtesy Donaldson Co.)

Dust Composition

The abrasiveness of dust, as measured by the rate of engine wear produced, depends on its origin. Also, dust fineness varies greatly among different locations around a tractor. To provide comparative test conditions, the SAE air-cleaner test code specifies two standardized grades of test dust: "fine" and "coarse."

Sizes of the standardized dusts are shown in table 7-1, along with dust sizes collected at a 1.5-m height during a Kansas dust storm. Although direct size increment comparisons are not possible, it is interesting to note the similarity between the actual field dust and the test dusts.

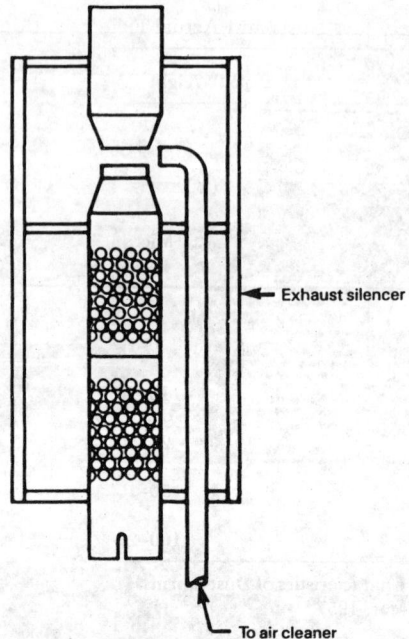

FIGURE 7-7   Aspirator used to remove dirt from the precleaner section of a dry-type air cleaner. Pipe to air cleaner would be attached to *d* on figure 7-6.

Air-Cleaner Tests

The primary objectives of an air-cleaner test are to determine (1) the efficiency in removing dust from the airstream and (2) the restriction to airflow. The Society of Automotive Engineers has adopted a standard procedure for the laboratory testing of air cleaners. The arrangement for the test apparatus is shown in figure 7-8.

Engine Cooling Systems

The *function* of the cooling system of an internal combustion engine is to maintain an optimum engine operating temperature. For this temperature

TABLE 7-1 Standardized Test Dusts and Actual Field Dusts

	Percent by Weight				
	SAE J726b			Kansas Dust Storm[a] 1.5-m Height	
Micron Range	Fine Grade	Coarse Grade	Micron Range	Sandy Soils	Silt Loam Soils
0–5	39 ± 2	12 ± 2			
			0–10	26.0	29.3
5–10	18 ± 3	12 ± 3			
10–20	16 ± 3	14 ± 3	10–20	5.0	6.1
20–40	18 ± 3	23 ± 3			
			20–50	29.7	43.7
40–80	9 ± 3	30 ± 3			
			50–100	37.3	19.5
80–200	—	9 ± 3			
			100–250	2.0	1.4

[a]W. S. Chepil, "Sedimentary Characteristics of Dust Storms: III Composition of Suspended Dust," *Amer. Jour. of Sci.*, vol. 255, Mar. 1957.

to be maintained, the rejected heat during the combustion process and the heat generated by engine friction and compression of gases must be removed.

The proper design and maintenance of a cooling system are extremely important because the amount of heat to be dissipated is great. The engine must be cooled to maintain proper lubrication, to prevent overheating of the engine parts, and to ensure proper combustion. On the other hand, the engine temperatures must be high enough to ensure vaporization of the fuel and prevent dilution of the oil. Cooling by air is used on some small tractors, but the majority are cooled by liquids.

Cooling Load

The heat balance in an internal combustion engine varies according to the efficiency, design, speed, load, and size. A typical heat-balance chart is shown in figure 7-9. At full load about 35 percent is converted into useful work, about 28 percent of the fuel energy is passed on to the cooling medium, about 26 percent is lost as exhaust heat, and the remainder is lost by radiation.

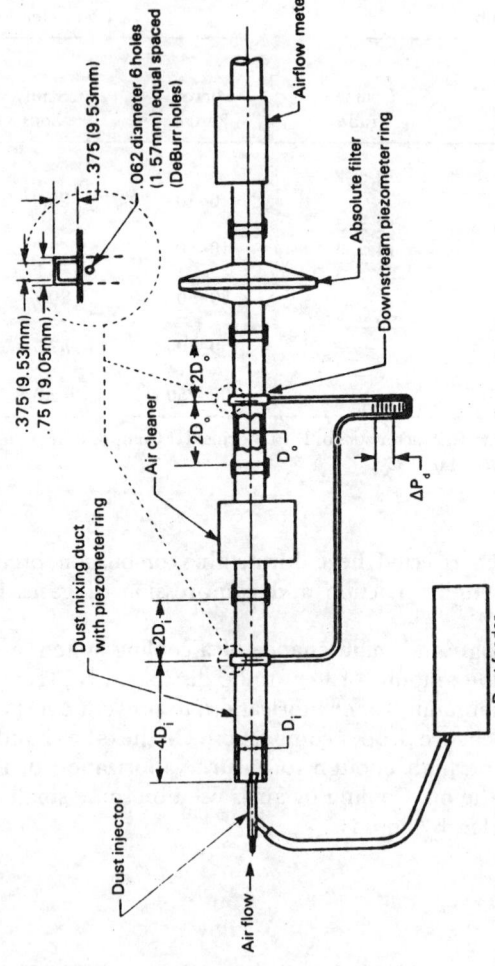

FIGURE 7-8 Equipment for testing air cleaners. (From SAE J726 [May 1981], "Air Cleaner Test Code.")

Airflow meter

Absolute filter

Downstream piezometer ring

.375 (9.53mm)

.062 diameter 6 holes
(1.57mm) equal spaced
(DeBurr holes)

.375 (9.53mm)

.75 (19.05mm)

Air cleaner

$2D_o$

$2D_o$

$2D_o$

D_o

ΔP_d

Dust mixing duct
with piezometer ring

$2D_i$

$4D_i$

D_i

Dust injector

Air flow

Dust feeder

168

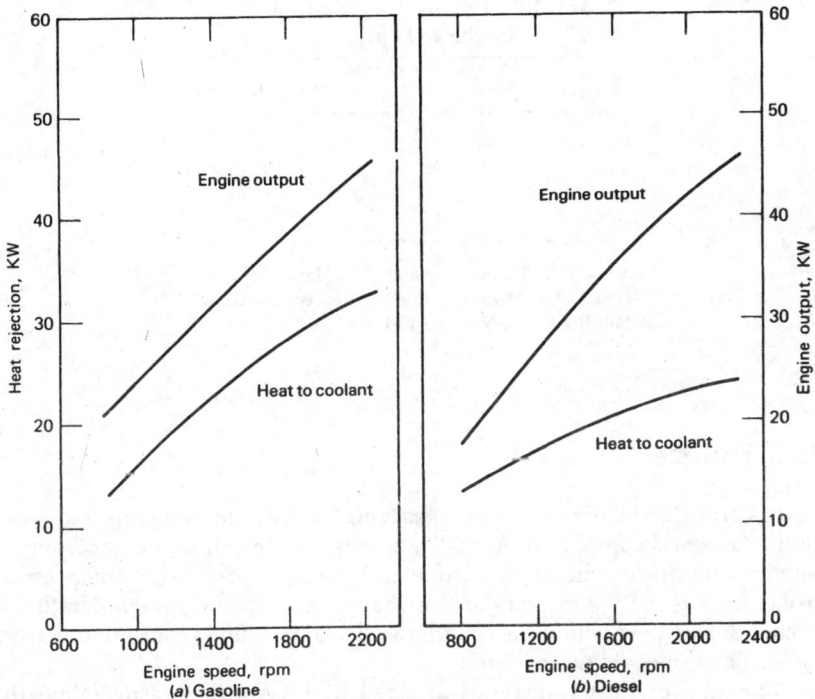

FIGURE 7-9 Heat rejected to coolant for (*a*) gasoline engine and (*b*) diesel engine with both having a 3.0-L displacement. (From J. T. Kulhavy, Paper 66A, Society of Automotive Engineers, Sept. 10–13, 1962.)

Although the heat loss from an engine seems excessively high, it should be remembered that the efficiency of the cycle cannot exceed that of the ideal air standard cycle.

The heat rejected to the cooling water for one specific basic engine is shown in figure 7-9. Note that at 2000 rpm and at full load the gasoline engine rejects approximately 0.75 kW for each kilowatt output. Since more of the energy from the fuel is transformed into useful work in a diesel engine than in a lower-compression gasoline engine, it is logical that less heat would have to be rejected to the coolant. Note that at 2000 rpm the diesel engine rejects approximately 0.58 kW for each kilowatt of output. The compression ratio of the diesel engine in figure 7-9 was 17.5:1, whereas the gasoline engine had a compression ratio of 7.3:1.

TABLE 7-2 Conductivity of Metals
kJ/(m² · C · m, thick · hour)

Material	Conductivity
Cast iron	1810
Steel	1740
Aluminum	8720
Brass	4090

SOURCE: T. Baumeister and L. S. Marks. *Standard Handbook for Mechanical Engineers*, 7th ed., Mc-Graw-Hill Book Co., New York, 1967.

Heat Transfer

Heat is transferred from the hot gases to the cylinder walls by radiation, conduction, and convection. Actually, heat is transferred to the gas from the cylinder wall during the intake stroke and the early part of the compression stroke. Because of the rapidity of this change, an equilibrium temperature is reached somewhere within the cylinder wall, and heat flows continuously from this point to the cooling medium.

The rate of heat transfer is affected by several factors including the material in the cylinder wall (table 7-2). Aluminum is used quite generally for heads and jackets in air-cooled engines, where high rates of heat transfer and light weight are required. Cast iron and steel are the predominant materials for engine cylinder walls because of wear resistance, stiffness, and strength rather than because of heat-transfer rates.

Air Cooling

The cylinder wall temperature of air-cooled engines is generally higher than that of water-cooled engines because of the low value of the heat-transfer coefficient between metal and air. The control of cylinder temperatures is more difficult in air-cooled engines. Considerable work has been done toward more efficient heat dissipation, but most of this has been in the aeronautical field. Special fans and baffles are used to direct the air around the engine so as to avoid hot spots. The hottest points in a cylinder head are the exhaust valve and the exhaust port, and special attention must be given to them whether it be by air or liquid cooling.

Air cooling has the advantage of eliminating water jackets, pumps, radiators, and water connections (fig. 7-10)

Radiator Design

Most systems are of the forced circulation type, which use an engine-driven water pump, as illustrated in figure 7-10. A thermostat is generally employed to maintain the desired temperature of the coolant.

Because a tractor usually operates at a high load factor, a relatively large cooling capacity must be supplied compared to that for an automobile. Heat-rejection rates may be calculated, but in the final form the values must be obtained from actual tests on given models of engines.

Other information necessary to the selection of an engine cooling system is the rate of water flow, the control temperature desired, and the temperature drop through the radiator. For tractor conditions, a flow of 0.16 l/(s·kW$_r$) with a temperature drop through the radiator of $5.5°$ to $8.5°$C may be assumed, where kW$_r$ is the heat flow through the radiator.

Airflow rates through a radiator core will usually be in the range of 5 to 10 m/s for an economical and efficient installation. The required heat-transfer surface can be determined from a typical radiator curve sheet (fig. 7-11). Curves such as these enable the design engineer to compare the performance of different types of surfaces against the same cooling fan.

Heat-rejection rates to the coolant have decreased as the efficiencies of the engines have increased. The quantity of water to be circulated depends on the initial and final temperatures of the water admitted to the engine; for the average engine cooled by a radiator, the temperatures are governed by the size of the radiator, the amount of water circulated, and the air temperature and velocity through the radiator. Tractor engines operating on gasoline are generally held at a water temperature of about 70°C; diesel engines are operated at a lower temperature.

To calculate the quantity of cooling water to be moved by the pump, one must make some assumption as to the desired temperature drop in the water as it moves from the top of the radiator to the bottom. This gives the temperature differential to be imparted to the water as it moves through the engine and should be assumed to be about $5.5°$ to $8.5°$C.

Example 7-1   A diesel tractor develops 100 kW, with a coolant temperature differential of 5.5°C. The quantity of water, Q, to be circulated is

$$Q(\text{L/s}) = \frac{100(\text{kW}) \cdot 0.58(\text{kW/kW}) \, 0.16l}{(\text{s}\cdot\text{kW}_r)} = 9.3 \text{ L/s}$$

Design practice varies according to the designer and the type of service, but

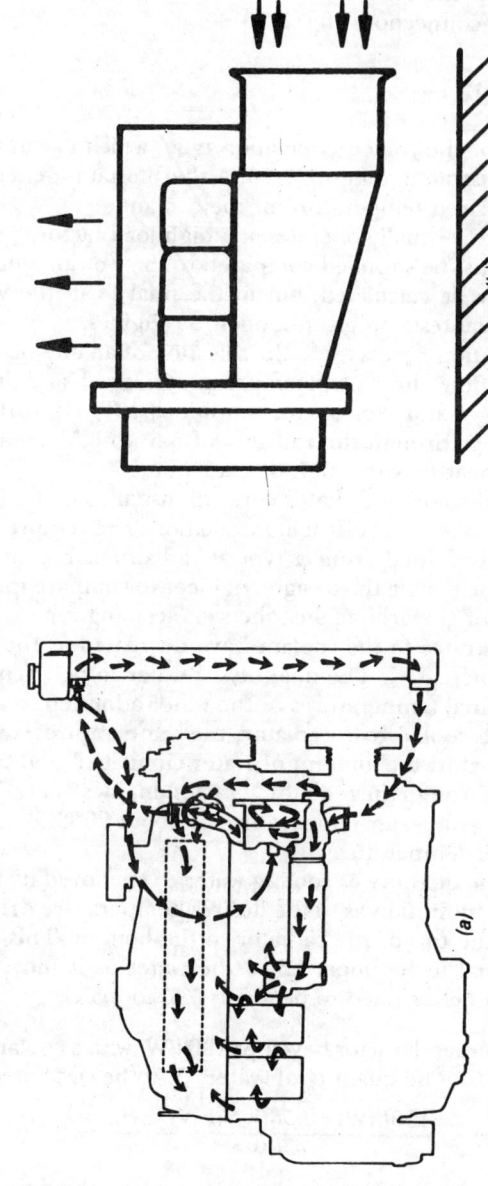

FIGURE 7-10 (a) Liquid-cooling media flow path. (b) Air-cooling media flow path. (From W. L. Sprick and T. H. Becker, "The Application and Installation of Diesel Engines in Agricultural Equipment," ASAE Distinguished Lecture Series No. 11, American Society of Agricultural Engineers, Dec. 17, 1985.)

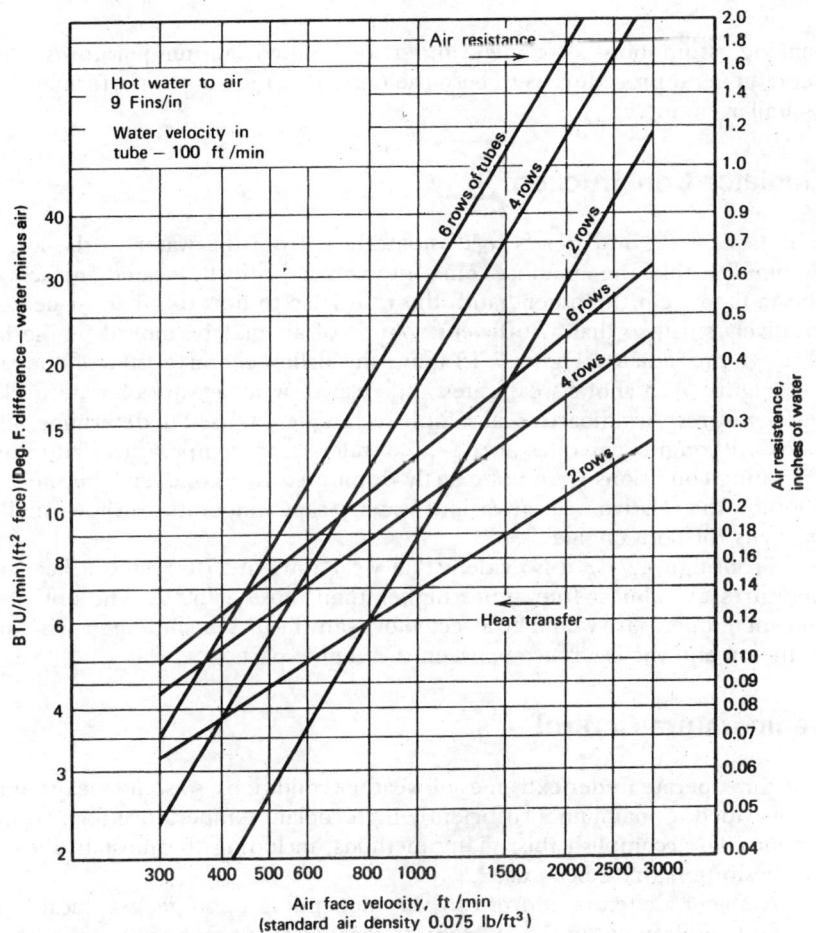

FIGURE 7-11 Performance of air-cooled water radiators. (Courtesy Young Radiator Co.)

pumps are usually selected to deliver 25 to 90 ml/s per kilowatt. For coolants with a lower specific heat than water, the flow would necessarily be increased.

The heat H to be dissipated to the air by the radiator for this example is

$$H = 0.58 \times 100 = 58 \text{ kW}$$

The heat to be dissipated from a given engine may be calculated by making assumptions as to the cylinder wall and coolant temperatures, but a more general procedure is to base the design on the known performance of a similar engine.

Radiator Construction

The task of the radiator is to dissipate heat from the water to the air; to accomplish this, the air must come into contact with the heated surfaces of the radiator. On the other hand, the resistance to flow of air must be kept relatively small so that a sufficient volume of air may be moved by the fan through the radiator. Figure 7-12 shows the airflow characteristics of one fan.

Figure 7-13 shows wear rates and relative wear at various water-jacket temperatures. A radioactive oil sample technique was used to determine wear rates. Although wear rates at the same water-jacket temperatures and same operating conditions were not exactly duplicated from one day's running to another, the relative wear at various water-jacket temperatures shows similar patterns for both engines.

From figure 7-13 it is evident that wear rates at 27°C water-jacket temperatures are almost four times higher than those at 60°C. The optimum coolant temperature would be affected by many factors, including the viscosity of the oil and the clearance between the engine parts.

Temperature Control

Tractors operate under extremes of weather conditions, so some means must be provided to maintain a sufficiently high coolant temperature for best operation. To accomplish this, many methods, including thermostats and radiator shutters, have been used.

Ambient air temperatures can vary depending upon geographical location and application (fig. 7-14). Many applications have ambient temperatures above 43°C. To allow for trash collection on the radiator air inlet, a temperature of 52° to 57°C above the ambient air should be used in the design process.

The temperature at which an engine operates affects both fuel economy and wear. Fuel consumption tests showed increasing specific fuel consumption as the jacket temperature was decreased from 93° to 38°C. A change in jacket temperature when the engine was operating at near maximum load affected the specific fuel consumption very little, whereas decreasing the temperature at light loads affected it considerably. The increase in specific fuel consumption with a decrease in engine temperature at low loads was attributed to lower manifold temperatures.

FIGURE 7-12 Tubular radiator construction. (Courtesy Young Radiator Co.)

Antifreeze Materials and Coolants

Water is most commonly used as a cooling medium because of its relatively high heat-transfer properties. The principal disadvantages are that it has a high freezing point and that it may cause corrosive action on the radiator and engine. Other coolants have been used primarily for antifreeze protection.

An antifreeze material should have the following characteristics:

1. Prevent freezing at the lowest temperature encountered
2. Not attack the materials of the cooling system
3. Be chemically stable under engine operating conditions

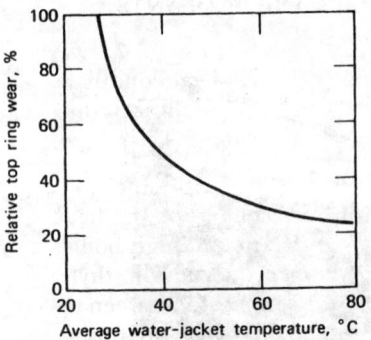

FIGURE 7-13 Effect of water-jacket temperature on the relative wear of the top ring. (From SAE Special Publication 194, June 1961.)

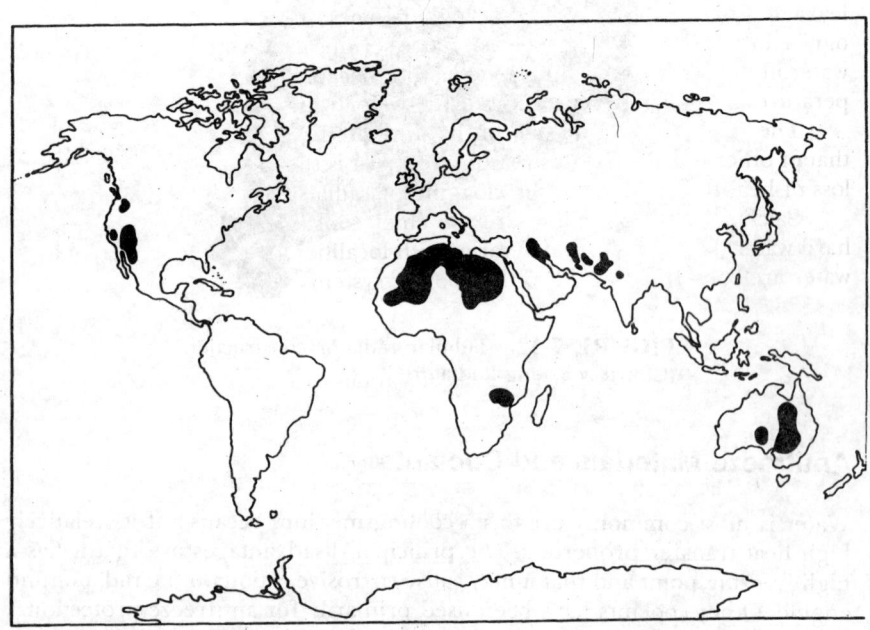

FIGURE 7-14 Geographical location of high ambient temperatures. (Courtesy Cummins Engine Co.)

4. Have a high specific heat and heat conductivity
5. Have a low coefficient of expansion to reduce overflow losses
6. Be nontoxic and nonflammable

The most successful antifreezes for automotive uses have been solutions of glycols (ethylene and propylene). For the heavy-duty operation that may be demanded of tractors, one of the high-boiling-point glycols is the most satisfactory. See table 7-3 for properties of ethylene glycol.

Figure 7-15 shows the relationship between the percent of ethylene glycol with water and the freezing point of the mixture of the antifreeze and coolant (AF & C). The eutectic point (or lowest freezing point) occurs at 68 percent.

Corrosion and Radiator Deposits

The most common water contaminants that corrode cooling systems are oxygen from the air, dissolved minerals in water, and acids from exhaust-gas leakage. Corrosion is increased at higher temperature, and it should be recognized that corrosion may be even more serious in warm weather (with only water in the radiator) than in cold weather because of higher operating temperatures and increased aeration of the water with greater turbulence.

The rate of corrosion of iron by untreated water is much higher than that of other metals. For this reason, iron rust is the principal problem in the loss of heat transfer and in the clogging of radiators and water jackets.

Calcium and magnesium are the chief constituents in scale formed by hard water. Hardness of water varies with localities, and if large quantities of water are constantly added to the cooling system, they cause the deposition

TABLE 7-3 Boiling Points of Various Concentrations of Ethylene Glycol

Ethylene Glycol Concentration by Volume	Boiling Point	
	Atmospheric Pressure	(103 kPa) System Pressure
44	107°C	128°C
50	108°C	129°C
60	111°C	132°C
70[a]	114°C	136°C

[a]Concentrations higher than 70 percent are not recommended; 68 percent provides maximum freezing protection, about −69°C.

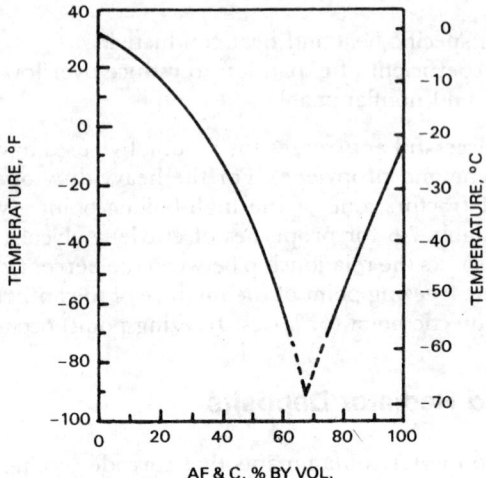

FIGURE 7-15 Freezing points of aqueous ethylene glycol AF&C solutions. A 68 percent solution gives maximum protection. (From *Lubrication*, Texaco, Inc., vol. 65, no. 3, 1979.)

of scale, which reduces heat transfer and leads to hot spots in the engine. The best way to avoid scale in the engine is by filling the radiator with soft water. If soft water is not available, periodic cleaning will be satisfactory. A good cooling-system cleaner combines oxalic acid with sodium bisulfate.

After cleaning, a corrosion inhibitor should be added. Corrosion inhibitors do not remove rust already formed in the system, and any necessary cleaning should be done before the inhibitor is installed. Corrosion inhibitors are available under a variety of trade names. They may be a sodium chromate or contain mercaptobenzothiazole (commonly called MBT) and sodium phosphite. The inhibitor prevents corrosion by forming a very thin film over all the interior surfaces of the cooling system.

Pressure Cooling

The practice of operating cooling systems under pressure has become common because of the reduction in evaporation losses, because of the increase in engine-operating temperatures possible without overflow loss from boiling, and because of the need to increase the radiator cooling capacity. The boiling point will decrease approximately 1.4°C for each 500 m above sea level.

Quantity of Air

The heat removed from the engine by the liquid must be given off to the air, and in the conventional system, this is accomplished by blowing air through the radiator by means of a fan. Expressed as an equation

$$H = A v M_a C_a \, \Delta t_a \tag{3}$$

where H = kilowatts
         A = frontal area of the radiator in m^2
         v = velocity of the air through the radiator in m/s
        M_a = mass of the air in kg/m^3
        C_a = specific heat of the air in kJ/kg per degree Celsius
        Δt_a = temperature rise of the air in degrees Celsius in passing
             through the radiator

The specific heat of the air is 1.0 kJ/kg°C; the weight of the air may be taken as 0.34 kg/m³.

The radiator selected must have a frontal area that will permit fan sizes and speeds proportional to the allowable power for driving the fan, which may be as much as 5 percent of the engine output. Regardless of the individual fan and radiator specifications, the combination selected must be such that the product of mass of air and its allowable temperature rise will be equal to the maximum required cooling load.

The design of the fan, radiator, engine compartment, chaff screen, and grill each accounts for a portion of the airflow restriction. Openings to both the upstream and downstream sides of the radiator should be at least 1.2 times the radiator core area. Fan power can be minimized by using a thin, large-frontal-area radiator with a large-diameter, low-speed fan. Fan power is a cubic power of the fan speed and directly proportional to its diameter.

Fans

The design and selection of a fan are important, since the air passing through the radiator is the primary means of removing the heat from the liquid. The volume of air required for any given installation is determined by test and may be calculated from equation 3.

The efficiency of a fan is the ratio of the power output to the power input. The air power is the power determined from the product of the volume of air, Q, (m³/s) and the pressure rise, p, (N/m²).

$$\text{Eff} = \frac{\text{air power}}{\text{input power}} \tag{4}$$
$$\text{air power [kW]} = pQ$$

where $Q = m^3/s$

p = total pressure, 10^3 Pa, above barometric pressure

The power requirement of a fan is generally assumed to vary as the cube of the speed. The discharge of the fan varies directly as the speed up to the point at which blade interference occurs; blade interference prevents the air from entering the spaces as rapidly as it is removed by the blades.

From these data it may be noted that the fan should be as large as possible and should run as slowly as possible to obtain the greatest efficiency.

Attempts to decrease the power requirement of fans have resulted in the development of thermostatically controlled fans. When the engine temperature reaches the desired setting, the pitch of the blades is decreased or the fan is disconnected, which results in a lowering of the power requirement at any given speed.

The fan can either pull air through the radiator with flow toward the engine or blow it in the opposite direction. The direction is selected on the basis of the application. Desired operator compartment comfort levels, radiator fouling source locations, and cleanability of the screens are some factors to be considered. Fan performance varies greatly with airflow restriction (fig. 7-16). The design of the fan, radiator engine compartment, chaff screen, and grill each accounts for a portion of the airflow restriction. Openings to both

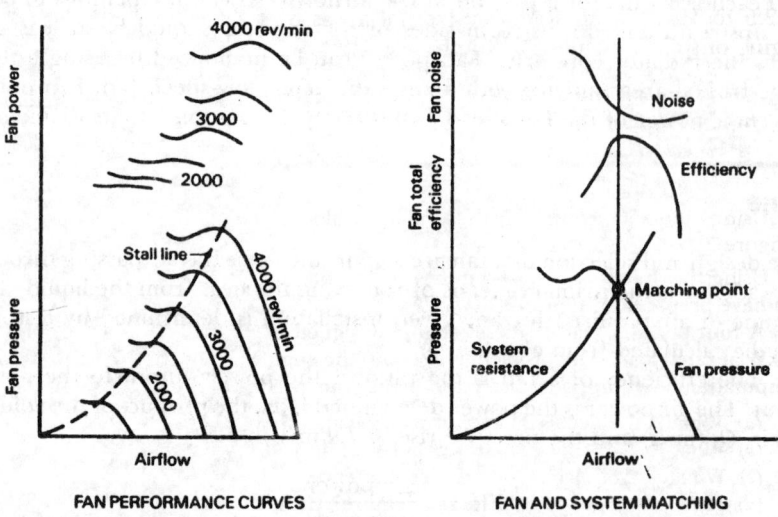

FIGURE 7-16   Cooling fan characteristics. (Courtesy Perkins Engines Ltd.)

the upstream and the downstream sides of the radiator should have at least 120 percent of the radiator core area.

Fan power can be minimized by using a thin, large frontal-area radiator with a large-diameter, low-speed fan that sweeps the maximum area of the core. Fan power is approximately a cubic function of its speed and a direct function of its diameter.

Fan shrouds with close fan blade tips to shroud clearance will provide the most efficient airflow through the radiator. The fan should ideally be at least 10 cm from the core. Suction fan blades should have two thirds of their projected width inserted into the shroud. Fans that blow the air should have two thirds of their projected width outside the shroud. Fan tip clearance should not exceed 1.2 cm. To avoid interference between the fan and the shroud, practicality dictates that this clearance should not be less than 0.6 cm. Recirculation of hot radiator exhaust air to the cold side of the radiator should be prevented by baffling, which seals leaks between the radiator and the engine compartment hood.

Engine Cooling Summary

Specific heat rejection by a diesel engine ranges from about 0.6 to 0.94 kW/kW of engine output, and water flow will range from 0.7 to 1.4 l/kW of engine output. Fan power is nominally 5 percent of gross engine power, with a range of 2.5 to 10 percent. Radiator frontal areas of 19 to 29 cm²/kW per gross engine output are rules of thumb.

PROBLEMS

1. Using automatic control analysis, make a block diagram of the governor shown in figure 7-1.

2. Using the dimensions shown in figure 7-1, fill in the boxes of the block diagram you have prepared of figure 7-1.

A four-cylinder tractor diesel engine designed to run at 2000 rpm will develop 75 kW at rated load. The problem is to equip the engine with a proper radiator. Show computations, assumptions, and reasons for your procedure.

(a) What is the necessary rate of heat dissipation to the air?

(b) What rate of cooling-water circulation do you recommend?

(c) What rate of airflow do you recommend?

(d) How much power will the fan require? If the fan is the only accessory, what percentage is the fan power of the total power that the engine would develop without accessories?

REFERENCES

Beranek, L. L. et al. *Noise and Vibration Control.* McGraw-Hill Book Co., New York, 1971.
Larson, R. E. "Advantages of a Well-chosen Air Cleaner Inlet System." SAE Preprint No. 798, Oct. 1952.
Siemens, J. C., and J. A. Weber, "Dry-Type Air Cleaners on Farm Tractors." SAE Preprint No. 77A, Oct. 1958.

SUGGESTED READINGS

Brooks, D. B., and R. E. Streets. "Automotive Antifreezes." *Natl. Bur Standards (U.S.) Circ.* 474, Nov. 10, 1948.
Chepil, W. S. "Sedimentary Characteristics of Dust Storms: III Composition of Suspended Dust." *Am. J. Sci.,* vol. 255, Mar. 1957, pp. 206–213.
Church, A. H., "Centrifugal Governors — Analysis of Properties with Design Procedure." *Product Eng'g,* vol. 12, Aug. 1941, pp. 409–412.
Millington, B. W., and E. R. Hartles. "Frictional Losses in Diesel Engines." *SAE Trans.,* vol. 77, 1968, pp. 2390–2406.
Millington, B. W. "Centrifugal Governors with Flyweights of Distributed Mass." *Engineering* (London), vol. 163, part 1, Mar. 28, 1947, p. 232.
SAE. *Heavy Duty Engine Cooling Systems.* Society of Automotive Engineers SP-24, Sept. 1982.
Sprick, W. L., and T. H. Becker. "The Application and Installation of Diesel Engines in Agricultural Equipment." *ASAE Distinguished Lecture Series,* no. 11. American Society of Agricultural Engineers, Dec. 17, 1985.
Vasey, G. H., and W. F. Baillie. "Some Experiences with Testing of Spark Arresters for Tractor Engines." *J. Agric. Eng'g Res.,* vol. 6, no. 1, 1961.

8
LUBRICATION

The primary objective of lubrication is to reduce friction, wear, and power loss. Lubrication accomplishes this requirement by interposing a film of oil between the sliding surfaces. Lubricating oils in internal combustion engines also function to cool surfaces, such as the pistons, by absorbing heat and dissipating it through cooling surfaces and radiators and to reduce compression losses by acting as a seal between the cylinder walls and the piston rings.

Much progress in lubrication technology has been made in recent years as modern machines are being designed to maintain higher loads and speeds. Because of this advance in bearing-load requirements, the proper design of bearings and lubrication systems and the selection of proper lubricants are increasing in importance.

Types of Lubricants

Lubricants commonly used in engines either are derived from mineral oils or are classified as synthetic. The synthetic oils available are based on esters derived from animal or vegetable oils, and another group is derived from hydrocarbons. However, lubricants that are derived from mineral oils predominate.

Mineral lubricants are obtained from crude petroleum and change very little on exposure to air. A great variety of lubricants are produced, ranging from light oils to heavy greases.

Properties of Lubricants

A lubricant must be able to perform certain tasks in order to accomplish its purpose satisfactorily. It must possess sufficient viscosity and oiliness to protect the mechanical devices at the necessary speeds, pressures, and temperatures. It must be of such a nature that it can be satisfactorily handled by the lubrication system, and finally it must be able to withstand service conditions. A

183

number of tests, both physical and chemical, have been developed to determine the characteristics of a lubricant and its suitability for a given purpose. Most of the tests are suitable only for the laboratory; the tractor owner must depend on recommendations of the petroleum and tractor companies.

Viscosity

The most important physical property of a lubricating oil is its viscosity. Viscosity is the internal resistance of a fluid as one layer is moved in relation to another layer.

The viscosity of an oil must be sufficient to support an oil film between a bearing and its journal. Excessive viscosity, however, causes unnecessary power consumption.

Because viscosity is such an important property of lubricating oils, much effort has been devoted to devising systems for measuring this property. Most of the many systems in use have been devised for a specific fluid and range of viscosity.

Sir Isaac Newton deduced the following relationship for a fluid being shear-stressed between two plates (fig. 8-1):

$$F = \mu A \frac{v}{h} \tag{1}$$

where F = force
         μ = absolute viscosity
         A = area
         v = velocity
         h = clearance between plates

A fluid is said to be Newtonian if the viscosity is constant, except for temperature changes, or

$$\mu = \frac{F/A}{v/h} = \frac{\text{shear stress}}{\text{rate of shear}} = \text{constant} \tag{2}$$

The absolute value of μ can be determined by several methods. The two most common methods are based on (1) laminar flow through some type of a capillary tube such as the Cannon-Fenske viscometer shown in figure 8-2 or (2) some type of rotational viscometer such as the MacMichael viscometer shown in figure 8-3. The latter instrument was designed for very viscous liquids, but variations of this type of instrument have been adapted to many kinds of fluids.

If SI units are used in equation 2, the coefficient of viscosity, μ, will be

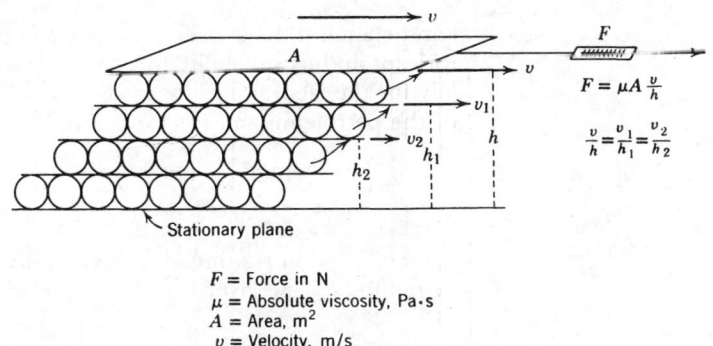

F = Force in N
μ = Absolute viscosity, Pa·s
A = Area, m^2
v = Velocity, m/s
h = Plane separation, m

FIGURE 8-1 Newton's theory of streamline or viscous flow. (From *Lubrication*, vol. 52, no. 3, Texaco, Inc. 1966.)

in pascal-second (Pa · s = N · s/m^2). If the English system of units is used in equation 2, the viscosity will be in reyns (lb · s/in.^{2}). Conversion to SI units can be made as follows:

$$1 \text{ Pa} \cdot \text{s} = 10 \text{ poise}$$
$$1 \text{ reyn} = 6.89 \times 10^4 \text{ poise}$$

In practice, determining the viscosity by Newton's method (fig. 8-1) is difficult, so several other methods have been devised, all based upon flow through a capillary tube or orifice. Flow through a capillary tube can be related to viscosity by *Poiseville's law* as follows:

$$\mu = \frac{\pi \, p r^4 t}{8Vl} \tag{3}$$

where p = pressure difference, dynes per cm^2
 r = radius of tube, cm
 t = time in seconds
 V = volume of liquid, cm^3
 l = length of the tube, cm
 μ = absolute viscosity, poises

A capillary tube apparatus lends itself to rapid determinations of viscosity; however, equation 3 is valid only for laminar flow.

All capillary tubes are seriously affected by emulsions and dirt so that in an effort to further simplify the determination of viscosity there are several short-tube, or orifice-type, viscometers in general use. Although the orifice type requires less time for a determination, it does not result in the precision of the capillary type.

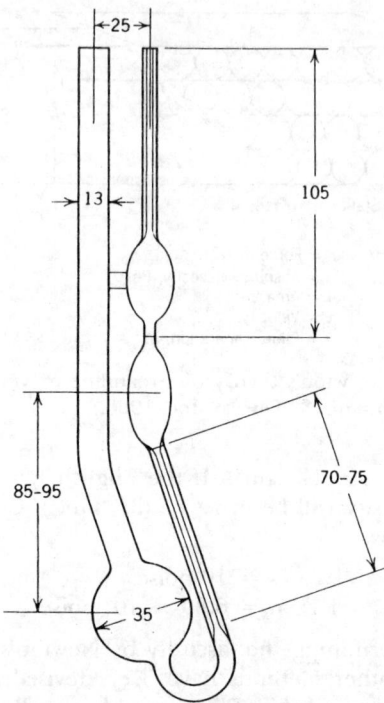

All dimensions are in millimeters

FIGURE 8-2 Cannon-Fenske viscometer for transparent liquids. (From
1976 Supplement to *Book of ASTM Standards,* Test for Kinematic Viscosity D445.)

An example of the orifice type is the Saybolt Universal Viscometer shown
in figure 8-4. The orifice-type viscometer is an *empirical* instrument in that
the viscosity cannot be computed directly. Instead, the viscosity is reported
as the time in seconds required for a known quantity of oil, at a constant
temperature, to flow through the orifice. Thus the viscosity of an oil, as
determined by the Saybolt Universal Viscometer, is reported in Saybolt Universal Seconds (SUS or SSU).

The viscosity of motor oil is usually reported in SUS. Even though a
capillary viscometer is used, its results will often be converted to SUS. For
more viscous fluids, a larger orifice is used and the results are reported in
Saybolt Furol Seconds.

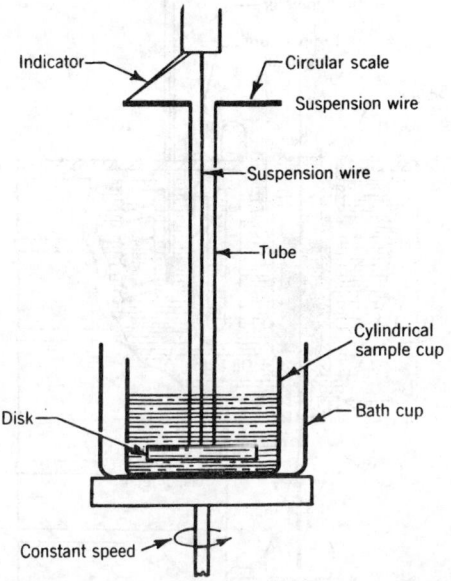

FIGURE 8-3 MacMichael Viscometer. (From *Lubrication*, vol. 52, no. 3, Texaco, Inc., 1966.)

Figure 8-5 is a nomograph for converting from one orifice system to another, or to centistokes, which is a measure of kinematic viscosity.

The apparent viscosity, as determined by an orifice-type or capillary instrument, is affected by the density of the liquid. To keep viscosity independent of the density, kinematic viscosity is computed as follows:

$$\text{Kinematic viscosity} = \frac{\text{absolute viscosity}}{\text{mass density}}$$

$$v = \frac{\mu}{\rho} \tag{4}$$

The relationship between SUS (or SSU) and kinematic viscosity can be determined from figure 8-5 or grossly from figure 8-6. Equations can be used to give more precise conversion. Figure 8-6 also introduces the SAE number system of expressing the viscosity of the oil, a method more easily understood by the consumer.

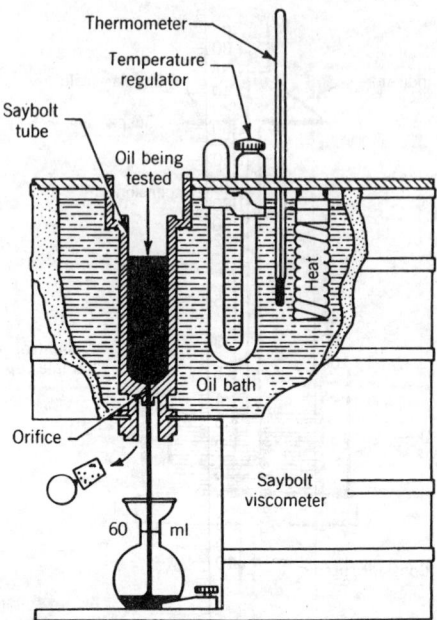

FIGURE 8-4 Saybolt Universal Viscometer. (From *Lubrication*, vol. 52, no. 3, Texaco, Inc., 1966.)

Classification of Oil by Viscosity

Oil, as used in the crankcase of most internal combustion engines, is normally classified in two ways. The crankcase oil will have an SAE number that corresponds to a range of the oil viscosities over a range of temperatures. It will also be classified as to quality. The SAE number (viscosity) system will be discussed first.

The SAE has adopted for convenience a series of numbers (table 8-1) that constitute a classification for crankcase lubricating oils in terms of viscosity only. Other factors of oil character or quality are not considered.

Viscosity numbers without an additional symbol are based on the viscosity at 99°C. Viscosity numbers with the additional symbol W(winter) are based on the viscosity at -18°C. The viscosity of crankcase oils included in this classification is not less than 39 s at 99°C, Saybolt universal.

Viscosity index expresses the variation in viscosity with a change in tem-

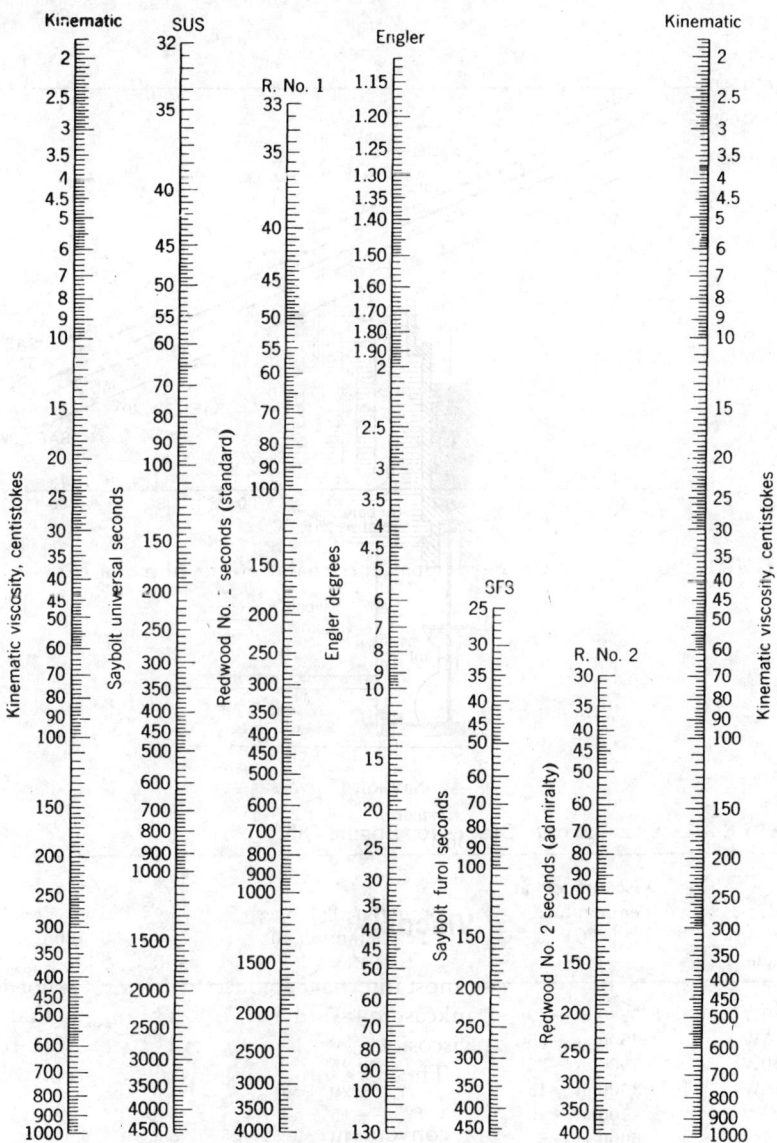

Line up straight edge so centistoke value on both kinematic scales is the same.
Viscosities at the same temperature on all scales are then equivalent.

To extend range of only the kinematic, saybolt universal, redwood No.1, and Engler
scales: multiply by 10 the viscosities on these scales between 100 and 1000 centistokes
on the kinematic scale and the corresponding viscosities on the other three scales. For
further extension, multiply these scales as above by 100 or a higher power of 10.

(Example: 1500 centistokes = 150 × 10 cs ≅ 695 × 10 SUS = 6950 SUS)

FIGURE 8-5 Nomograph for conversion of viscosity units. (From *Lubrication*, vol. 52, no. 3, Texaco, Inc., 1966.)

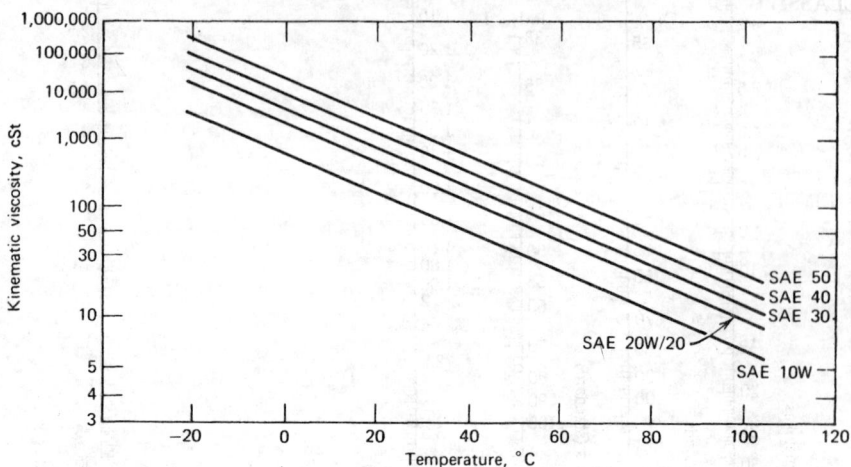

FIGURE 8-6 Typical viscosity-temperature characteristics of motor oils.

TABLE 8-1 SAE Viscosity Grades for Engine Oils

SAE Viscosity Grade	Viscosity (cP) at Temperature (°C) Max	Borderline Pumping Temperature (°C) Max	Viscosity (cSt) at 100°C Min	Viscosity (cSt) at 100°C Max
0W	3250 at −30	−35	3.8	—
5W	3500 at −25	−30	3.8	—
10W	3500 at −20	−25	4.1	—
15W	3500 at −15	−20	5.6	—
20W	4500 at −10	−15	5.6	—
25W	6000 at −5	−10	9.3	—
20	—	—	5.6	Less than 9.3
30	—	—	9.3	Less than 12.5
40	—	—	12.5	Less than 16.3
50	—	—	16.3	Less than 21.9

NOTE: 1 cP = 1 mPa·s; 1 cSt = 1 mm²/s
SOURCE: SAE J300 (Sept. 1980), "Engine Oil Viscosity Classification."

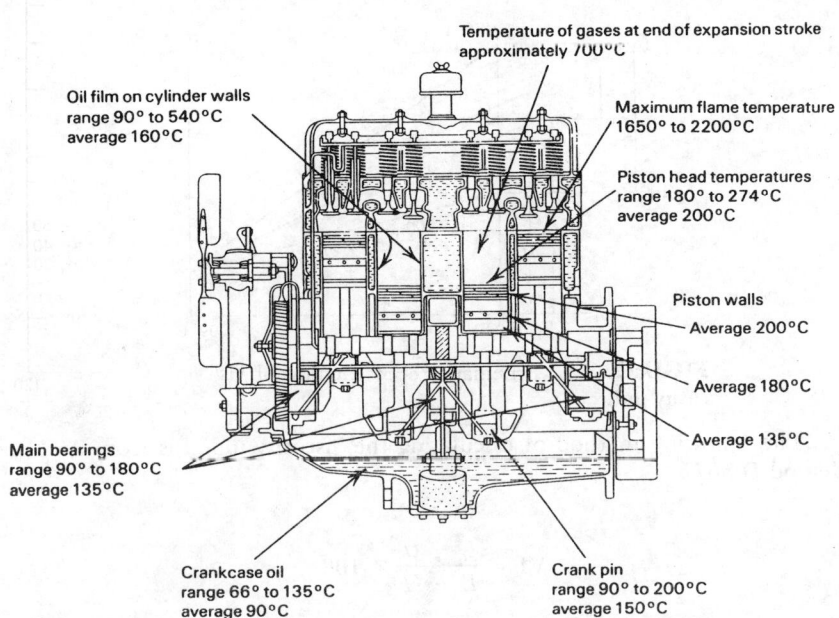

Temperature of gases at end of expansion stroke approximately 700°C

Oil film on cylinder walls range 90° to 540°C average 160°C

Maximum flame temperature 1650° to 2200°C

Piston head temperatures range 180° to 274°C average 200°C

Piston walls
Average 200°C

Average 180°C

Average 135°C

Main bearings range 90° to 180°C average 135°C

Crankcase oil range 66° to 135°C average 90°C

Crank pin range 90° to 200°C average 150°C

FIGURE 8-7 Operational temperatures of typical tractor engine in reasonably hard service, but not overloaded, at 75 percent of rated drawbar power and at an average speed of 4.8 km/h. (Courtesy Standard Oil Co. of Indiana.)

perature. An oil with a high viscosity index (VI) has less change in viscosity with a change in temperature than an oil with a low VI. Refineries have no problem producing oil of the correct viscosity for a constant operating temperature, but the necessity of starting an engine at a low temperature and running it at a higher temperature complicates the problem. The variation of temperature within the engine also adds to the problem (fig. 8-7). Therefore, an oil with a high viscosity index is desirable.

When the system of calculating the viscosity index was first developed, it was based upon a crude oil "L" (naphthenic), which was assigned an index of 0, and a crude oil "H" (paraffinic), which was assigned an index of 100. It was assumed that the "L" oil possessed the absolute maximum limit of viscosity temperature sensitivity and that the "H" oil was the least sensitive.

Although the system is still in use, the upper and lower limits of 100 and 0 no longer contain all of the oils. The improvements in refining, the use of additives to improve the viscosity index, and the manufacture of synthetic oils have resulted in lubricating oils that are far outside the scale in either direction.

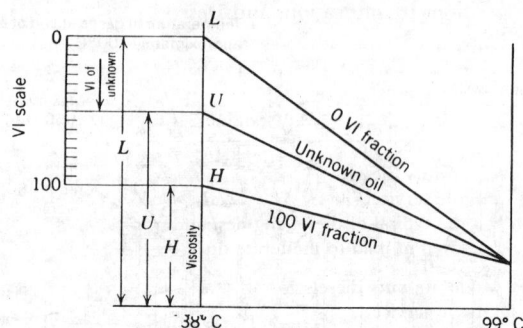

FIGURE 8-8 Schematic representation of viscosity index.

The following method of calculating the viscosity index is from ASTM Method D 567

$$VI = \frac{L - U}{L - H} \times 100$$

where L is the viscosity at 38°C of the "L" oil that has an arbitrary viscosity index of zero; H is the viscosity at 38°C of the "H" oil that has an arbitrary viscosity index of 100; and U is the viscosity at 38°C of the oil whose index is unknown.

The two reference lubricants and the unknown oil all have the same viscosity at 99°C. The values of L and H may be read from a set of previously prepared tables of viscosity that can be either Saybolt or kinematic. Figure 8-8 is a schematic representation of viscosity index.

Multigrade motor oils have been especially compounded to have a high viscosity index. They are compounded in such a way that they will have a low-temperature (−18°C) viscosity of one SAE viscosity number, whereas at a higher temperature (99°C) they will have a viscosity higher than would be expected. One multigrade oil is shown in figure 8-6. In general, a multigrade oil of, for example, 5W-20 will have a low-temperature viscosity of an SAE 5W oil and at high temperatures it will have a viscosity of SAE 20.

Engines subjected to wide temperature variations would normally benefit from using a multigrade oil, whereas engines subjected to relatively constant temperatures would not be benefited. The extra cost of multigrade oils as compared with single-grade oils suggests that the multigrade oil be used only under winter conditions or other variable temperature conditions.

TABLE 8-2 Designation, Identification, and Description of Engine Oil Categories for Diesel Engines

Letter Description	API Engine Service Description
CA	Light-duty diesel Engine service Service typical of diesel engines, and occasionally gasoline engines, operated in mild to moderate duty with high-quality fuels.
CB	Moderate-duty diesel Engine service Service typical of diesel engines operated in mild to moderate duty but with lower-quality fuels that necessitate more protection from wear and deposits. Occasionally has included gasoline engines in mild service.
CC	Moderate-duty diesel and gasoline engine service Service typical of lightly supercharged diesel engines operated in moderate to severe duty and includes certain heavy-duty gasoline engines.
CD	Severe-duty diesel Engine service Service typical of supercharged diesel engines in high-speed, high-output duty requiring highly effective control of wear and deposits.

SOURCE: Condensed from SAE J183 (Feb. 1980)—SAE Recommended Practice.

Classification by Service

In general, there are only two specifications for motor oil: the *viscosity* (SAE number) and the *service* classification. As a practical means of matching the oil to the engine, a system has been devised by the American Petroleum Institute that allows oil manufacturers to classify their oils in such a way as to "fit" oils to working conditions. The system considers two things: (1) the type of service and (2) the type of engine and fuel it uses. Table 8-2 describes the service classifications for diesel engines.

The oil companies now use these service classifications. Nothing prohibits an oil from being recommended for more than one service classification. Thus, for example, it is possible that an oil will be classified for service CB and CC. The tractor and engine manufacturers also recommend the use of a service classification of oil for each tractor.

Universal Tractor Oils

The development of a lubricating oil used for several purposes has led to so-called *universal tractor oils*. They are used for the most lubricated parts of a tractor, including the engine, transmission, hydraulic system, and brake system. This type of oil has been popularized in Europe and has been adopted by other countries. Universal tractor oils vary in quality and specifications according to the manufacturer. Some questions have been raised concerning the relative merits of these oils. The relatively high price and the difficulty of maintaining required performance under varying temperature and operating conditions seem to be limiting factors at present. However, the development of universal tractor oils will likely be a worldwide trend, and further improvements will be made by the tractor and oil manufacturers.

Oil Additives

The properties of lubricating oils can be altered considerably by the use of additives.

The trend toward continuing higher-performance engines has placed a greater demand on the performance of the lubricating oil. For example, a low-compression spark ignition engine demands much less from the oil than a high-speed, high-compression diesel engine. The performance of the latter engine is more seriously affected if the piston rings should become stuck in the ring grooves of the piston. Since the diesel engine depends upon a high-compression pressure for ignition, and also for its high efficiency, it is obvious that such an oil must also keep the piston rings free. Hence, there is a need for a *detergent* in the oil.

High-compression, high-speed engines require bearings that are resistant to high temperatures, higher pressures, and higher speeds. The result has been the use of alloy bearings that meet the requirements; but unfortunately, alloy bearings are more susceptible to corrosion than were the babbit bearings formerly used in engines. Thus, the demand for a *corrosion preventive* in the oil.

Table 8-3 lists the most important oil additives, their purpose, and the mechanism by which the additive performs its purpose. Obviously, no one oil contains all the compounds listed. As a general rule, the more additives contained in an oil, the better its service classification will be and the more it will cost to produce.

It is not feasible to discuss all of the oil additives currently being used. For the student interested in pursuing the subject further, the suggested readings at the end of the chapter will be of value.

TABLE 8-3 Lubricating Oil Additives

Type	Typical Compounds	Reason for Use
Detergents, dispersants	Succinimides, neutral metallic sulfonates, phenates, phosphates, polymeric detergents, amine compounds	Keep sludge, carbon, and other deposit precursors suspended in the oil.
Oxidation inhibitors	Zinc dialkyl dithiophosphates Compounds of nitrogen and sulfur Hindered phenols Bis-Phenols	Prevent or control oxidation of oil, formation of varnish, sludge, and corrosive compounds. Limit viscosity increase.
Alkaline compounds	Overbased metallic sulfonates and phenates	Neutralize acids, prevent corrosion from acid attack.
Extreme pressure (EP) antiwear, friction modifiers	Zinc dialkyl dithiophosphates Tricresyl phosphates Organic phosphates Chlorine compounds	Form protective film on engine parts. Reduce wear, prevent galling and seizing.
Rust inhibitors	High-base additives, sulfonates, phosphates, organic acids or esters, amines	Prevent rust on metal surfaces by forming protective surface film or neutralizing acids.
Metal deactivators	Zinc dialkyl dithiophosphates, metal phenates, organic nitrogen compounds	Form film so that metal surfaces do not catalyze oil oxidation.
Viscosity index improvers	Polyisobutylene, methacrylate, acrylate polymers, may incorporate detergent groups	Reduce rate of viscosity change with temperature. Reduce fuel consumption. Maintain low oil consumption. Allow easy cold starting.
Pour point depressants	Methacrylate polymers	Lower "freezing" point of oils, ensuring free flow at low temperatures.
Antifoamants	Silicone polymers	Reduce foam in crankcase and blending.

SOURCE: Courtesy Chevron Oil Company.

There is little advantage to using an oil that has a higher *service classification*—that is, better—than the conditions require.

Journal Bearing Design

The coefficient of friction of a journal bearing is a function of several variables. For any given bearing, the only three variables that may be changed are the

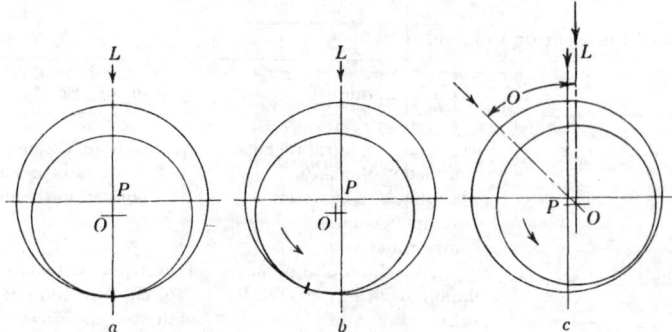

FIGURE 8-9   Action of journal bearing (exaggerated). (*a*) At rest.
(*b*) Partial lubrication, just starting. (*c*) Fluid film lubrication.

absolute viscosity of the lubricant, μ, the speed n of the journal, and the
bearing pressure p (force/projected area). These three factors are commonly
combined into the nondimensional quantity $\mu n/p$.

The action of a bearing in starting from rest and passing through the
region of unstable or thin-film lubrication is shown in figure 8-9. When at
rest, the journal is as shown in figure 8-9(*a*). The clearance space is filled with
oil. At the beginning of operation, the shaft tends to roll up one side of the
bearing. However, since some oil adheres to the journal and is carried between
the load-carrying surfaces, the rotation tends to build up a supporting oil
film, after which the journal moves to the right, as shown in figure 8-9(*c*).
Note that the shaft during operation is displaced to the side opposite the one
that it tends to roll up during starting.

So that no metal-to-metal contact may occur between the journal and
bearing, a film of oil sufficiently thick must be kept between the two. The
thickness of the film required will depend on how smooth the surfaces are,
since no bearing is perfectly smooth. When the layer of oil is thick enough
to prevent any metal-to-metal contact, the bearing is said to be operating in
the region of thick-film or stable lubrication. Under conditions of thin-film
or unstable lubrication, the projections touch so that the bearing materials
and the conditions of the surfaces affect the frictional loss.

Referring to figure 8-9(*c*), note that the oil flows toward the point of
minimum thickness. The oil film is built up by the rotation of the journal,
and since the film thickness increases with the speed of rotation, the load that
the bearing can carry also increases with the speed of rotation.

The coefficient of friction has been found to be a straight-line function

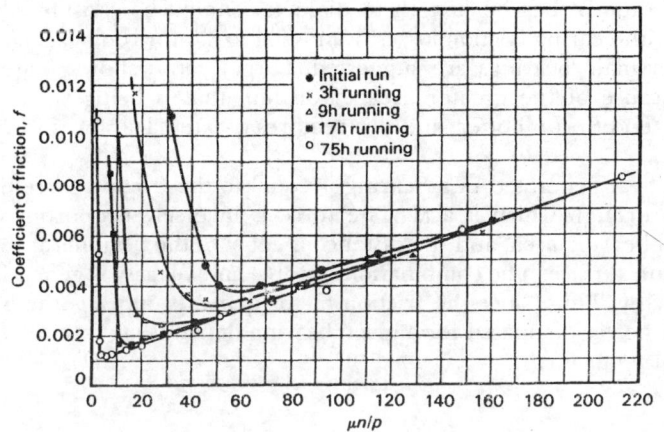

FIGURE 8-10 Journal friction curves for bronze bearings, showing the progressive amounts of running in.

of the quantity $\mu n/p$ except at low values, which usually indicate either high loads or low speeds (fig. 8-10).

Operation of the bearing to the right of the point of minimum coefficient of friction tends to be stable as an increase in bearing temperature lowers the oil viscosity and decreases the coefficient of friction. Viscosity is the only oil property that influences friction in this region. There is no measurable effect as a result of oiliness.

When the value of $\mu n/p$ becomes sufficiently small to pass to the left of the lowest point on the curve, it is impossible for a fluid film to be maintained, and lubrication is brought about largely by the ability of the lubricant to adhere to the surfaces. This is the region of thin-film or unstable lubrication, because an increase in the bearing temperature results in a further decrease in the viscosity of the oil and a further increase in the coefficient of friction. There is then, for any given combination of bearing and operating conditions, a point at which a transition occurs from stable to unstable lubrication. The location of this critical point is affected by the bearing material, the speed, the load, and the absolute viscosity of the lubricant.

Design factors play an important part in the power consumed by bearings. The important relationship between the coefficient of friction and the dimensionless coefficient $\mu n/p$ shows how the power required to turn a shaft varies with the shaft speed, the bearing pressure, and the viscosity of the oil. Film thickness may become so small that irregularities of the shaft or bearing may break through the oil film and result in partial lubrication conditions with a larger increase in friction loss.

The ratio of bearing length to diameter should be considered in connection with bearing friction losses. Ratios 0.4 to 0.6 for rod bearings and 0.6 to 0.8 for main bearings are suggested. Larger ratios show greater power losses because of the greater area of oil film that is being sheared. Such bearings tend to run hotter, and this requires greater clearances for more oil flow.

The heat generated in a bearing is equal to the friction force times the distance through which it acts. The force is inversely proportional to the projected bearing area, and the distance is equal to the peripheral velocity of the bearing surface. The combination of force and velocity is generally called the pv factor. Thus, since the friction factor depends on $\mu n/p$, the heat generated or power dissipated for a given bearing depends on both the pv factor and the $\mu n/p$ factor.

Oil Contamination

Oil in an internal combustion engine and other lubricated parts does not wear out or break down but becomes unfit for further use through contamination and through loss of additives. Oil contamination may consist of solid particles, such as dirt, grit, and other abrasives; it may be water; or it may be of a chemical nature.

Water enters the engine in several ways. Water is formed during the combustion of the fuel, and some of this may enter the crankcase by flowing by the pistons. Some water is condensed in the fuel tanks and brought in with the fuel. Under very wet field conditions, water sometimes enters hydraulic and brake systems. In whatever way it may occur, water in the lubricating oil gives rise to the possibility of emulsions and sludge.

Low-temperature sludge is largely the result of contaminating the lubricating oil with extraneous matter originating from incomplete combustion of the fuel-oxidation products. Detergent oils, which are capable of keeping contaminants in finely divided suspension, help alleviate the trouble, but they do not offer a complete solution. Operating jacket and crankcase temperatures should be kept in the 80° to 100°C range at all times.

Engine deposits can be reduced by adequate ventilation and scavenging of combustion products from blow-by and oil oxidation vapors. Pressure ventilation has been found to be helpful in engines, especially in those operating under intermittent and light loads.

Crankcase ventilation, if it is exhausted into the engine intake manifold, reduces engine emissions considerably. In the future, agricultural tractors may have to meet the same EPA (Environmental Protection Agency) requirements now required of highway vehicles. See SAE J900 (Nov. 1980) (Crank-

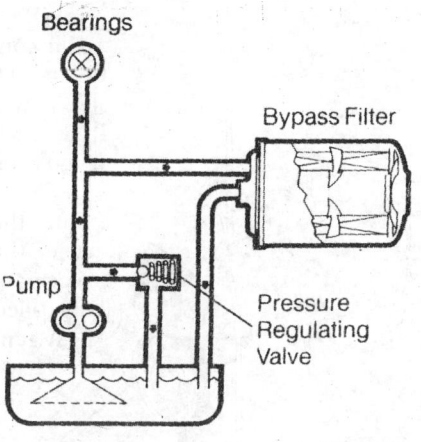

Bearings

Bypass Filter

Pump

Pressure
Regulating
Valve

FIGURE 8-11 Bypass engine oil filtering system. (Courtesy Fram Corp.)

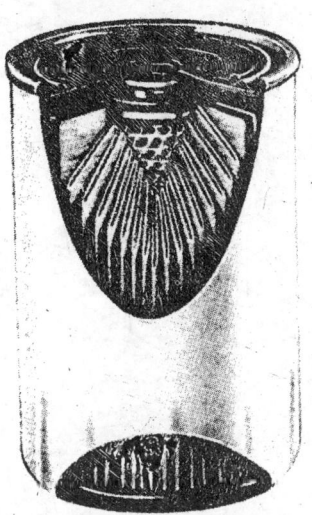

FIGURE 8-12 Disposable type of oil filter made of pleated resin-impregnated paper. (Courtesy Fram Corp.)

case Emission Control Test Code) for the standard method of measuring the crankcase emissions.

Oil Filters

Lubricating oil in an engine becomes contaminated with various materials: dirt, metal particles from the engine, carbon from fuel and oil that have burned, water, and fuel dilution. An ideal filter should remove all contaminants larger than engine clearances.

The majority of oil filters are classified as bypass filters (fig. 8-11) in that only a percentage of the oil that is pumped by the oil pump passes through the filter. The remainder bypasses the filter and goes directly to the oil distributor lines. The rate of flow through a filter depends on several factors, the most important being the pressure on the filter, the viscosity and temperature of the oil, the size of orifice, and the area of the filter.

Rather extensive tests (*Univ. Nebraska Agr. Expt. Sta. Bull.* 334, 1941) reveal some interesting information about oil filters. For instance, most filters tend to lower the acidity of the oil passing through them. Filters offering high

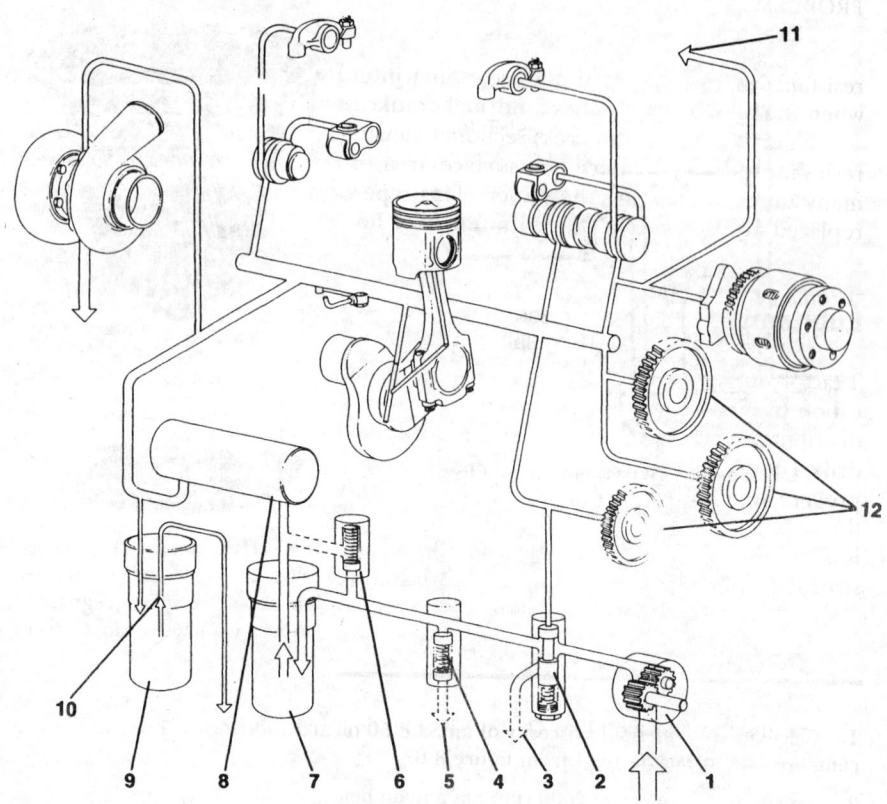

FIGURE 8-13 Lubrication flow schematic on a modern diesel engine. (Courtesy Cummins Engine Co., Inc.)

Key:

1. Oil Pump
2. Pressure Regulator Valve
3. Oil Return to Pan
4. High Pressure Relief Valve
5. Oil Return to Pan
6. Full Flow Filter Bypass Valve
7. Full Flow Filter
8. Oil Cooler
9. Bypass Filter
10. Bypass Filtered Oil Return
11. Accessory Drive/Air Compressor
12. Idler Gears

resistance to flow are practically useless in winter for short periods of operation when the oil seldom reaches a normal crankcase temperature.

Figure 8-12 shows a cross-sectional view of an oil filter made of pleated resin-impregnated paper. The surface area of the filter has been increased many times by pleating the paper. This type of filter is not reusable but is replaced after a predetermined number of hours of tractor engine use.

Lubricating Oil Systems

Tractor engines are equipped with an internal pressurized system of lubrication (fig. 8-13). A pump takes the oil from the sump and delivers it to a distributor duct that connects with all the main bearings. The crankshaft is drilled to provide an oil passage to the connecting rod, which is drilled to provide an oil passage to the piston pin. A pressure-regulating valve controls the pressure of the oil at the desired level. The system shown in figure 8-13 is a "full-flow" type. In other words, this system continually filters the entire stream of oil being pumped to the lubricated parts.

PROBLEMS

1. Calculate the kinematic viscosity of an SAE 50 oil at 60°C. How does your answer compare with the value read from figure 8-6?

2. An engine operating at 2000 rpm has a main bearing 57.2 mm long and 36.3 mm in diameter, carrying an estimated load of 1335 N. The engine crankcase is filled with SAE 30 oil at a temperature of 121°C. Assuming that the bearing follows the curves of figure 8-10 and that the oil has a specific gravity of 0.92 and follows the curves of figure 8-6, what is the coefficient of friction of the bearing?

3. Considering only the friction in the connecting rod bearings, how much power is required to motor the following engine having (a) four crankpin (connecting rod) bearings 63.5 mm diameter by 42.2 mm long? Bearing clearance 0.076 mm. Assume journal and bearing are concentric and use SAE 20 oil. Engine rotates at 2200 rpm. Mean oil temperature 121°C. (b) What horsepower is required if SAE 5W is used at the same temperature? Assume specific gravity is 0.9 at 20°C.

4. A capillary tube has a diameter of 3 mm and is 100 mm long. A head of 10 mm of oil above the inlet of the capillary maintains a pressure difference across the capillary. The specific gravity of the oil is 0.92. A volume of 200 cm³ of oil passes through the tube in 10 s.
   (a) What is the pressure difference across the tube?
   (b) What is the dynamic viscosity of the oil?
   (c) What is the kinematic viscosity of the oil?

REFERENCE

Univ. Nebraska Agr. Exp. Sta. Bull. 334, 1941.

SUGGESTED READINGS

ASTM. *Single-Cylinder Engine Tests for Evaluating the Performance of Crankcase Lubricants* (Abridged Procedure). ASTM Special Publication 509, Mar. 1972.

Chevron Oil Company. "Automotive Engine Oils, What They Are and How They Work." *Chevron Research Bulletin,* 1975.

Cummings, W. M. "Fuel and Lubricant Additives—I." *Lubrication* vol. 63, no. 1, 1977.

Fitch, E. C. "The Evaluation of Anti-Wear Additives in Hydraulic Fluids." *Wear,* vol. 36, 1976, p. 255.

Fuller, D. D. *Theory and Practice of Lubrication for Engineers.* John Wiley & Sons, New York, 1956.

John Deere. *Fuels, Lubricants and Coolants, Fundamentals of Service,* FOS-5805B. Deere & Co., 1985.

Kreuz, K. L. "Diesel Engine Chemistry." *Lubrication,* vol. 56, no. 6, 1970.

McLain, James A. "Diesel Engine Lubrication." SAE Paper No. 740S16 (also SP-390), Society of Automotive Engineers, 1974.

Rein, S. W. "Viscosity—I." *Lubrication,* vol. 64, no. 1, 1978.

Rein, S. W. "Viscosity—II." *Lubrication,* vol. 64, no. 2, 1978.

SAE. "Engine Lubrication." SP 640, Society of Automotive Engineers, 1985.

SAE. "Bearing Design for Off-Highway Equipment." SP-521, Society of Automotive Engineers, 1982.

SAE. "Multigrade Oils for Diesel Engines." SP-472, Society of Automotive Engineers, 1980.

SAE. "The Relationship Between Engine Oil Viscosity and Engine Performance." SP-416, Society of Automotive Engineers, 1977.

Note: *Lubrication* is a publication of TEXACO, Inc.

9
HUMAN FACTORS IN TRACTOR DESIGN

I've tried to show the endless opportunity and need for improvements and innovations in this field that are going to frustrate and excite the next generation of design engineers, human factors engineers and industrial designers.

WILLIAM F. H. PURCELL

The design of the modern tractor includes considerations of human factors. These factors, when properly incorporated in design, allow the operator to perform many complex tasks with efficiency, safety, and a minimum of fatigue. In general, human factors include such items as riding comfort, visibility, location and arrangement of controls, ease of operating controls, design for thermal comfort, and sound control. A typical work-space control center for a modern tractor is shown in figure 9-1.

This chapter briefly discusses and summarizes the main human factors that are considered in the design of the tractor operator's work space. This subject could occupy an entire book, and in fact, much has been written in recent years by researchers and designers whose interests were quite varied. For the benefit of the student or the designer interested in further exploration of this topic, numerous references and suggested readings are provided at the end of the chapter.

Operator Exposure to Environmental Factors

Tractors are used under varied geographical and climatological conditions. Direct exposure to temperature, humidity, wind, thermal radiation, dust, and

FIGURE 9-1   Major controls and instruments for a modern tractor. (Courtesy Deere & Company.)

Key:

1. Clutch pedal or inching pedal
2. Light switch
3. Fuel shutoff valve (diesel)
4. Air restriction indicator
5. Transmission oil indicator light
6. Power takeoff clutch lever
7. Engine oil pressure gauge

8. Engine temperature gauge
9. Steering shaft adjusting knob
10. Steering wheel
11. High-beam indicator light
12. Speed hour meter
13. Shift levers
14. Hand throttle

15. Remote-hydraulic operating levers
16. Rockshaft control lever
17. Brake pedals
18. Differential lock pedal
19. Light dimmer switch
20. Steering wheel tilt lock
21. Ignition key and starter switch

chemicals is encountered. Design of a suitable enclosure for the tractor operator minimizes the effects of the extremes that these environmental parameters can generate.

Table 9-1 defines comfort and bearable zones for four of these parameters as they apply to humans. Temperature, humidity, ventilation, and thermal radiation zones are interrelated.

Research has been done in regard to dust effects. Permissible guideline limits for dust concentrations on combines in the Netherlands is 15 mg/m³. Zander (1972) reported greater levels on combines without cabs.

Noren (1985) reported peak dust concentrations of 577 mg/m³ occurring during soil tillage operations. Mean values were 146 mg/m³. Tractor cabs that were pressurized at 50 Pa and fitted with air filters of "fine" quality reduced the dust concentration to a mean value of 24.7 mg/m³ in the operator's breathing zone.

TABLE 9-1 Environmental Zones for Selected Parameters

	Comfort Zone		Bearable Zone	
Environmental Parameter	Lower Limit	Upper Limit	Lower Limit	Upper Limit
Temperature, °C	18	24	−1	38
Humidity, % RH	30	70	10	90
Ventilation, m³/min	.37	.57	.14	1.4
Ultraviolet radiation	Unknown		Unknown	

Source: Henry Dreyfus. *The Measure of Man*. Whitney Library Design, New York, 1959, 1960.

Greater use of chemicals in agriculture and forestry and increased human protection requirements suggest the importance of a tractor cab as a protective device. Mick (1973) at the University of Iowa has reported preliminary results indicating that cabs offer a certain degree of protection during pesticide application. Akesson et al. (1974) have measured the health hazards to workers applying certain chemicals commonly used in California agriculture.

Miller et al. (1979) studied the potential contamination of the operator's environment in an air-conditioned tractor cab when applying toxic pesticides. Pesticides were applied to cotton and soybeans with rear-mounted spray booms and to pecan groves with an air-blast sprayer. The tractor cab was equipped with a standard roof-mounted fresh-air filter, an evaporator core, a heater core, and a recirculating air filter. Recirculation-to-fresh-air ratio was 3 : 1. Conclusions were that concentrations of methyl parathion, toxaphene, and xylene inside the operator's enclosure were well within acceptable limits and at most were 15 percent of acceptable limits.

Thermal comfort is defined in the *ASHRAE Handbook of Fundamentals* as the state of mind that expresses satisfaction with the thermal environment.

The thermodynamic process of heat exchange between humans and the environment can be described by the general heat balance equation of Gagge et al. (1941).

$$S\text{(storage)} = M\text{(metabolism)} - E\text{(evaporation)}$$
$$\pm R\text{(radiation)} \pm C\text{(convection)} \qquad (1)$$
$$- W\text{(work accomplished)}$$

S is the amount of heat gained or lost. If the body is in a state of thermal balance, S becomes zero. Positive values for storage will cause the mean body temperature to rise; negative values will cause it to fall. It is most convenient to express the preceding terms as power per unit of body surface (W/m^2 or kcal/hr-m^2), since heat exchange is always related in some way to the body's surface area. The metabolic rate for a tractor driver will be in the range of

70 to 174 watts/m². Average body surface area of a man will be 2 m², resulting in 140 to 348 watts. An enclosure environment must be maintained that will balance the metabolism so that thermal equilibrium is maintained.

Extensive research has been devoted to developing indices to evaluate thermal comfort. Much of this work has been under the sponsorship of the American Society of Heating, Refrigerating and Air-conditioning Engineers (ASHRAE). An empirical index—effective temperature (ET*)—was first developed by Houghton and Taylor. This has been upgraded over the years, and in 1972, a new comfort chart, shown in figure 9-2, was published by ASHRAE. The new *effective temperature* scale (ET*) is based on a simple model of human physiological response. Most individuals will be comfortable when the ET* is between 24° and 27°C (see fig. 9-2). One sees that various combinations of dry- and wet-bulb temperatures satisfy the "comfortable" zone. The ET* scale applies to altitudes from sea level to 2134 m and to the special case for indoor thermal environments in which the mean radiant temperature is nearly equal to dry-bulb air temperature and the air velocity is less than 14 m/min.

Caution must be used in applying the (ET*) criteria to tractor cab enclosures, since the mean radiant temperature may be 2° to 5°C greater than or 2° to 20°C less than the dry-bulb, and air velocities may be two to three times greater than 14 m/min.

Additional discussion relative to thermal comfort in design for operator enclosures is covered in the section Thermal Comfort in Operator Enclosures later in this chapter.

Operator Exposure to Noise

Noise-induced hearing loss does not occur in a sudden, traumatic manner unless the noise exposure is extremely severe. This type of hearing loss is insidious because its occurrence is imperceptibly slow and painless. The official guideline for noise exposure in the United States is based on the Walsh-Healy Act. Table 9-2 shows current acceptable levels. These criteria were developed on the basis of factory worker environments. Direct application to farm tractor environments cannot be made because of intermittent, variable-intensity noise and seasonal variations in exposure. However, these criteria do provide a starting point. The interested reader may want to peruse SAE Research Project R-4, 1969, "A Study of Noise Induced Hearing Damage Risk for Operators of Farm and Construction Equipment."

Manufacturers and farmers recognize that tractors have been too noisy. Engineers made the first official reports on tractor noise levels at the Nebraska Tractor Test Laboratory in 1971. Average noise level for all tractors tested

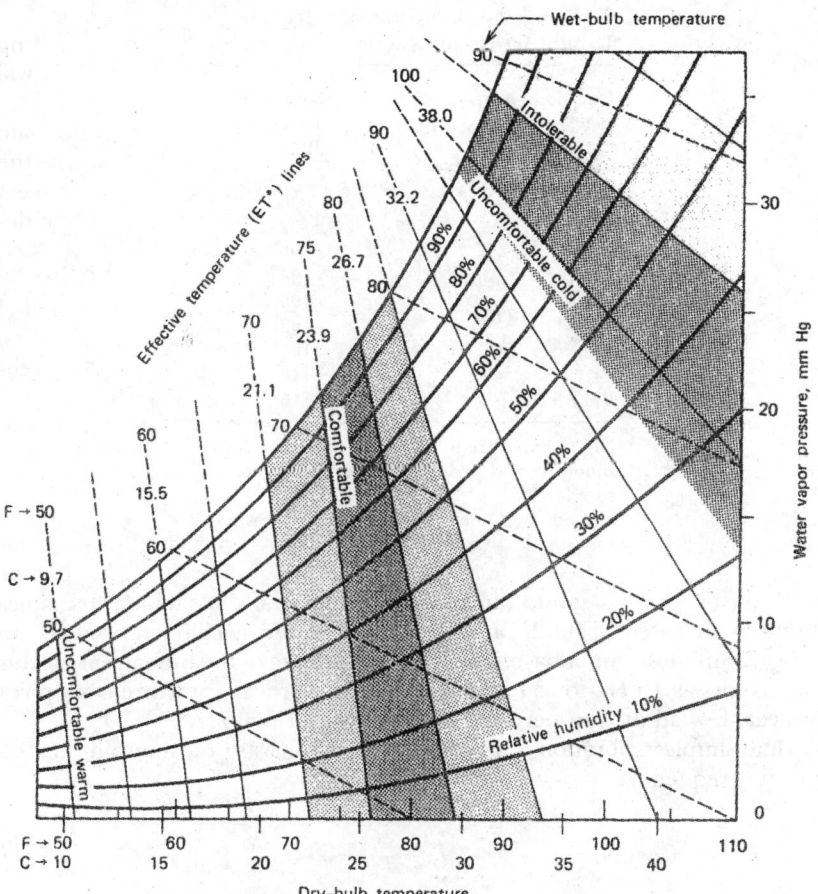

FIGURE 9-2 New effective temperature ET* scale. Each ET* line represents combinations of dry-bulb temperature and relative humidity that generally produce the same level of skin "wettedness" as caused by regulatory sweating. (From *ASHRAE Handbook of Fundamentals*, 1972.)

without cabs at 75 percent pull was 95.2 dB(A) at the operator's site. Tractors tested with cabs averaged 91.4 dB(A). Tractors tested in 1984 without cabs at 75 percent pull averaged 94.0 dB(A) at the operator's site. Tractors with cabs averaged 79.6 dB(A). It is evident from these figures that manufacturers have chosen to reduce noise levels at the operator's site primarily by engineering sound control in cabs.

TABLE 9-2 Occupational Safety and
Health Act Noise Criteria

Duration per Day, Hours	Sound Level, dB(A)[a]
8	90
6	92
4	95
3	97
2	100
1 $^1/_2$	102
1	105
$^1/_2$	110
$^1/_4$ or less	115

NOTE: Exposure to impulsive or impact noise should not exceed 140-dB peak sound pressure level.
[a]Sound level meter using A-scale, slow response.

Acoustic noise is sound, and sound is, physically speaking, mechanical vibrations in gaseous, fluid, or solid media. Sound is characterized by frequency, amplitude, and phase. The frequency range of the human ear extends from as low as 16 Hz to as high as 20,000 Hz. From a practical standpoint, however, few adults can perceive sound above 11,000 Hz.

The simplest vibration is a pure tone that consists of a sinusoid (fig. 9-3) with the frequency:

$$f = \frac{1}{T} \tag{2}$$

where T = period in seconds

The magnitude is most commonly expressed as the RMS (root mean square) value because of its direct relation with the energy content of the signal in linear systems. For a simple, pure tone

$$A_{RMS} = \frac{1}{\sqrt{2}} A_{PEAK} \tag{3}$$

Airborne sound is a variation in the normal atmospheric pressure. Most sound-measuring instruments are calibrated to read RMS airborne sound pressures on a logarithmic scale in decibels. Sound pressure can be expressed

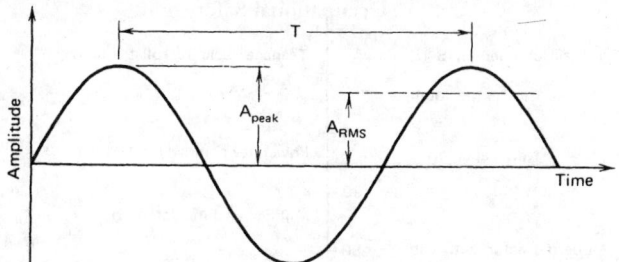

FIGURE 9-3 A sinusoidal vibration with peak and the RMS value of the signal.

as a sound pressure level (SPL) relative to a reference sound pressure. For airborne sounds, this reference sound pressure is 0.00002 N/m². Mathematically,

$$SPL = 20 \log \frac{P}{P_0} \, dB \qquad (4)$$

and

$$SPL = 20 \log \frac{P}{0.00002} \qquad (5)$$

where
SPL = sound pressure level, decibels
P = measured RMS sound pressure, N/m²
P_0 = reference sound pressure, N/m²
log = logarithm to the base 10

Figure 9-4 shows some typical values for sound pressure levels. Sound pressure doubles with each increase of 6 dB.

There are situations in which several noise levels must be combined to predict the overall effect. A convenient method for combining decibels is by use of the chart in figure 9-5. For example, if tractor noise were the result of two 80-dB components, for example, transmission and engine, an overall level of 83 dB would result. Combining 75 dB and 80 dB would result in 81.2 dB.

General-purpose sound-measuring instruments are normally equipped with three frequency-weighting scales, A, B, and C. These scales approximate the ear's response characteristics at different sound levels. Nebraska tractor

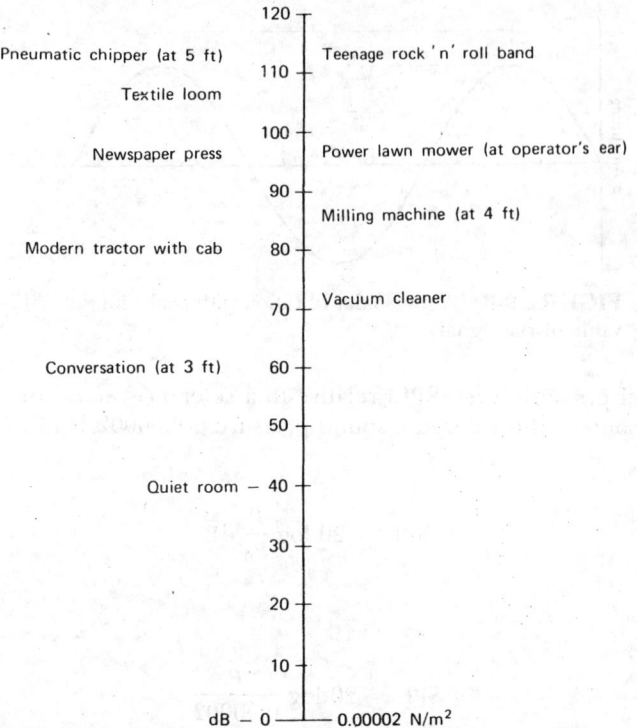

FIGURE 9-4   Typical airborne sound levels.

test data are reported in decibels, using the A weighting scale, and written as dB(A). Design for noise control, however, requires more detailed analysis of sound spectrums. Sound analyzers measure sound levels in bandwidths ranging from octaves to only a few hertz in width.

Operator Exposure to Vibration

Exposure of the human body to vibrations can result in biological, mechanical, physiological, and psychological effects. Ride vibration intensities are normally positively correlated with ground speed and often become intolerable as speed is increased. Matthews (1972) reported in a survey that the majority of large tractors were found to be operating at less than two-thirds maximum power because of the operator's inability or unwillingness to withstand the ride at full speed. The Rosseggers (1960) have conducted research that suggests that

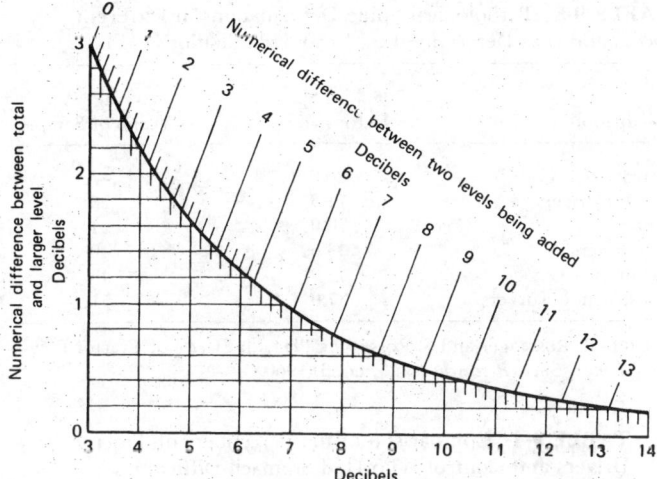

FIGURE 9-5   Chart for combining or subtracting decibels. (From *Handbook of Noise Measurement*, 7th ed., GenRad, Inc., 1974.)

Example: Subtract 81 dB from 90 dB. The difference is 9 dB. The 9-dB vertical line intersects the curved line at 0.6 dB on the vertical scale. Thus the unknown level is 90 − 0.6 or 89.4 dB.

Example: Add 75 dB and 80 dB. The difference is 5 dB. The 5-dB line intersects the curved line at 1.2 dB on the vertical scale. Thus the total value is 80 + 1.2 or 81.2 dB.

ride-induced vibrations do have ill effects on the operator. The results of their work are shown in table 9-3, which emphasizes the importance of proper seating for tractor drivers. Of all the occupations listed, it was found that only truck drivers had more difficulties with their spines than tractor drivers. As a general rule, the percentage of the normal population with symptoms of spine degeneration increases with age, so if a correction for age were made in table 9-3, the seriousness of the problem for tractor drivers would be even greater than shown.

The Rosseggers also studied the gastrointestinal difficulties of tractor drivers as compared to a control group. The results are shown in table 9-4.

The Rosseggers conclude:

The investigation has revealed that tractor driving may have considerable ill effects on the health of the operator. This is largely due to the effects of vibration and shocks continuously acting upon the human body and setting up harmful stimuli, and partly to the need to keep the body in a cramped condition and

TABLE 9-3 Pathological Spine Deformations in Different
Occupations as Determined by X-ray Examinations

Occupation	% with Spine Deformations	Average Age (Years) of Sample
Truck drivers	80.0	—
Tractor drivers	71.3	26
Miners	70.0	51
Bus drivers	43.6	40
Factory workers	43.0	45
Construction workers	37.0	51

SOURCE: R. Rossegger and S. Rossegger. "Health Effects of Tractor Driving." *J. Agr. Engr. Research*, vol. 5, no. 3, 1960.

TABLE 9-4 Comparison of the Percentage of Tractor
Drivers and Control Who Had Stomach Difficulties

	Tractor Drivers (322)	Control (37)
Stomach complaints	76%	46%
X-ray findings	52%	30%

SOURCE: R. Rossegger and S. Rossegger. "Health Effects of Tractor Driving." *Jour. Agr. Engr. Research*, vol. 5, no. 3, 1960.

unhealthy posture for long periods. The response of the body in an effort to counteract these effects and to maintain equilibrium imposes an additional strain upon the tractor driver, thus increasing fatigue.

Care should, therefore, be taken in designing tractors, and particularly tractor seats, to reduce vibration and shocks to a minimum by appropriate suspension and shock absorption and to arrange the tractor controls in a manner to insure a comfortable posture and minimum effort.

Effects reported by the Rosseggers are caused by low-frequency vertical vibration, that is, frequencies up to 20 Hz. This low-frequency vibration results in whole-body excitation. Suggs (1973) has observed the 4- to 8-Hz range as being critical. In this range resonance occurs in parts of the human body producing discomfort. Suggs states:

One of the more interesting resonances which occurs is that of the viscera. In vertical vibration the viscera resonates at about 4 to 5 Hz and forces the diaphragm at the same frequency. This produces a mechanical pumping action in the chest which overventilates the lungs. Carbon dioxide, a component of the plasma acid balance is flushed out of the blood stream and respiratory alkalosis results. The drowsiness associated with alkalosis may be implicated in some tractor accidents.

Suggs further states:

Hand coordination, foot pedal control, reaction time, visual acuity and tracking are all adversely affected by low frequency vibration. While these effects in themselves do not constitute a health or safety problem, they do produce a decrement in the ability of the operator to control a vehicle which implies a higher accident probability.

Low-frequency vertical vibration is present during normal field operations. Amplitude of vibration is, in part, dependent on roughness of the field. The undamped natural frequencies of wheel tractors commonly lie in the 3- to 5-Hz range.

Higher frequency vibration (30 Hz and up) results in part-body vibration. Although it is not important with regard to whole-body vibration, it is the source of foot and hand-arm excitation. The higher frequency vibrations can be present at the steering wheel, gear shift levers, control levers, and floor panels.

Vibration can be described as an oscillatory motion of a mechanical system and can generate sound in a gas, fluid, or solid as previously discussed. In our present discussion, vibration will be confined to the mechanical context of a solid medium. Many important mechanical vibrations in tractors lie in the frequency range of 1 to 100 Hz (corresponding to rotational speeds of 60 to 6000 rpm). The oscillatory motion can be simple harmonic or extremely complex. Like sound, mechanical vibration is characterized by frequency, amplitude, and phase.

In the simplest case, mechanical vibration can be derived from a simple harmonic wave form as shown in figure 9-3 and mathematically described by the displacement function:

$$X = A \sin \omega t \tag{6}$$

where $\quad X$ = displacement, m
$\quad A$ = peak amplitude, m
$\quad \omega$ = frequency of vibration, rad/s
$\quad t$ = time, s

$$\dot{X} = \omega A \cos \omega t \tag{7}$$

where $\quad \dot{X}$ = velocity, m/s

$$\ddot{X} = -\omega^2 A \sin \omega t \tag{8}$$

where $\ddot{X}$ = acceleration, m/s^2

One readily notes that the acceleration is proportional to the displacement and to the square of the frequency.

Vibration is usually measured with an accelerometer, which senses $\ddot{X}$ in equation 8. An accelerometer is an electromechanical transducer that produces an output voltage signal proportional to the acceleration to which it is subjected. Readout is normally expressed as acceleration, RMS, which mathematically is

$$\ddot{X}_{RMS} = \frac{1}{\sqrt{2}}(\omega^2 A) \tag{9}$$

It is also convenient to express acceleration in terms of g's. This is achieved by dividing $\ddot{X}$ or $\ddot{X}_{RMS}$ by 9.8 m/s^2. Quite frequently, vibration levels are expressed in decibels using the form of equation 4. A convenient reference level for vibration is 1 m/s^2 RMS. Equation 10 expresses it mathematically as

$$VAL = 20 \log V/1, dB \ re \ 1 \ m/s^2 \tag{10}$$

where   VAL = vibration acceleration level, dB

V = measured RMS acceleration, m/s^2

Engineers and scientists have been working for more than 50 years to quantify human vibration tolerance to vehicle ride vibration. One such effort is represented by International Standards Organization ISO 2631, *Guide for the Evaluation of Human Exposure to Whole Body Vibrations*. This guide is the result of some 10 years of work. Figure 9-6 shows the format of the boundary curves for vertical vibration as experienced by the operator. It is important to note that in the 4- to 8-Hz range, human tolerance to vibration is a minimum.

The preceding standard has raised many questions and concerns as to its validity. The reader is encouraged to refer to Janeway (1975), Gerke and Hoag (1981), Griffen (1978), and Stikeleather (1976 and 1981).

Discussion has been limited to vertical vibration effects with respect to a seated operator. However, horizontal (lateral and longitudinal) and angular (pitch, roll, and yaw) acceleration inputs are also experienced by the seated operator. Design for vibration isolation and control requires detailed analysis of complex vibrations. Vibration analyzers (similar to sound analyzers) are used to determine vibration levels in bandwidths ranging from octaves to only a few hertz.

The Operator-Machine Interface

Every time a person operates a tractor, the sensing, decision-making, and muscular powers of the operator are joined to an engineering system. The

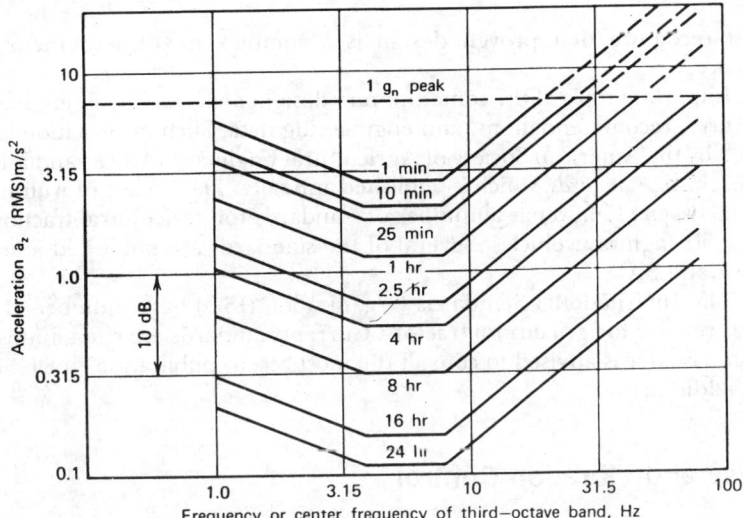

FIGURE 9-6 Fatigue-decreased proficiency boundaries for vertical acceleration (per ISO 2631).

operator uses hearing, sight, and feel to interpret inputs and to interface with the control-instrumentation components to achieve the desired output of the tractor. Human engineering data defining acceptable ranges for environmental factors, noise, and vibration must be incorporated in the design to ensure operator safety and comfort. This is a challenging task because many gaps exist in human engineering data. Although significant progress has been made in the last 10 years, additional research, testing, and experience are needed to develop more useful and reliable levels of design technology.

Regarding design of the operator's workplace, Hansson et al. (1970) state:

Safety, comfort and convenience should be considered in the design, location and construction of the operator's work place. The work place should be located on the machine so that visibility in the driving position is good without requiring the operator to work in an awkward, tiring position. Levers, pedals and instruments should be conveniently and logically located and the work place should fit both tall and short operators. In addition, the operator should be able to change his working position easily and the work area should be free of sharp edges and obstructions such as transmission cases.

The remainder of this chapter will introduce the reader to elementary aspects of tractor design for operator safety, comfort, and convenience. One must recognize that proven design is a complex mixture of theory and practice.

Important tools of the contemporary designer are standards, engineering practices, recommendations, and engineering data. Such information is published by the American Society of Agricultural Engineers (ASAE) and is found in the *ASAE Standards,* which is published annually. The Society of Automotive Engineers (SAE) also has a number of standards for agricultural tractors and other off-highway vehicles. Several of the standards are published jointly by ASAE and SAE.

The International Standards Organization (ISO) has a number of standards relative to agricultural tractors. Current standards are cited in this text, but the reader is advised to consult the most recent publications for revisions and additions.

Noise and Vibration Control

The overall mechanical design of a tractor, including the operator enclosure, has a direct bearing on the noise and vibration levels at the operator's site. Noise levels are the summation of rotating and reciprocating parts, structural vibrations, and gas and fluid flows, all of which are transmitted either through the tractor's structure or through the air. Vibration levels are the summation of components that are similar to noise source components plus ground inputs from wheels. Energy from noise and vibration sources is a result of time-dependent forces that reach the operator's site by structural and airborne transmission. Thus the human operator experiences various levels of acoustic energy and tractor chassis motion. Figure 9-7 shows a schematic of the situation.

The design engineer has the challenge to specify the design for mechanical construction of the tractor to limit the levels of acoustic energy and tractor chassis motion to acceptable values. The design process calls for determining the magnitude, frequency, and phase of the time-dependent forces and for subsequent application of the proper engineering technique to reduce them.

Sound and vibration are generated in the various rotating, reciprocating, and vibrating components of the tractor. Fans, blowers, gears, and internal combustion engines generate sound at a fundamental frequency, respectively, of rpm/60 times the number of blades, impellers, teeth, and number of cylinders. Also, components of higher harmonics are generated. Design techniques to control structure and airborne noise must determine proper stiffness of stationary and moving parts and proper use of absorbing, barrier, and damping materials. For example, a six-cylinder tractor engine rated at 2200

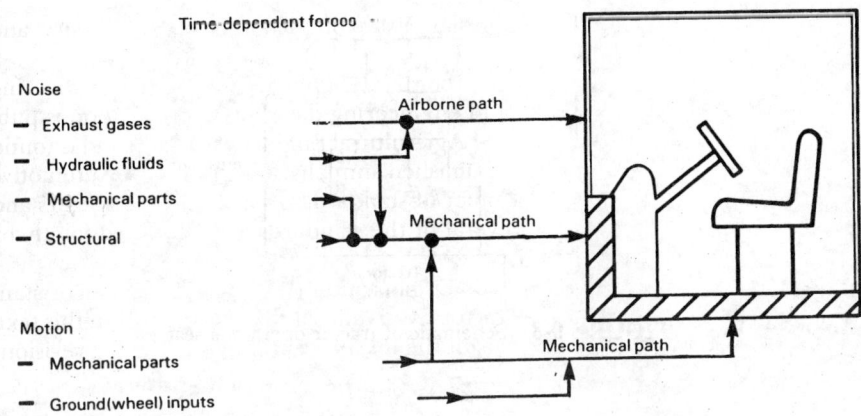

FIGURE 9-7 Schematic diagram of noise and motion transmission.

rpm will generate a fundamental sound frequency of 220 Hz. This will produce airborne noise at the operator's site.

The designer may choose to use an operator enclosure for sound control. If a 6.3-mm single glass window is used, approximately a 25 percent transmission loss can be achieved for the 220-Hz frequency. Use of a 12.7-mm single glass would achieve approximately a 30 percent transmission loss into the enclosure. Higher frequency components near 5000 Hz would be attenuated approximately 36 percent and 44 percent, respectively.

Control of motion experienced by the operator requires proper seat suspension design and operator platform isolation and suspension.

Operator Seating

A tractor can be considered as a spring-mass-damper system vibrating with 6 degrees of freedom, as described in Chapter 11. Displacement and acceleration inputs to the base of an operator's seat will consist of vertical, lateral, longitudinal, roll, pitch, and yaw components. The research of Raney et al. (1961), Pradko and Lee (1968), and Janeway (1975) shows that the predominant vibrational motion of a wheel tractor is vertical (Z_s) and that the seated operator is most sensitive to vertical acceleration.

If we consider only vertical motion imparted to the base of the operator's seat, the 2-degree-of-freedom system described in Chapter 11 under Longitudinal Stability applies. Referring to figure 11-7, the vertical motion imparted to the tractor seat is approximately

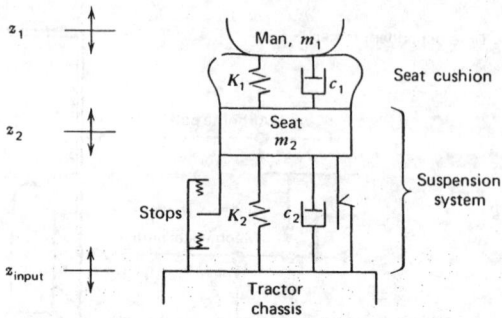

FIGURE 9-8   Schematic of tractor operator's seat.

$$Z_{input} = z_t + R\theta \tag{11}$$

where Z_{input} = vertical motion imparted to base of seat, m
   z_t = vertical motion of tractor center of gravity, m
   θ = rotation or pitch about tractor center of gravity, radians
   R = longitudinal distance from operator's seat to the tractor center of gravity, m

For small θ, the term $R\theta$ represents the vertical motion resulting from the pitch motion of the tractor.

Then

$$\ddot{Z}_{input} = \ddot{z}_t + R\ddot{\theta} \tag{12}$$

where $\ddot{Z}_{input}$ = vertical acceleration imparted to base of operator's seat, m/s^2
   $\ddot{z}_t$ = vertical acceleration of tractor center of gravity, m/s^2
   $R\ddot{\theta}$ = vertical acceleration resulting from pitch, m/s^2

The operator's seat and suspension are represented schematically in figure 9-8. For suspension-type seats the cushion is mounted to a guided linkage incorporating some type of spring and damper (shock absorber) along with travel limit stops. Friction is also inherent. This seat suspension system is a vibratory mechanical system subjected to the tractor chassis input $\ddot{Z}_{input}$. Seat motion is a function of seat spring rate, mass, damping, and frequency content of the input vibration. For a given input vibration $\ddot{Z}_{input}$ to the seat base, the seat develops a corresponding output vibration.

If we grossly simplify the system in figure 9-8 to consist of a rigid mass, $M = m_1 + m_2$, coupled to the tractor chassis via a linear spring K_2 and a viscous damper C_2, we can simplify the following discussion about reducing the vibration intensity transmitted to the operator.

The relationship between the output (seat) vibration for a given input

(tractor chassis) vibration as a function of frequency is known as the transfer
function or the frequency response function of the system. The ratio of output
vibration intensity to input vibration intensity is called the transmissibility, as
expressed by equation 13:

$$\text{Transmissibility} = \frac{\text{output vibration intensity}}{\text{input vibration intensity}} \tag{13}$$

For proper seat design, values of C_2 and K_2 must be judiciously selected
so that the transmissibility is less than 1. This will reduce the vibration intensity
imparted to the seat by the tractor chassis. A transmissibility in the range of
0.5 to 0.65 is considered satisfactory.

The transmissibility function (equation 13) for a single degree of freedom
linear spring-mass-damper system is derived by Den Hartog (1956) in Chapter
2. The resulting equation for a steady-state input $\ddot{Z}_{\text{input}}$ is

$$T = \left[\frac{1 + 4\zeta^2 \left(\dfrac{\omega_t}{\omega_s}\right)^2}{\left[1 - \left(\dfrac{\omega_t}{\omega_s}\right)^2 \right]^2 + 4\zeta^2 \left(\dfrac{\omega_t}{\omega_s}\right)^2} \right]^{0.5} \tag{14}$$

where T = transmissibility
        ζ = damping ratio
        ω_t = tractor chassis frequency, rad/s
        ω_s = undamped natural frequency of the seat, rad/s

Equation 14 is shown in figure 9-9. For a clearer understanding of equation 14 some additional details are as follows:

$$\zeta = c/c_c \tag{15}$$

c = seat suspension damping rate, N/m · s
c_c = critical damping rate, N/m · s

where

$$c_c = 2M\omega_s \tag{16}$$

M = mass of seat and operator, kg
ω_s = undamped natural frequency of the operator seat, rad/s

where

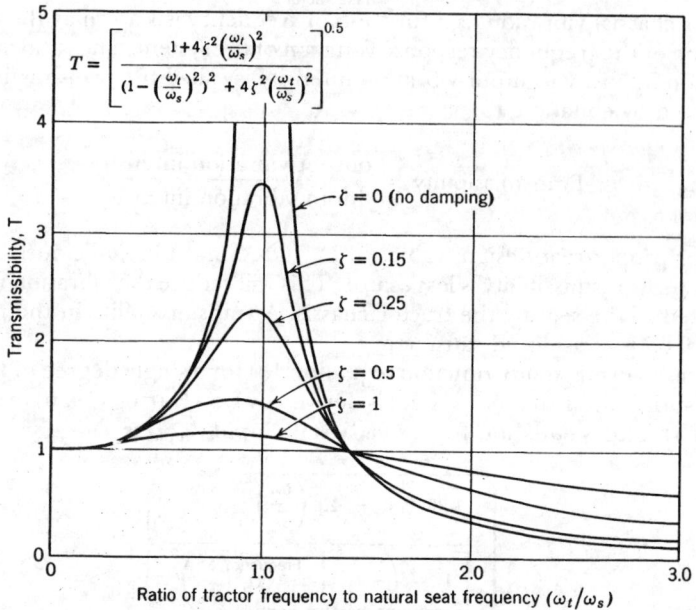

FIGURE 9-9 Transmissibility as a function of frequency ratio for various amounts of viscous damping.

$$\omega_s = \sqrt{\frac{k}{M}} \tag{17}$$

$k = K_2 =$ spring rate of seating system, N/m
$M =$ mass of seat and operator, kg

It is important that the student thoroughly understand the contents of figure 9-9 and equation 14. In proper seat-suspension design, the transmissibility should be less than unity, which occurs at ω_t/ω_s ratios of greater than 1.414. It should be apparent that the frequency ω_t of the input vibration must be known in order to design a seat-suspension system for the tractor. The tractor frequency in general will be in the 3- to 10-Hz range. Because occasionally a ω_t/ω_s ratio near unity will be experienced, the importance of damping should be apparent. Finally, it should be pointed out that transmissibility is independent of input amplitude for linear systems.

The true transmissibility of a seat is much more complex than equation 14 indicates because of nonlinear cushion characteristics and nonlinearities in the suspension such as inherent friction and limit stops. In practice, the

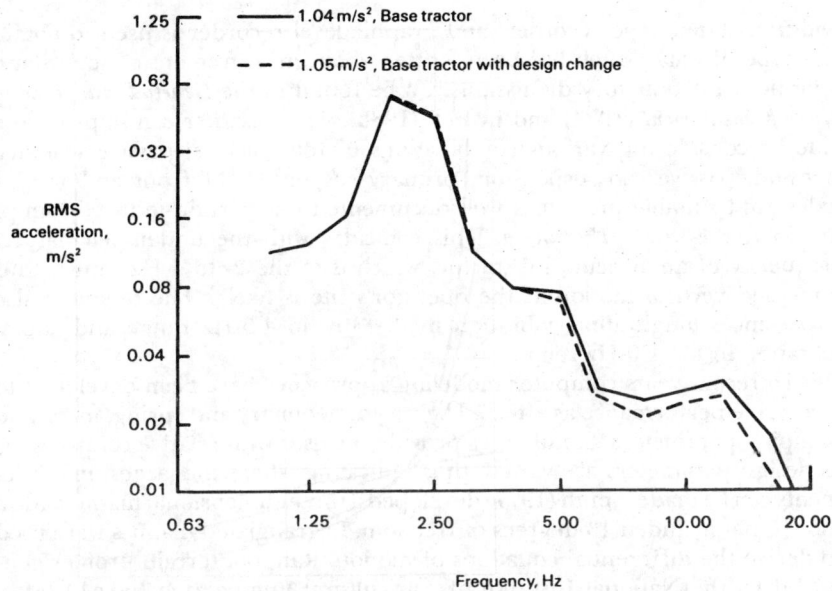

FIGURE 9-10 Third-octave spectra of the vertical seat acceleration (From Smith 1977.)

transmissibility of a seat is determined experimentally by laboratory procedures.

Once the seat transmissibility is determined, the ride quality can be related to vehicle ride criteria. For example,

$$a_z = (\text{transmissibility}) (\ddot{Z}_{input}) \qquad (18)$$

where a_z = vertical acceleration, (RMS) m/s²
$\ddot{Z}_{input}$ = RMS value of the acceleration defined in equation 12, m/s²

a_z corresponds to the acceleration used in the acceleration criteria of ISO 2631 shown in figure 9-6.

In the foregoing discussion, a wheeled agricultural tractor will have a vertical acceleration frequency content in the 1.0- to 20.0-Hz range. It is typical to represent vibration data for tractor ride comfort by a one-third octave band rms acceleration versus frequency plot. Some representative data from Smith (1977) are shown in figure 9-10. Similar measurements for horizontal and lateral components of vibration can be made.

A measurement system consisting of accelerometers, third-octave bandwidth analyzer, tape recorder, and graphic level recorder is used to obtain this type of data. Much has been written about measurement of sound and vibration. Introductory discussions can be found in the *GenRad Handbook of Noise Measurement* (1974) and in Bell (1982). The designer can improve the ride by considering various combinations of four factors: passive seat suspension, passive cab suspension, primary suspension of front and/or rear axles, and suitable tires. It is well documented that terrain-induced bumps are in the 1- to 10-Hz range. This coincides with the undamped natural frequency of an agricultural tractor, which is in the 2- to 4-Hz range: thus amplified vertical motion at the operator's site is likely. The operator also experiences longitudinal vibration in the 1.5- to 4.5-Hz range and lateral vibration in the 1.0-Hz range.

In recent years computer modeling approaches have been developed to study ride performance as affected by tractor geometry and spring, mass, and damping parameters. Simulations provide the user with relative comparisons as design parameters are varied, thus indicating where the largest improvements can be made. Smith (1977) developed a three-dimensional mathematical model that included 13 degrees of freedom. Lagrangian dynamics were used to derive the differential equations of motion. Random terrain profiles generated by the National Institute of Agricultural Engineering (NIAE) (Matthews 1966), for smooth and rough test tracks at 12 km/h were used as the excitation for the model developed. Results of the simulation indicated a vertical motion frequency of 1.99 Hz of the chassis and cab. The longitudinal motion of the chassis and cab had a fundamental frequency of 2.22 Hz, with higher order resonance at 4.75 and 10.17 Hz. Also, lateral motion of the chassis and cab occurred at frequencies of 1.21, 1.51, 2.67, 5.97, 6.48, and 12.4 Hz. Because it is difficult to correlate acceleration amplitude levels with operator-perceived ride comfort, the simulation was largely used for providing a relative rather than an absolute evaluation of the effect of a design change.

Claar and Buchele (1980) utilized a mathematical computer model, Integrated Mechanism Program (IMP), which is capable of analyzing two- and three-dimensional, motion-constrained, rigid-body, and closed-loop kinematic chain mechanisms. The IMP system was used to evaluate the effect of operator cab position and suspension system parameters on characteristics of tractor rides. Nine model tractor configurations were studied. Each model had one of three operator cab positions and one of three suspension systems. Cab positions were (1) conventional; (2) mid-chassis; and (3) behind the front wheels. The three suspension systems were (1) unsprung front and rear axles; (2) unsprung rear axle and sprung front axle; and (3) sprung front and rear axles. A tractor with and without a semimounted moldboard plow was modeled. Results indicated that (1) for unsprung front and rear axles, the con-

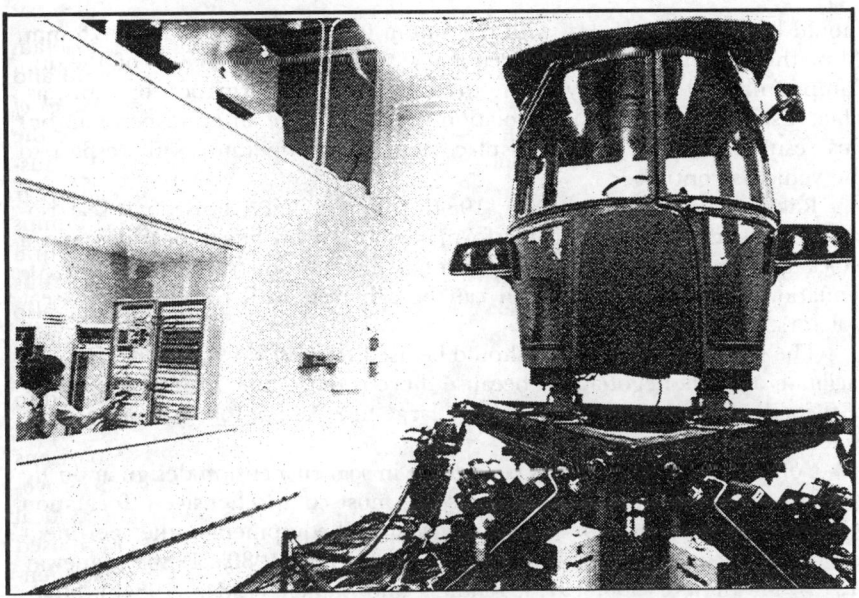

FIGURE 9-11   Off-road vehicle motion simulator with tractor operator enclosure. (Courtesy Deere & Company.)

ventional cab location gave the best operator ride comfort; (2) with suspended front axle and rigid rear axle, the forward cab location gave the best ride comfort; and (3) with both front and rear axles suspended, the conventional and mid-chassis cab location provided the best operator ride comfort.

Duncan and Wegscheid (1982) have developed an off-road vehicle motion simulator. This laboratory device, shown in figure 9-11, is used to determine effects of vibration, noise, and task complexity on operator and system performance. The simulator can be used to improve suspension systems for seats and operator enclosures. Six degrees of freedom with motion in x, y, z, roll, pitch, and yaw coordinates are possible. Ride programs drive the simulator from data files made from recordings of vehicle accelerations collected from field operations. Data files constructed from vehicle computer simulations can also drive the simulator.

There are basic guidelines governing overall design for ride comfort. For passive seat suspension, components used are springs with constant or variable rate, shock absorbers to dampen low-frequency vibrations, and torsion bar mechanisms or pneumatic cylinders to compensate for static deflection. The undamped natural frequency of the seat suspension system should be less than 1.5 Hz, with maximum travel of approximately 76 mm. Also, the undamped natural frequency should be 0.4 to 0.5 of the undamped natural frequency of the tires. Isolation performance, or transmissibility, should be 0.4 to 0.5. Isolation improves with softer suspension, but this results in larger relative displacement of the operator with respect to the vehicle's controls.

Rakheja and Sankar (1984a, 1984b) suggest that a horizontal seat isolator consisting of a sprung platform supported on linear bearings and a shock absorber can attenuate horizontal, direction-of-travel vibration. Attenuation of rotational vibration can be achieved with roll and pitch seat isolators.

The largest-diameter tires should be used for best ride comfort. This will facilitate obstacle negotiations because there is more time to raise the center of the axle to the bump height. Also, a large tire will not follow rough terrain exactly and will act as a filter.

Considerable research has been done in seat-suspension design and ride comfort. Stikeleather (1981) provides the most comprehensive information to date relative to operator seats for agricultural equipment. Other pertinent information can be found in Bell (1982), Claar et al. (1980a, 1980b), Janeway (1975), Matthews (1966, 1973), Rakheja and Sankar (1984a, 1984b), Stikeleather and Suggs (1970), Stikeleather (1973, 1976), and Young and Suggs (1973).

Sound Control in Operator Enclosures

The first generation of tractor cabs came on the market in quantity in the mid 1960s. Many farmers complained that the cabs were noisy. It was not uncommon for noise levels to be greater at the operator's ear with a cab than without one. This was documented by Ryland and Turnquist (1970). Using a 75 percent engine load, they reported a mean sound pressure level of 88.2 dB for a tractor without a cab. Installation of a commercial cab increased the noise level to 94.1 dB. Ryland and Turnquist then insulated the cab, which lowered the noise level to 84.3 dB.

Reduced noise levels have been primarily achieved by incorporating sound control measures in the operator enclosures. This design approach includes isolation mounts for the cab and suitable insulating materials for ceiling, walls, and floors. Sellon (1976) and Zitko (1977) report design practice for proper sound control.

Sound reaching the operator is structure-borne, airborne, or a combination of both. Structure-borne sound results from vibrations transmitted from the vehicle through the cab attaching points. Airborne sound is trans-

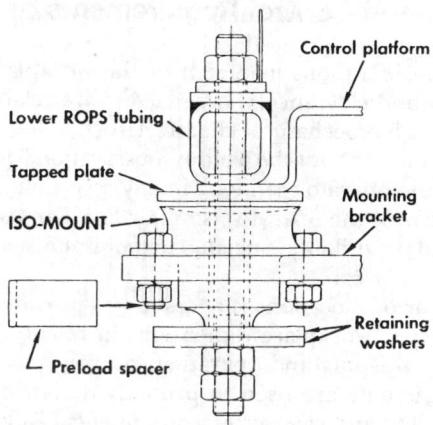

FIGURE 9-12 Typical vibration isolation mount for an operator enclosure. (From R.F. Zitko, "Control Center Design Concepts Series 86 Tractors," ASAE Paper 77-1049, 1977.)

mitted through air and enters the operator area through holes or through the enclosure walls.

One type of cab isolation mount currently in use is shown in figure 9-12. Vibration isolation is achieved by a rubber interfacing between the mounting bracket and the control platform. The magnitude of isolation is dependent on the transmissibility of the rubber being used (curves similar to those in fig. 9-9) and the input frequency of the vibration.

The floor of the cab is treated with a barrier material. A rubber material then overlies the barrier. Surface areas above the floor are treated with noise absorption materials that are effective in the 125- to 2000-Hz range. The predominant noise in tractor and other off-highway equipment is in the frequency range of 125 to 500 Hz.

Large sheet-metal areas tend to resonate and generate airborne sound in the operator enclosure. Constrained layer damping consisting of two layers of steel laminated with a resilient mastic controls the resonance effect.

Proper measurement of sound is necessary to evaluate the effect of design changes on sound control. A current recommended practice is specified by SAE J1008 (Sept. 1980) "Sound Measurement—Self Propelled Agricultural Equipment—Exterior." Fundamental principles of noise and vibration control are covered by Bell (1982).

Spatial, Visual, and Control Requirements of the Operator

Two functional considerations in design of the workplace for a tractor operator are visibility and clearance. These factors are related to the operator's anthropometric and biomechanical characteristics.

Primarily visibility, or "out the window observation," for a tractor requires provisions so that the operator can look in any direction. Near-ground vision to the front and rear of the operator is important. Far vision in all directions is necessary. Secondary visibility is needed to monitor instruments and controls inside the workplace or cab.

Clearance at various locations is necessary to provide access to and from the workplace. Proper workplace dimensions in relation to the seat are important for ease in grasping and operating controls.

Anthropometric data are used to properly design the operator's workplace to meet visibility and clearance requirements. Table 9-5 gives selected anthropometric data for U.S. adult males. The data shown have been adopted for use in SAE J833 as a recommended practice. For example, height inside a tractor cab should be 185 cm to ensure that 95 percent of male tractor drivers will not bump their heads when standing. Use of 185-cm standing height would also ensure satisfactory design for adult women (95 percent is 170 cm).

TABLE 9-5　Selected Anthropometric Dimensions for Male Adults

	Dimension, cm		
Body Measurement	5th Percentile	50th Percentile	95th Percentile
1. Height	162	173	185
2. Sitting height, erect	84	91	97
3. Sitting height, normal	80	87	93
4. Knee height	49	54	59
5. Popliteal height	39	44	49
6. Elbow-rest height	19	24	30
7. Thigh-clearance height	11	15	18
8. Buttock-knee length	54	59	64
9. Buttock-popliteal length	44	50	55
10. Elbow-to-elbow breadth	35	42	51
11. Seat breadth	31	36	40
12. Mass, kg	58	75	98

SOURCE: E. J. McCormick. *Human Factors in Engineering and Design*, 4th ed. McGraw-Hill Book Co., New York, 1976.

It is general practice to use the 95th and 5th percentile values in workplace design. A minimum dimension in some cases should be based on an upper percentile value. Typically, minimum dimensions are used in establishing clearance in doors and escape hatches. On the other hand, maximum dimensions are established from lower percentiles to ensure that a short functional arm can reach a control or steering wheel. Purcell (1980) suggests that dimensions for the 2.5 percentile to 97.5 percentile be used to include 95 percent of the population. Considering the current worldwide marketing of many given tractor models, the designer needs to consider the use of ethnic anthropometric data. A comprehensive source of data can be found in NASA Reference Publication 1024, *A Handbook of Anthropometric Data,* Volume II, 1978.

Purcell (1980) and Stikeleather (1981) discuss in depth the functional and dimensional requirements for the design of seat cushions, back cushions, and armrests relative to a seated operator. Also addressed are functional and dimensional requirements for proper seat location relative to the controls. Purcell (1980) provides functional and dimensional visual requirements for instrumentation location and readout.

Graphical layout and use of anthropometric data are the basic tools for seating placement and location of controls. Figure 9-13 shows a typical layout. Figure 9-14 is a top view of figure 9-13.

The weight of the thighs and upper body should be supported by a seat cushion with a 5-degree seat angle from horizontal (front of cushion elevated). The angle between the backrest and seat should be approximately 95 degrees, with a minimum of 90 degrees. The backrest must support the lumbar back.

Armrests are necessary to provide support and comfort to the shoulders. Also, they must be adjustable vertically to accommodate varying sizes of persons for freedom of movement of the forearm and elbow when operating hand controls. The leg-to-thigh angle should be 110 to 120 degrees for pedal operation, with a minimum foot-to-leg angle of 90 degrees. Seat adjustments should provide at least 150 mm of fore and aft movement and 100 mm of vertical movement. SAE Recommended Practice J899 (Oct. 1980) provides operator's seat dimensions, and SAE J1163 (Jan. 1980) provides dimensions for seat location relative to platform and controls. SAE J898 (July 1982) defines zones of comfort and reach for control location.

In regard to accessing the operator's platform, SAE J185 (June 1981) provides dimensional recommendations for steps, handrails, and handholds. SAE J154a is a design guide for determining the minimum normal operating "space envelope" around the operator in a cab.

Procedural efficacy in operator workplace design concerns the logical arrangements of task elements to enhance operator performance and minimize error. Proper association between controls and displays along with logical grouping of instruments and controls is important in a compatible man-ma-

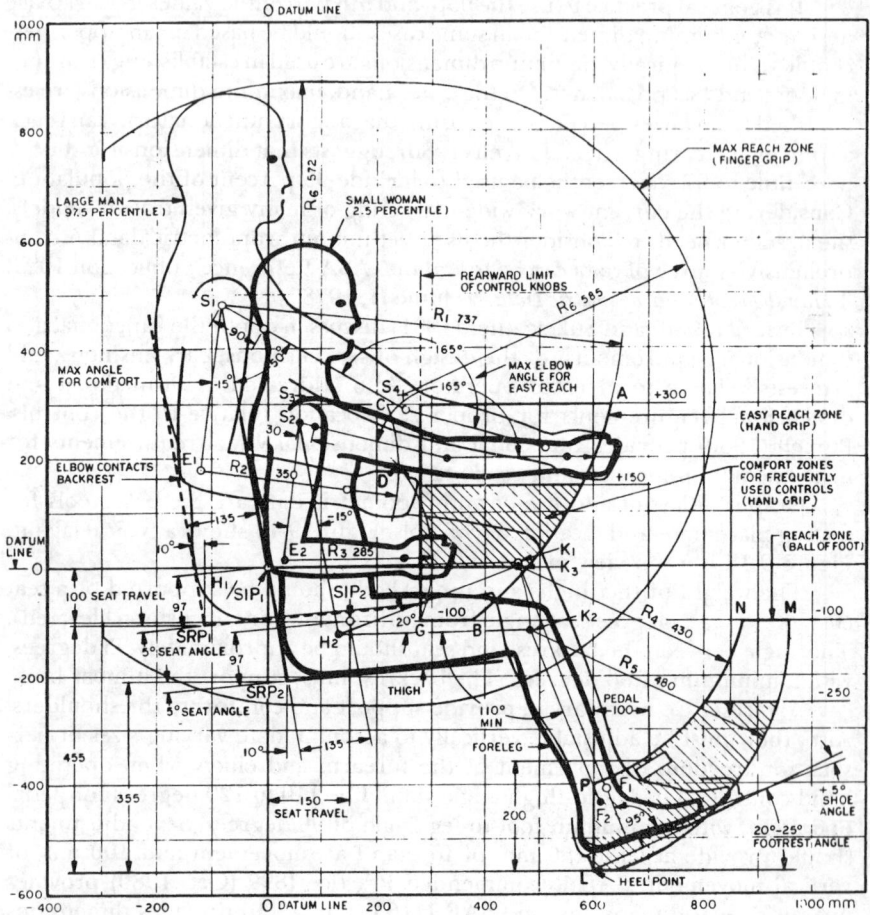

FIGURE 9-13 Graphic solution for finding optimum hand and foot control positions. (From Purcell 1980.)

chine system. A number of standards and engineering practices are used to achieve proper location of operator controls and instruments. ASAE Standard ASAE S335.3 ("Operator Controls on Agricultural Equipment") provides guidelines for the uniformity of location and direction of motion of operator controls to improve operator efficiency and convenience. The controls considered are brake, clutch, engine speed, ground speed (transmission), lift, and steering. For example,

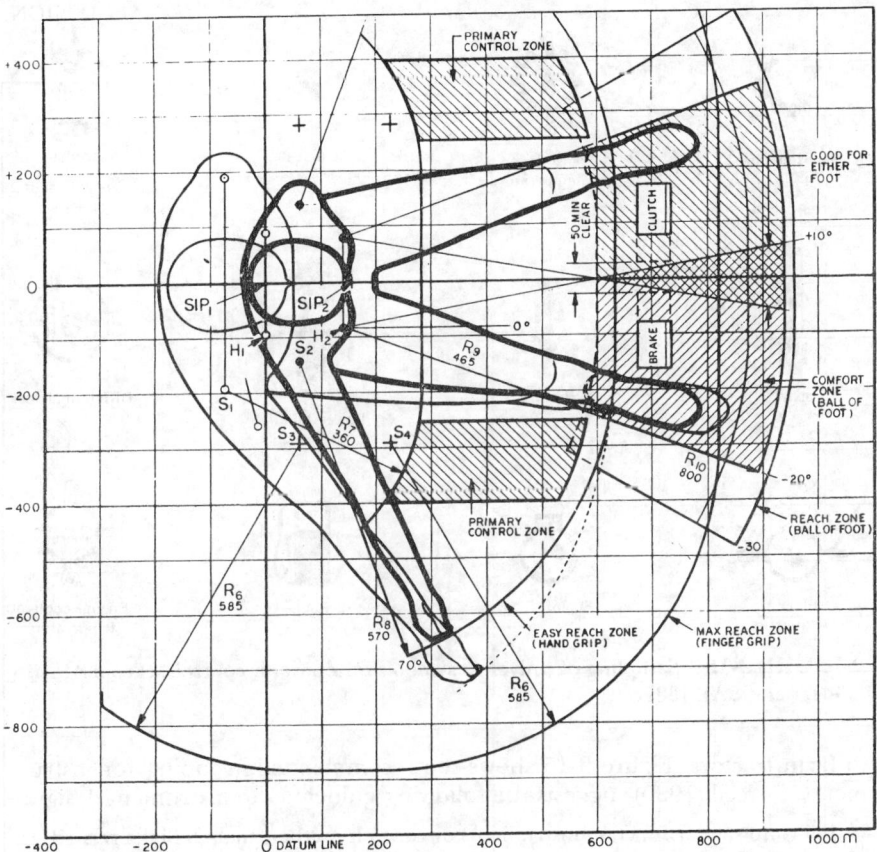

FIGURE 9-14 Optimum hand and foot control positions (top view of fig. 9-13). (From Purcell 1980.)

Clutch Control 5.1.1. When a foot pedal control is provided, it shall be actuated by the operator's left foot with the direction of motion forward and/or downward for disengagement.

To achieve proper association between controls and displays, universal symbols for operator controls have been developed and are published in the standard ASAE S5304.5/SAE J389b. A few selected symbols are shown in figure 9-15. SAE Recommended Practice J209 (Apr. 1980) ("Instrument Face Design and Location for Construction and Industrial Equipment") is applied

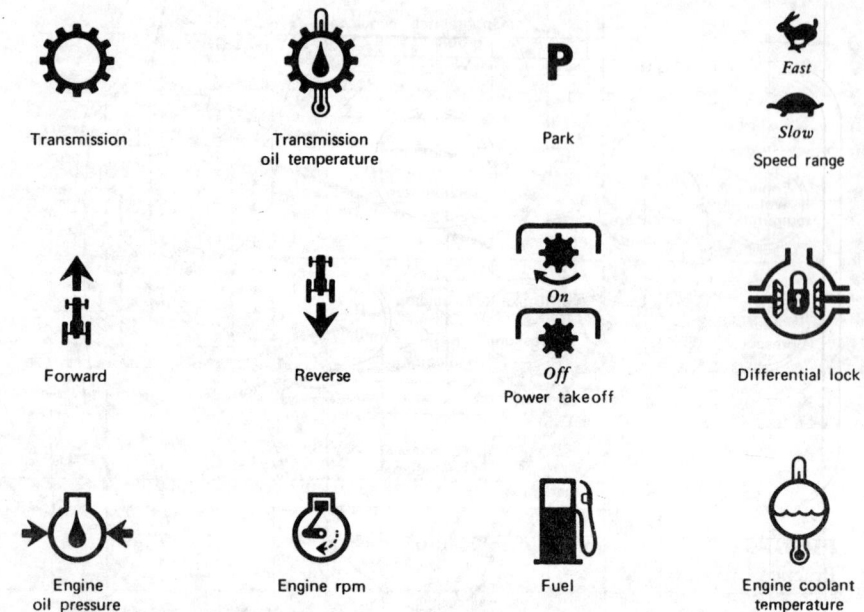

FIGURE 9-15 Sampling of universal symbols for operator controls. (From ASAE S304.5 and SAE J389b.)

to farm tractors. Figure 9-16 shows the recommended grouping for instruments. Purcell (1980) suggests the following guidelines to instrument design:

1. Group similar function instruments or controls on the panel; 2. Make related groups of instruments equal in size; 3. Design the panel as symmetrically as possible; 4. Provide a central zone on the panel for the highest priority instruments. This is the arc lying within the 30 degree cone for easy eye movement, which is 380 mm wide at 700 mm from the operator's eye; 5. Group the gauges in the priority zone horizontally and according to function with engine gauges to the left of the panel center and transmission gauges to the right. The panel center should be marked by either the steering column, or by the tachometer, or by a group of indicator lights; 6. Place all remaining instruments on either side of the priority group and keep their relative positions the same on all vehicles in the line; 7. Controls on the instrument panel should also follow a standard for the whole line of vehicles.

Electronic monitoring and digital readout of tractor performance functions are replacing dials and gages. A digital instrument cluster used by one manufacturer is shown in figure 9-17. Details of the electronics are discussed in chapter 6.

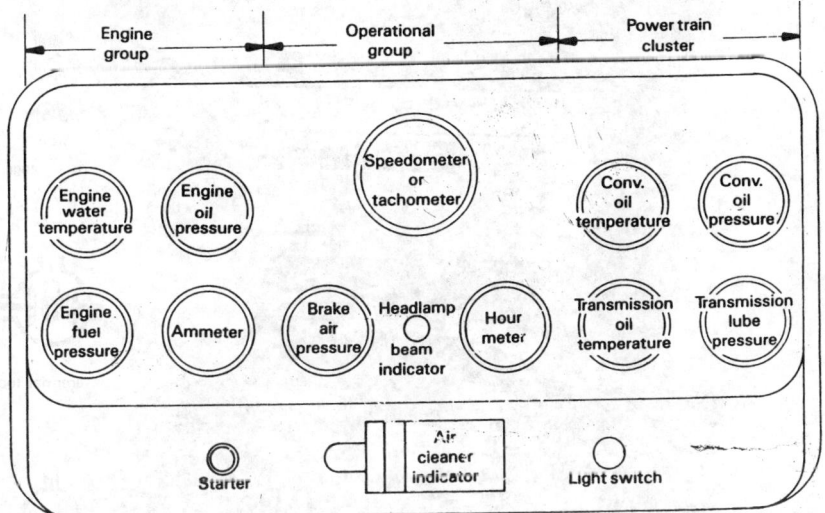

FIGURE 9-16   Instrument panel grouping as specified by SAE Recommended Practice SAE J209.

Primary hand controls that are used continuously such as hand throttle, shift levers, and rock shaft control lever should be placed within the cross-hatched area shown in figure 9-13 and to the right of the operator as shown in the cross-hatched area of figure 9-14, leaving the left hand available for steering at all times. The less frequently used controls such as PTO clutch, parking brake, and differential lock can be allocated to left hand or foot operation outside of the cross-hatched areas indicated previously. Color coding of hand controls, as recommended in ASAE EP443, will aid the operator in control identification.

Rollover Protection for Wheeled Agricultural Tractors

The National Safety Council reported in the 1983 edition of *Accident Facts* that for 1982, overturns accounted for 49 percent of all on-farm tractor fatalities. Twenty percent of fatalities were due to the victim's being run over by the tractor, after having first fallen from the tractor in more than half of the cases. These data cover one third of farm tractors in the United States. During the 14-year period from 1969 to 1982, deaths per 100,000 tractors have decreased from 17.6 in 1969 to 8.1 in 1982. These figures include overturn, runover, power takeoff, and other.

FIGURE 9-17   Digital instrumentation cluster. (Courtesy Case-International.)

The preceding figures emphasize the importance of rollover protection and seat belt use. As the percentage of tractors equipped with rollover protective structures (ROPS) increases along with the operator usage of seat belts, fatality rates due to overturn and runover should decrease. There are three basic types of ROPS: 1. skeleton frame, 2. frame with overhead roof, and 3. enclosed cab. ROPS protect the operator from tractor overturn and/or falling objects. With the enclosed cab, environmental and sound control can also be provided.

The engineering design of ROPS is a complex problem and beyond the scope of this text. Sophisticated computer techniques using the finite element method are currently being used. These techniques will identify and predict the mode and sequence of yielding that takes place during a rollover situation. It also aids in analysis of welds and joints. Yeh et al. (1976) outline an analytical procedure and computer program description for ROPS design.

Much effort has been expended to develop standards for ROPS that establish test and performance requirements. Current standards are ASAE S383.1 ("Roll-Over Protective Structures [ROPS] for Wheeled Agricultural

Tractors") and ASAE S310.3 ("Overhead Protection for Agricultural Tractors—Test Procedures and Performance Requirements"). The reader is encouraged to consult the most recent *ASAE Standards* for revisions or additions to those mentioned.

Testing of ROPS consists of either a static or dynamic loading test in the laboratory, a crush test, a field upset test, and a temperature-material test. After the static or dynamic test, the same ROPS is subjected to a static crush test. The field upset test may be omitted if the laboratory test (static or dynamic) indicates certain compliances. Special steels that must meet certain impact strength have been developed and certified for ROPS.

Thermal Comfort in Operator Enclosures

Operator enclosure design must include cab pressurization, filtration, air movement, heating, cooling, and window defrosting. These factors must be considered in order to provide clean air and proper air velocity, temperature, and humidity for human thermal comfort. Design values reported by Zitko (1977) to meet a wide range of climatic conditions are given in table 9-6. Cab pressurization is necessary to prevent dust from entering the enclosure. Proper measurement of cab pressurization is outlined in SAE J1012 (March 1980) ("Agricultural Equipment Enclosure Pressurization System Test Procedure").

Fanger (1970) was the first to generalize the physiological basis of comfort so that for any activity level (metabolism) and clothing values, it is possible to analytically predict comfort in terms of combinations of air temperature, mean radiant temperature, air humidity, and relative air velocity. Fanger derived a comfort equation that takes the form:

$$F(H/A_{Du}, I_{cl}, t_a, t_{mrt}, P_a, v) = 0 \qquad (19)$$

TABLE 9-6 Design Parameters for Maintaining Thermal Environment in a Tractor Cab

Parameter	Rating or Capacity
1. Heating	8.2 kW at 66°C and water flow 11.4L/min
2. Cooling	7.0 kW at 32°C and 60% relative humidity
3. Air movement	Three-speed blower rated at .236 m³/s at 50 Pa
4. Cab pressurization	50–100 Pa above outside of cab
5. Fresh air filter	1.92 m² pleated paper, self-cleaning

SOURCE: R. F. Zitko. "Control Center Design Concepts Series 86 Tractor." ASAE Paper 77-1409, 1977.

where $\dfrac{H}{A_{Du}}$ = internal heat production per unit body surface area

$(A_{Du}$ = DuBois area)

I_{cl} = thermal resistance of clothing

t_a = air temperature

t_{mrt} = mean radiant temperature

P_a = pressure of water vapor in ambient air

v = relative air velocity

A tractor operator will expend 60 to 150 kcal/h · m². Sedentary persons will average 50 kcal/h · m². Light, medium, and heavy clothing will have values of 0.5, 1.0, and 1.5 clo respectively. On a bright, sunny day, the mean radiant temperature, t_{mrt} can be 5° to 10°C above the cab air temperature t_a. Cab air velocities should be in the 1- to 3-m/s range.

The solution to equation 19 is complex. Fanger (1970) solved the equation on a digital computer and presents the solutions in terms of a family of graphs for practical engineering use. The reader is encouraged to consult these graphs for application and use.

Kaufman et al. (1976) concluded that Fanger's approach to thermal comfort was effective in predicting comfort in a tractor cab for summer conditions in South Dakota. Eriksson and Domier (1975) have done work on the comfort of subjects in heated tractor cabs. They compared the cab temperature chosen by the subject to a calculated comfortable temperature based on Fanger's comfort equation and found a large spread in the data. In 1975, Domier reviewed the research relating to thermal comfort in tractor cabs. Other research of interest is reported by Turnquist and Thomas (1976) and Eriksson (1974).

Safety

Safety is improved by education, by regulation or laws, and by design. Although all of these methods are effective, the most positive method is to design the product or system in such a way that the hazard is removed. The engineer's responsibility and expertise are to design safety into the machines and systems for which he or she is responsible.

Designing for safety is not a simple process. The engineer must consider standards, laws, court decisions, human factors, function, user reaction, and sales. There are conflicts in this process, unfortunately. The engineer's job is to design the product so that there is a minimum of conflict among the preceding considerations. For example, it does little good to design a safety shield that is soon discarded because it is inconvenient to use. On the other hand, if the function of a machine is seriously hampered by a safety device, the machine will not sell. The engineer must not ignore the safety aspect of a machine because the problem is difficult.

PROBLEMS

1. Using figure 9-2, define two sets of dry-bulb temperature and relative humidity that would be in the comfortable zone. Define one set of conditions for an uncomfortably cold environment and an uncomfortably warm environment. How would you propose to make measurements inside a cab to relate cab conditions to the comfort criteria of figure 9-2?

2. A test engineer has isolated a pure-tone noise in a tractor and determined its frequency to be 8000 Hz and peak-to-peak amplitude to be 0.01 Pa. Determine the period, RMS sound pressure, and resulting sound pressure level. Should the engineer be concerned about eliminating or reducing this source of noise? Why?

3. Show through calculations that doubling the sound pressure results in a 6-dB increase.

4. Sound was measured at 80 dB in the operator's station on a tractor. What is the RMS sound pressure in newtons per square meter (N/m^2)? If the sound pressure increased eight times, determine the resulting sound pressure in decibels. Why is the decibel scale used in sound measurement rather than a sound pressure scale?

5. Two components of sound have been identified in a tractor and measured at 74 dB and 81 dB. What is the resulting noise level?

6. What is the maximum theoretical sound pressure in decibels that can exist? Hint: Let P equal 1 atm.

7. An accelerometer mounted to the waist of a seated tractor operator records a vertical RMS acceleration of 1.5 m/s^2 at 8 Hz. What is the most likely source of this vibration? Would it be desirable for the operator to experience this level of vibration? Why? What is the amplitude of this vibration in millimeters?

8. If the RMS acceleration in problem 7 were decreased by 10 dB because of improved seat design, what would be the resulting RMS acceleration? How many hours could the operator tolerate this level of vibration?

9. Using the anthropometric data in table 9-5, determine the operator workplace dimensions for (a) seat height range, (b) seat width, (c) seat depth (front-rear), (d) arm rest width and height, and (e) height inside a cab.

10. Design a seat suspension for (1) seat and operator mass of 90 kg, (2) transmissibility of 0.315, (3) tractor chassis frequency of 4 Hz, and (4) static spring deflection of seat and operator of 10 cm.

(a) Determine undamped natural frequency of the seat.

(b) Determine spring and shock absorber design values.

(c) Determine the decibel reduction in RMS acceleration that the seat suspension provides.

(d) Using ISO 2631 criteria, how long should a tractor driver be exposed to a tractor chassis input of 1 RMS m/s^2 for the suspension you designed?

(e) If the tractor chassis frequency decreases, will the seat suspension be more or less effective? Why?

11. Inlet air velocity to a cab is 300 m/min. How much louver area, in square centimeters, is needed if the fan is delivering 0.236 m^3/s?

REFERENCES

Akesson, N. B. et al. "Health Hazards to Workers from Application of Pesticides." ASAE Paper 74-1007, 1974.

Bell, L. H. *Industrial Noise Control Fundamentals and Applications,* 1st ed. Marcel Dekker, Inc., New York, 1982.

Claar II, P. W., W. F. Buchele, and S. J. Marley. "Agricultural Tractor Chassis Suspension System for Improved Tractor Ride Comfort." ASAE Paper 80-1565, 1980.

Den Hartog, J. P. *Mechanical Vibrations,* 4th ed. McGraw-Hill Book Co., New York, 1956.

Duncan, J. R., and E. L. Wegscheid. "Off-Road Vehicle Simulation for Human Factors Research." ASAE Paper 82-1610, 1982.

Eriksson, Hans-Arne. "Climatic Requirements in Tractor Cabs." Report No. 9, Swedish Institute of Agricultural Engineering, Uppsala, Sweden, 1974.

Eriksson, H., and K. W. Domier. "Heating and Ventilating of Tractor Cabs." ASAE Paper 75-1516, 1975.

Fanger, P. O. *Thermal Comfort: Analysis and Applications in Environmental Engineering.* Danish Technical Press, Copenhagen, 1970.

Gagge, A. P., A. C. Burton, and H. C. Bazett, "A Practical System of Units for the Description of the Heat Exchange of Man with His Environment." *Science,* vol. 94, 1941, pp. 428–430.

GenRad, Inc. *Handbook of Noise Measurement,* GenRad, Inc., West Concord, MA, 1974.

Gerke, F. G., and D. L. Hoag. "Tractor Vibrations at the Operator's Station." *Trans. of ASAE,* vol. 24, no. 5, 1981, pp. 1131–1134.

Griffen, M. J. "The Evaluation of Vehicle Vibration and Seats." *Applied Ergonomics,* vol. 9.1, 1978, pp. 15–21.

Hansson, J. E., Lars Sjøflot, and C. W. Suggs. "Matching the Farm Machine to the Operator's Capabilities and Limitations." *Implement & Tractor,* Aug. 21, 1970.

Janeway, R. N. "Human Vibration Tolerance Criteria and Applications to Ride Evaluation." SAE Paper 750166, 1975.

Kaufman, K. R., P. K. Turnquist, and R. N. Swanson. "Physiological Responses and Thermal Comfort of Subjects in a Tractor Cab." ASAE Paper 76-1577, 1976.

Matthews, J. "The Ergonomics of Tractor Design and Operation." *Proceedings of the XVI CIOSTA Congress,* Wageningen, The Netherlands, 1972.

Matthews, J. "The Measurement of Tractor Ride Comfort." SAE Paper 730795, 1973.

Matthews, J. "Ride Comfort for Tractor Operators. IV. Assessment of the Ride Quality of Seats." *J. Agric. Eng'g Res.* (England), vol. 2, no. 1, 1966, pp. 44–57.

Mick, D. L. "The Tractor Cab as a Protective Device during Pesticide Applications." ASAE Paper 73-1553, 1973.

Miller, M. L., A. W. Eissler, and J. W. Ackley. "Tractor Operator Enclosure Environment During Pesticide Application Operations." ASAE Paper 79-1009, 1979.

National Safety Council. *Accident Facts 1983 Edition,* National Safety Council, Chicago, 1983.

Noren, O. "Dust Concentrations During Operations with Farm Machines." ASAE Paper 85-1055, 1985.

Pradko, F., and R. Lee. "Analysis of Human Vibration." *SAE Trans.,* vol. 77, 1968.

Purcell, W. F. H. "The Human Factor in Farm and Industrial Equipment Design." ASAE Distinguished Lecture Series, no. 6, 1980.

Rakheja, S., and S. Sankar. "Suspension Designs to Improve Tractor Ride: I. Passive Seat Suspension." SAE Paper 841107, 1984a.

Rakheja, S., and S. Sankar. "Suspension Designs to Improve Tractor Ride: II. Passive Cab Suspension." SAE Paper 841108, 1984b.

Raney, J. P., J. B. Liljedahl, and R. Cohen. "The Dynamic Behavior of Farm Tractors." *Trans. of ASAE,* vol. 4, no. 2, 1961, pp. 215–218, 221.

Rossegger, R., and S. Rossegger. "Health Effects of Tractor Driving." *J. Agric. Eng'g Res.,* vol. 5, no. 3, 1960, p. 241.

Ryland, D. W., and P. K. Turnquist. "Effect of Cab, Soundproofing, and Exhaust-Control Methods on Tractor Noise at Operator's Site." *Trans. of ASAE,* vol. 13, no. 1, 1970, p. 148.

Sellon, R. N. "Design of Operator Enclosures for Agricultural Equipment." ASAE Distinguished Lecture Series, no. 2, 1976.

Smith, D. W. "Computer Simulation of Tractor Ride for Design Evaluation." SAE Paper 770704, 1977.

Stikeleather, L. F. "Operator Seats for Agricultural Equipment." ASAE Distinguished Lecture Series, no. 7, 1981.

Stikeleather, L. F. "Evaluating the Vibration of Shock Isolation Qualities of Operator Seats for Construction Machinery." SAE Paper 730823, 1973.

Stikeleather, L. F. "Review of Ride Vibration Standards and Tolerance Criteria." SAE Paper 760413, 1976.

Stikeleather, L. F., and C. W. Suggs. "An Active Seat Suspension System for Off-Road Vehicles." *Trans. of ASAE,* vol. 13, no. 1, 1970, p. 99.

Suggs, C. W. "Agricultural Machinery Noise and Vibration Levels in Comparison to Human Comfort and Safety Limits." ASAE Paper 73-524, 1973.

Turnquist, P. K., and J. C. Thomas. "The Subjective Response of Males to Comfort Under Controlled Tractor Cab Environment." *Trans. of ASAE*, vol. 19, no. 3, 1976, p. 402.

Yeh, R. E., Y. Haung, and E. L. Johnson. "An Analytical Procedure for the Support of ROPS Design." SAE Paper 760690, 1976.

Young, R. E. and C. W. Suggs. "Seat Suspension System for Isolation of Roll and Pitch in Off-Road Vehicles." *Trans. of ASAE*, vol. 16, no. 5, 1973, p. 876.

Zander, J. *Ergonomics in Machine Design (A Case-Study of the Self-Propelled Combine Harvester).* Agricultural University, Wageningen, The Netherlands, 1972.

Zitko, R. F. "Control Center Design Concepts Series 86 Tractor." ASAE Paper 77-1049, 1977.

SUGGESTED READINGS

Claar II, P. W., and P. Sheth. "Off-Road Vehicle Ride: Review of Concepts and Design Evaluation with Computer Simulations." SAE Paper 801023, 1980.

Cooper, W. A., "The Ear, Hearing, Loudness and Hearing Damage." In Malcomb J. Crocker (ed), *Reduction of Machinery Noise*. Proceedings of Short Courses at Purdue University. Purdue University Press, West Lafayette, IN, 1974 and 1975.

Crocker, M. J. "Noise Control by Use of Enclosures and Barriers." In Malcomb J. Crocker (ed), *Reduction of Machinery Noise*. Proceedings of Short Courses at Purdue University. Purdue University Press, West Lafayette, IN, 1974 and 1975.

Deere & Co. *Fundamentals of Machine Operation—Tractors.* John Deere Service Publications, Moline, IL, 1974.

Domier, K. W. "A Review of Research Relating to Thermal Comfort of Cab Operation." Report No. 15, Swedish Institute of Agricultural Engineering, Uppsala, Sweden, 1975.

Dreyfus, Henry. *The Measure of Man*. Whitney Library Design, 1960.

Hamilton, J. F. "Fundamentals of Vibration and Noise Control by Vibration." In Malcomb J. Crocker (ed), *Reduction of Machinery Noise*. Proceedings of Short Courses at Purdue University. Purdue University Press, West Lafayette, IN, 1974 and 1975.

"A Handbook of Anthropometric Data, Volume II." NASA Reference Publication 1024, 1978.

Human Engineering Guide to Equipment Design, rev. ed. Sponsored by Joint Army-Navy-Air Force Steering Committee, 1972.

Lane, R. S. "Sources and Reduction of Diesel Engine Noise." In Malcomb J. Crocker (ed), *Reduction of Machinery Noise*. Proceedings of Short Courses at Purdue University. Purdue University Press, West Lafayette, IN, 1974 and 1975.

McCormick, E. J. *Human Factors in Engineering and Design*, 4th ed. McGraw-Hill Book Co., New York, 1976.

"A Study of Noise Induced Hearing Damage Risk for Operators of Farm and Construction Equipment." Technical Report, SAE Research Project R-4, Society of Automotive Engineers, 1969.

Vaidya, R. G. "Sound Propagation Outdoors." In Malcomb J. Crocker (ed), *Reduction of Machinery Noise*. Proceedings of Short Courses at Purdue University. Purdue University Press, West Lafayette, IN, 1974 and 1975.

Woodson, W. E. *Human Engineering Guide for Equipment Designers*. University of California Press, Berkeley, CA, 1964.

10
TRACTION

Transmitting engine power to the drawbar in agricultural and other off-highway vehicles is achieved through traction devices, namely wheels and tracks. Of the three principal ways of transmitting tractor-engine power into useful work—power takeoff, hydraulic, and drawbar—the least efficient and most used method is the drawbar. Basic configurations of wheels and tracks are modified to meet special operating conditions in various parts of the world. The predominate traction device is the pneumatic tire. The majority of this chapter deals with the theory and operating parameters of pneumatic tires for agricultural tractors. Terminology used in this chapter conforms, when possible, to ASAE Standard: ASAE S296.2.

Traction Mechanics

Traction is developed by the interaction of mechanical devices with soil. Theory, experimental studies, and field tests provide the general nature of these interactions for analysis and design of tractive systems.

Central to design of traction systems is the development of performance equations. Freitag (1985) indicates that performance equations have developed by use of theoretical, analog, and empirical approaches. Owing to the complexities of soil-traction device interactions, the theoretical approach has not yielded readily usable solutions for the designer. However, it does provide general performance response for a particular combination of soil-vehicle parameters.

Mohr-Coulomb Failure Criteria

Application of the Mohr-Coulomb criteria to a soil-plate(lug) situation provides insights into the nature of plate(lug) loading and soil shear behavior. If a plate of width b and length l is equipped with lugs sufficiently long, such that an area $A = bl$ shears off as in figure 10-1(a), we find that the force required is usually dependent upon both the normal force and the area.

240

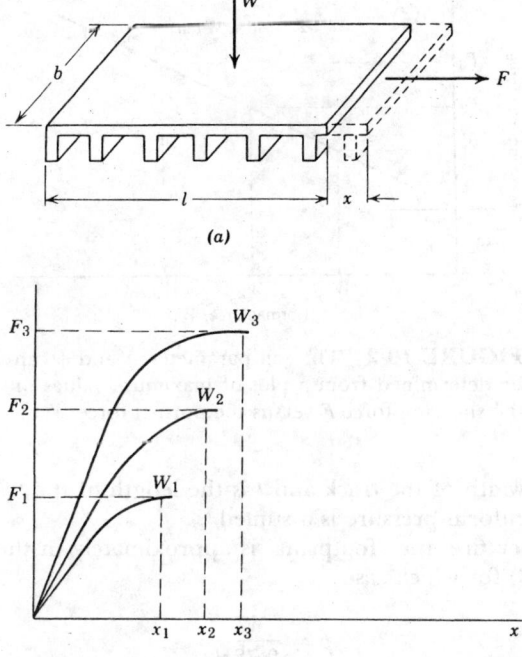

FIGURE 10-1 Method of determining maximum value of shearing force for various levels of vertical loading.

If we plot maximum values of F against W (as in figure 10-2), we find for soils having some cohesion that F does not approach zero as W approaches zero. If the maximum values are plotted for a soil that has both cohesion c and internal friction ϕ, the result will be similar to figure 10-2. The equation for such a curve is

$$F = Ac + W \tan \phi \qquad (1)$$

or

$$F = A(c + p \tan \phi) \qquad (2)$$

where $p = W/A$ is the average normal soil pressure and A is the area.

For a track

$$p = \frac{W}{bl} \qquad (3)$$

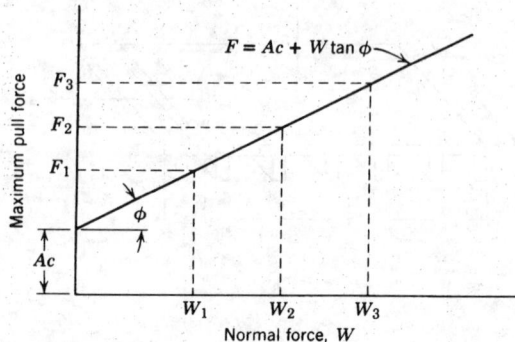

FIGURE 10-2 The soil parameters c and ϕ can be determined from a plot of maximum values of the shearing force F versus the normal force W.

where b is the width of the track and l is the length of the track in contact with the soil. Uniform pressure is assumed.

For a rubber tire, the "footprint" is approximately in the shape of an ellipse (fig. 10-3) for which case

$$p = \frac{W}{0.78bl} \tag{4}$$

If we know the soil values c and ϕ, the maximum soil thrust can now be approximated by equation 2.

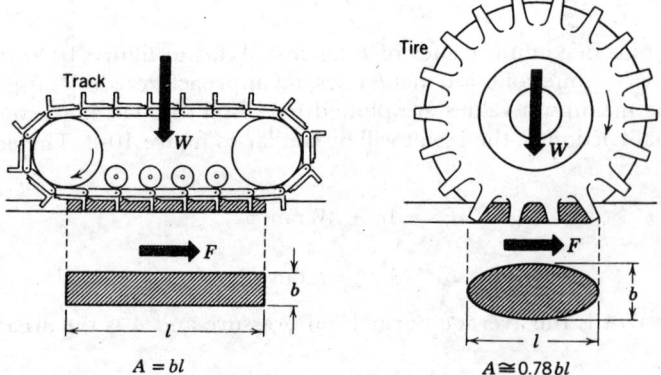

FIGURE 10-3 Soil thrust from a traction member is, in most cases, affected by both shear area and weight.

This simple model is not very useful because we seldom know c and ϕ, and p is not uniform. In addition, the model does not predict the magnitude of the motion resistance force acting on the traction device.

Traction Performance Equations

Dwyer (1984) provides a good overview in the development of analytical and empirical relationships for tractive performance of wheeled vehicles. The mobility number concept was first derived by Freitag (1965) and further developed by Turnage (1972) using dimensional analysis and a series of performance measurements over a wide range of tire size and shape. Wismer and Luth (1974) further developed the utility of this approach for predicting tractive performance.

Traction is the term applied to the driving force developed by a wheel, track, or other traction device.

Tractive efficiency (TE) is defined as the ratio of output power to the input power for a traction device. It is the measure of the efficiency with which the traction device transforms the torque acting on the axle into linear drawbar pull. Several factors lower the tractive efficiency; among these are steering, rolling resistance, slip, and friction in and deflection of the traction device.

Net traction coefficient (μ) is defined as the ratio of the net pull produced to the dynamic normal load on the traction device. The difference between traction efficiency and coefficient of net traction should be recognized.

Motion resistance ratio (ρ) is defined as the rolling resistance force divided by the normal load on the traction device.

The tractive ability is affected by the vertical soil reaction against the traction wheels. Weight transfer, from drawbar pull, decreases the soil reaction against the front wheels by an amount ΔR_f and increases the reaction against the rear wheels by an amount ΔR_r, thus adding to the maximum drawbar pull for a two-wheel-drive tractor. The symbols in the "weight transfer" equations are defined in figure 10-4. See chapter 11 for the derivation of these equations.

$$\Delta R_r = \frac{Py_f}{L_1} \tag{5}$$

and

$$\Delta R_f = \frac{Py_r}{L_1} \tag{6}$$

Any means of increasing the rear wheel reaction will increase the traction of the rear wheel if the soil has sufficient strength and if sinkage does not limit the traction.

Figure 10-5(a) illustrates one of several alternative methods of describing

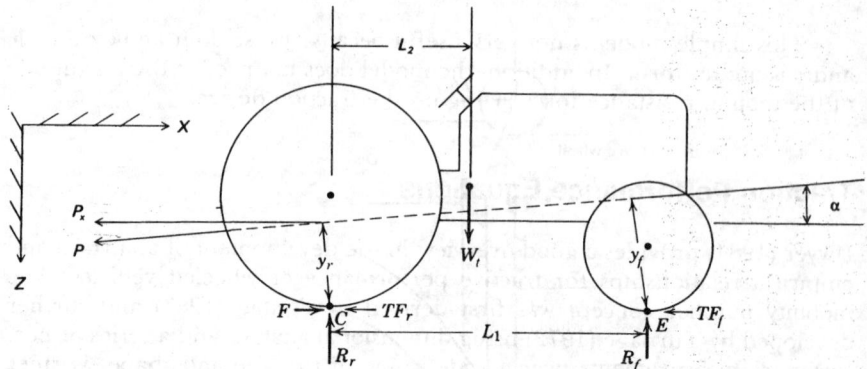

FIGURE 10-4 Steady-state free-body diagram of drawbar pull and soil reactions on a tractor.

the forces acting on a wheel. The method illustrated in figure 10-5(a) and used in this book is condensed from a publication by Wismer and Luth (1974).

Figure 10-5 is divided into three distinct force states: braked, driven, and driving. The transition point between the braked and driven force states is the towed wheel condition. A towed wheel is unpowered: axle torque is zero, neglecting bearing friction. The transition point between the driven and driving force states is the self-propelled wheel condition. For a self-propelled wheel, pull is zero, with the applied torque simply overcoming the motion resistance of the wheel.

The curves presented in figure 10-5(a) represent a given soil strength, tire size, and load. As soil strength increases, the curves move upward to the left; as soil strength decreases, they move downward to the right.

In figure 10-5(a), both the axle torque and pull are plotted as functions of wheel slip. These reactions develop from soil stresses resulting from slippage (motion loss) of the wheel. Slip is defined as

$$S = 1 - \frac{V_a}{V_t} \tag{7}$$

where S = wheel slip or travel reduction
V_a = actual travel speed
V_t = theoretical wheel speed = $r\omega$
r = rolling radius of wheel on hard surface
ω = angular velocity of wheel

The term "rolling radius" is defined in ASAE S 296.2 (ASAE Standards 1985) as the distance traveled per revolution of the traction device divided by 2π when operating at the specified zero condition. The zero condition selected here is the vehicle operating in a self-propelled condition on a hard surface, such as a smooth road, with zero drawbar load. This differs from another

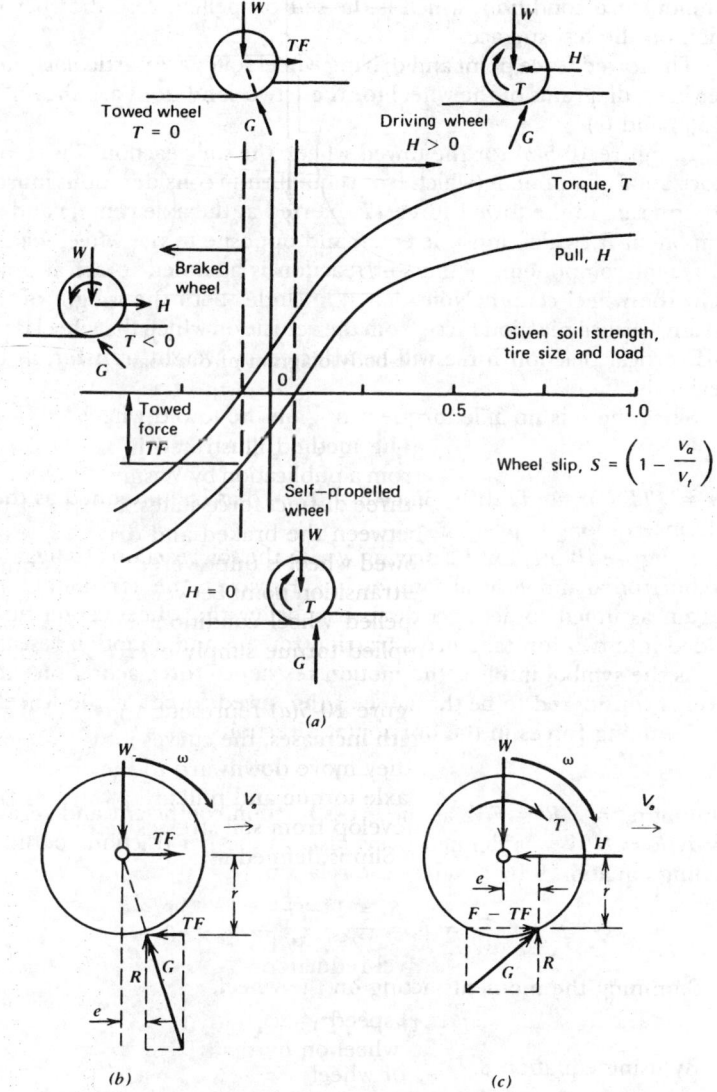

FIGURE 10-5 (*a*) Pull-torque-slip relation for wheels on soil. (*b*) Free-body diagram of a towed wheel. (*c*) Free-body diagram of a driving wheel.

common zero condition, which is the self-propelled, zero drawbar load condition on the test surface.

The towed force point and driving wheel states are particularly important. Free body diagrams of the wheel for these two conditions are shown in figures 10-5(b) and (c).

In figure 10-5(b) for the towed wheel, the soil reaction G is resolved into a horizontal component (which from equilibrium considerations must be equal and opposite to the towed force TF exerted at the axle center) and a vertical component R (which must be equal and opposite to the wheel load W). The horizontal component of the soil reaction is assumed to act at a distance r below the wheel center. Note that W includes both the weight of the wheel and any vertical reaction force from the vehicle on which the wheel is mounted. This vertical reaction force will be affected by weight transfer, as discussed previously.

Since there is no axle torque acting on the towed wheel,

$$TFr - Re = 0 \tag{8}$$

or $e = (TF/R)r = (TF/W)r$. Since $\rho = TF/W$ has been defined as the motion resistance ratio, $e = \rho r$.

In figure 10-5(c) for the driving wheel, the soil reaction G is again resolved into horizontal and vertical components. However, the horizontal component is again assumed to act at a distance r below the wheel center and is now divided into two forces: a gross traction force F and a motion resistant force TF. As the symbol implies, the motion resistance force acting on the driving wheel is considered to be the same as the towed force for the wheel.

Summing forces in the horizontal direction

$$H = F - TF \tag{9}$$

Defining $\mu_g = F/R = F/W$ as the gross traction coefficient and recalling that $\mu = H/R = H/W$ was defined previously as the net traction coefficient, and dividing equation 9 by W, gives the relation:

$$\frac{H}{W} = \mu = \frac{F}{W} - \frac{TF}{W} = \mu_g - \rho \tag{10}$$

Summing the moments acting on the wheel,

$$T - (F - TF)r - Re = 0 \tag{11}$$

By using equation 8,

$$T = Fr \tag{12}$$

Thus the wheel torque T is assumed equal to the gross tractive force F acting at a moment arm equal to the rolling radius r.

TABLE 10-1 Wheel Soil Model Parameters

Parameter	Symbol	Dimensions
Soil:		
Cone index	CI	FL^{-2}
Wheel:		
Tire section width	b	L
Overall tire diameter	d	L
Tire rolling radius	r	L
System:		
Load	W	F
Towed force	TF	F
Pull	H	F
Gross tractive force	F	F
Slip	S	—

Traction Prediction from Dimensional Analysis

Equation Development

Use of analytical equations for traction prediction has been limited because convenient evaluation of the soil properties for cohesion and internal friction is difficult. Dimensional analysis is used to simplify the prediction equations for the multivariable system. The variables considered in the following equations are presented in table 10-1.

As shown, nine pertinent variables are involved in the traction equations. Seven dimensionless ratios are needed to formulate a prediction equation.* An adequate set of dimensionless ratios relating the variables is

$$\rho\left(= \frac{TF}{W}\right), \mu\left(= \frac{H}{W}\right), \mu_g\left(= \frac{F}{W}\right) = f\left(\frac{CIbd}{W}, \frac{b}{d}, \frac{r}{d}, S\right) \qquad (13)$$

However, since $\mu = \mu_g - \rho$, experimental relations only need be developed for ρ and μ_g.

Towed Force

The towed force or motion resistance of a pneumatic tire is dependent on load, size, and inflation pressure, as well as soil strength. For soils that are

---

*For further information, see D. R. Freitag, "A Dimensional Analysis of the Performance of Pneumatic Tires on Soft Soils." Technical Report No. 3-688, USAE Waterways Experiment Station (1965), or a good textbook on similitude or model analysis.

not very soft and tires that are operated at nominal tire inflation pressures,*
the towed force can be predicted from

$$\rho = \frac{TF}{W} = \frac{1.2}{C_n} + 0.04 \tag{14}$$

where C_n = wheel numeric = $\dfrac{CIbd}{W}$

(Dimensions must be selected such that the wheel numeric
is dimensionless.)

CI = cone index measured with a cone penetrometer as in
ASAE S 313.2

It should be noted that for a firm surface such as compacted dry clay, the C_n
value would be very large and the towed force would be equal to 0.04 times
the wheel load. This rolling resistance is attributed to tire flexing and scrub-
bing. Equation 14 was developed for tires with a tire width/diameter b/d ratio
of approximately 0.3. Any large deviation from this width/diameter ratio can
be expected to change the quantitative relation of the towed force function.
The towed force number defined by equation 14 is presented graphically in
figure 10-6.

Gross Tractive Force

The variations of the gross tractive force with soil strength and slip have been
incorporated into a relation including the effect of wheel load and tire size:

$$\mu_g = \frac{F}{W} = \frac{T}{rW} = 0.75(1 - e^{-0.3c_n s}) \tag{15}$$

where e = base of natural logarithms.

Pull

Substituting equations 14 and 15 into equation 10, the net traction coefficient
μ is given by

$$\mu = \frac{H}{W} = 0.75(1 - e^{-0.3c_n s}) - \left(\frac{1.2}{C_n} + 0.04\right) \tag{16}$$

A practical restriction of $b/d \approx 0.30$ is imposed on the final equation along
with a tire deflection/section height ratio (δ/h) limitation of 0.20. The restriction on δ/h is associated with an $r/d \approx 0.475$. This reduces the pull relation to

---

*Normal tire inflation pressures in agriculture, earth-moving, and forestry applications produce
static tire deflections of approximately 20 percent of the undeflected section height. The traction
equations are based on this deflection value.

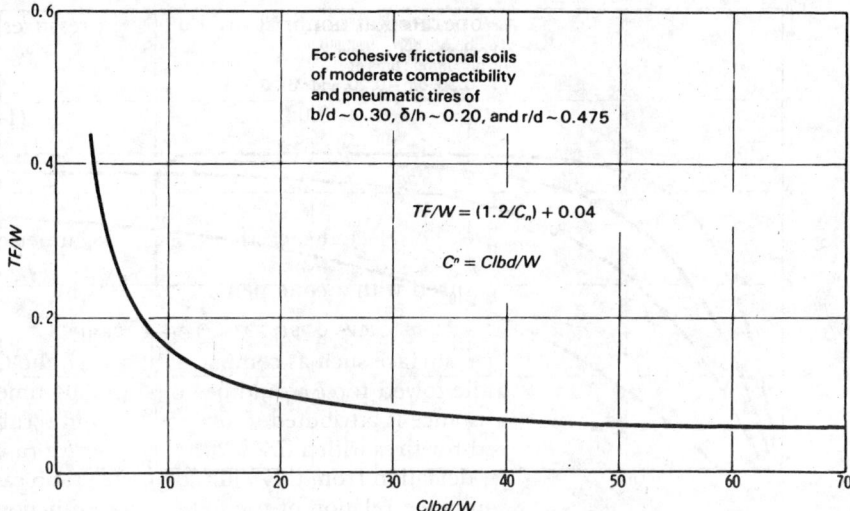

FIGURE 10-6 Towed wheel performance relation. (From Wismer and Luth 1974.)

one dependent (H/W) and two independent $(C_n$ and $S)$ dimensionless ratios resulting in equation 16, which is presented graphically in figure 10-7.

Prediction equations 14, 15, and 16 give good estimates of the performance of a single tire except where the surface of the soil is weak. However, the effect of tread differences cannot be predicted with the equations.

Tractive Efficiency

The pull, torque, and slip characteristics of a driving wheel define both the magnitude and efficiency of tractive performance. The "pull/weight" or "net tractive coefficient" is an accepted term for defining performance level. Similarly, the term "tractive efficiency" (TE) has been adopted to define efficiency. Tractive efficiency of a wheel is defined as:

$$TE = \frac{\text{output power}}{\text{input power}} \tag{17}$$

which can be expressed as

$$TE = \frac{HV_a}{T\omega} = \frac{HV_a}{T\left(\dfrac{V_t}{r}\right)} = \frac{\left(\dfrac{H}{W}\right)}{\left(\dfrac{F}{W}\right)}(1 - S) \tag{18}$$

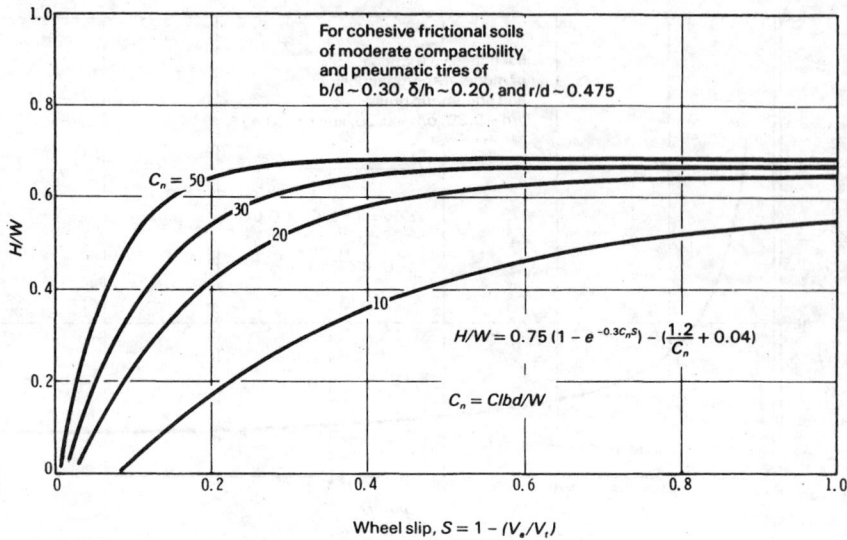

Wheel slip, $S = 1 - (V_a/V_t)$

FIGURE 10-7   Driving wheel performance relation. (From Wismer and Luth 1974.)

The variation of tractive efficiency and the pull/weight (H/W) ratio of a driving wheel with slip are shown in figure 10-8. It is readily observable that TE reaches a maximum at a relatively low slip and then decreases with increasing slip. Also note that the maximum TE occurs at lower slip values for the large C_n values that are associated with higher soil strengths or lower wheel loadings. Maximum power output of a wheel occurs at the wheel slip of maximum TE. However, the H/W ratio is not close to its maximum value at this slip. The requirement for a large drawbar pull necessitates that the design slip be selected to the right of the slip corresponding to the peak of the TE curve. A typical design TE curve is shown in figure 10-8. From this curve, the design TE, H/W, and slip for a variety of soil strengths and wheel loading combinations, in terms of C_n, can be determined: for example, for $C_n = 30$, TE $= 0.72$, $H/W = 0.51$, slip $= 0.16$. This approach permits balancing the design of the vehicle over the range of soil strengths it will probably encounter in its operational life.

Burt et al. (1979) conducted tests to determine drawbar pull, input power, output power, and tractive efficiency as a function of travel reduction and dynamic load. Figures 10-9, 10-10, and 10-11 show the results for a 12.4–28 bias four-ply tire having an R-1 tread. Soil type was a Decatur clay loam, and mean cone index was 252.5 N/cm for 0–15 cm depth.

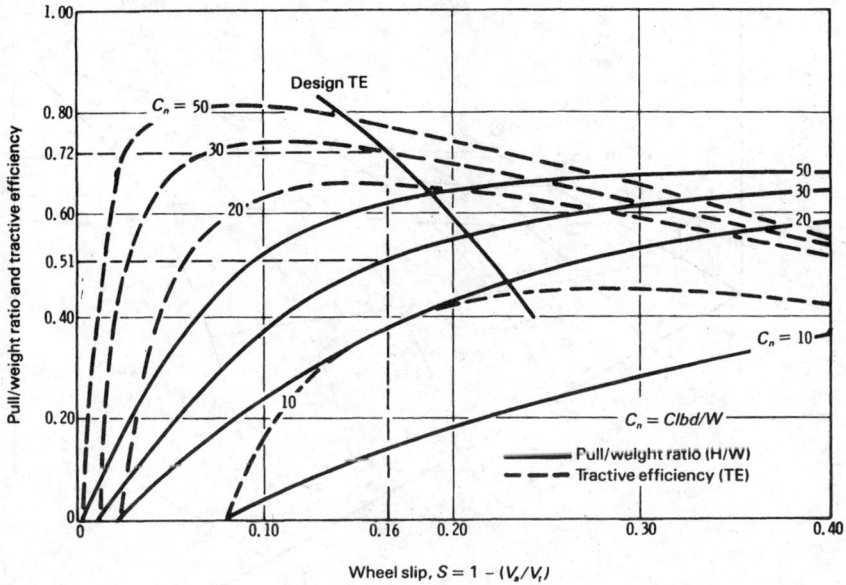

FIGURE 10-8   Tractive performance of wheels on soil. (From Wismer and Luth 1974.)

The Wismer-Luth equations discussed earlier have restrictions on the tire section width to tire diameter ratio (b/d), the tire deflection to tire section height ratio (δ/h), and the tire rolling radius to diameter ratio (r/d). The preceding ratios were held constant at values of approximately 0.30, 0.20, and 0.475, respectively. Ashmore et al. (1985) developed equations similar to the Wismer-Luth equations for log-skidder tires by using two skidder tire sizes of 18.4–34, 10-ply, and 24.5–32, 12-ply, both with steel-reinforced carcasses and special logging LS-2 lugs. Ashmore introduced an additional dimensionless ratio of dynamic load to rated load of tire. Prediction equations for towed force, wheel torque, and net traction as a function of travel reduction, dynamic load, soil strength, and tire size were developed.

Cone Index
Cone index is used as the measure of soil strength in the traction equations. Cone index is the average force per unit base area required to force a cone-shaped probe into soil at a steady rate. The design and use of the cone penetrometer are discussed in ASAE S 313.2 (ASAE Standards 1985).

Cone index characteristically varies with depth of penetration (see figure

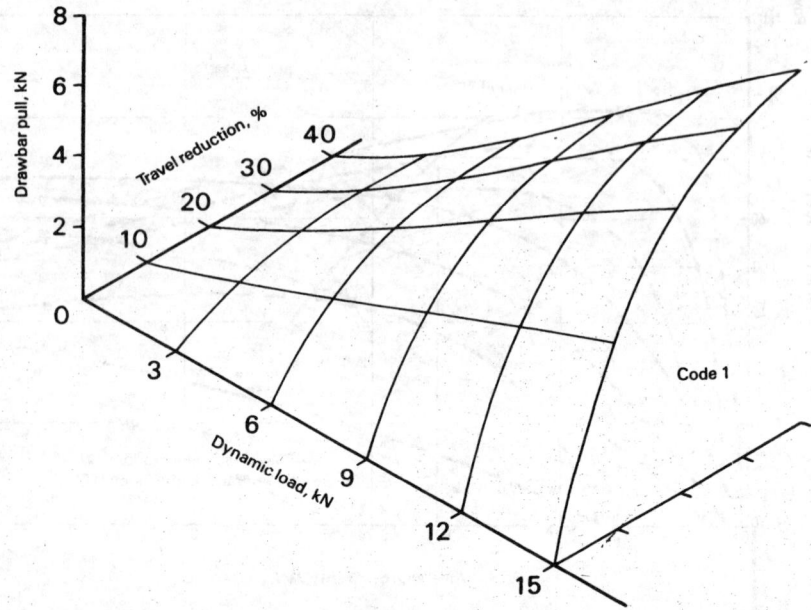

FIGURE 10-9   Drawbar pull surface for 12.4–28 tire on firm base conditions. (NTML Photo No. P-10, 278a.) (From Burt et al. 1979.)

10-12). Thus the question arises as to what cone index value should be used. For the traction equations, the 0- to 6-in. (15-cm) average cone index has produced the best correlations for machines with tire sinkages of less than 3 in. (7.5 cm). However, if the tire sinkage is greater than this value, the cone index should be averaged over the 6-in. (15-cm) layer, which includes the maximum sinkage of the tire. In general, cone index should be measured before the soil is subjected to wheel traffic.

Highly compactible soils, such as freshly tilled soils, present a special problem in predicting tractive performance. The soil tends to compact and increase in strength under heavy tire loads. Cone index measured after traffic may be several times the value measured before traffic. Best results to date have been accomplished by using after-traffic cone index values in the developed equations for highly compactible soils. No satisfactory method has been devised for predicting after-traffic cone index from before-traffic measurements.

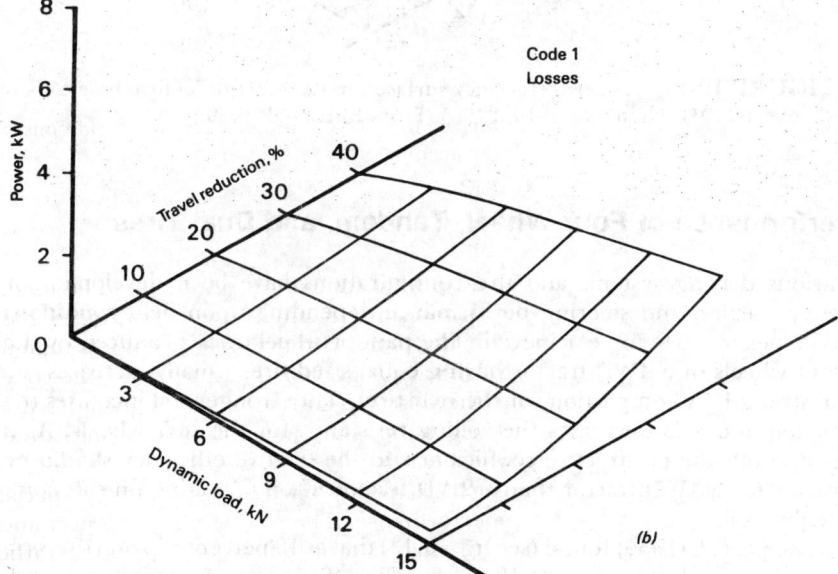

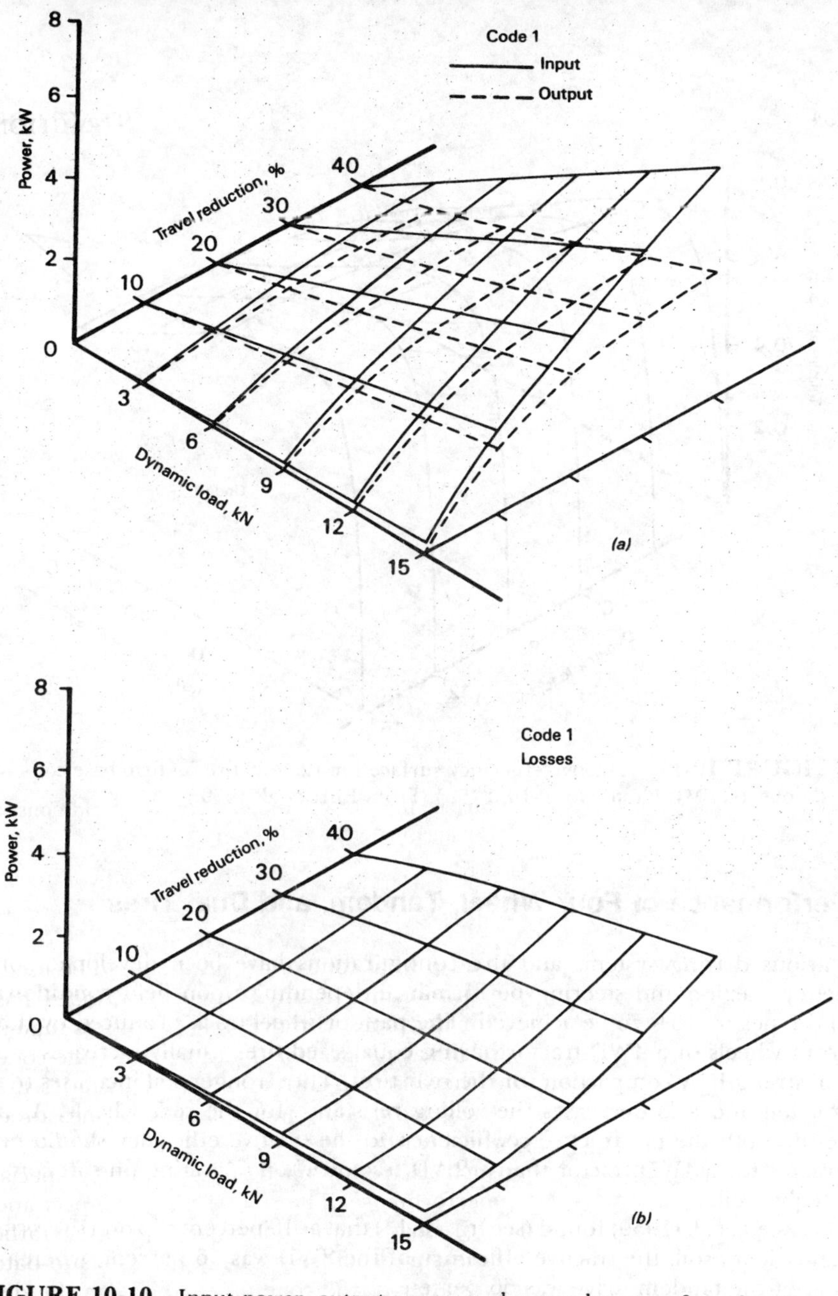

FIGURE 10-10 Input power, output power, and power losses surfaces for 12.4–28 tire on firm base conditions. (NTML Photo No. P-10, 275d, e.) (From Burt et al. 1979.)

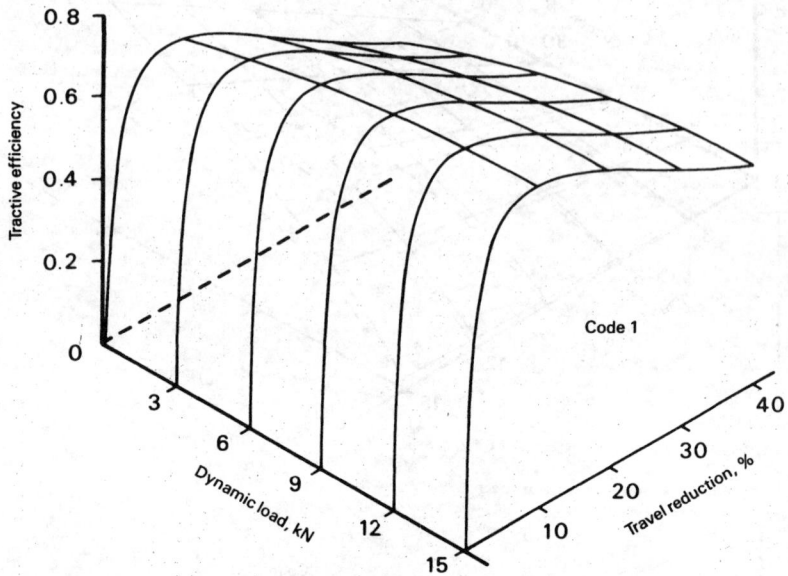

FIGURE 10-11   Tractive efficiency surface for 12.4–28 tire on firm base conditions. (NTML Photo No. P-10, 275f.) (From Burt et al. 1979.)

Performance of Four-Wheel, Tandem, and Dual Tires

Various driving systems and tire configurations have been developed for better traction and steering performance depending upon field conditions and types of work to be achieved. The path or wheel track produced by the front wheels of a 4WD tractor having equal-sized tires usually increases the soil strength by compaction for the rear tires. The stronger soil increases the traction and also decreases the rolling resistance for the rear wheels. As a result, both the net tractive coefficient and the tractive efficiency should be greater for a 4WD tractor than a 2WD tractor when operating on soft compactible soil.

Reed et al. (1959) found (see fig. 10-13) that at 10 percent slip on Hiwassee sandy loam soil, the tractive efficiency of the 2WD was 56 percent, whereas that of the tandem drive was 66 percent.

Dwyer and Pearson (1975) also found that the traction of the rear wheels (second pass) was greater, especially in soft soil. Dwyer found that the rolling resistance of the second pass was less than on the first pass. The average

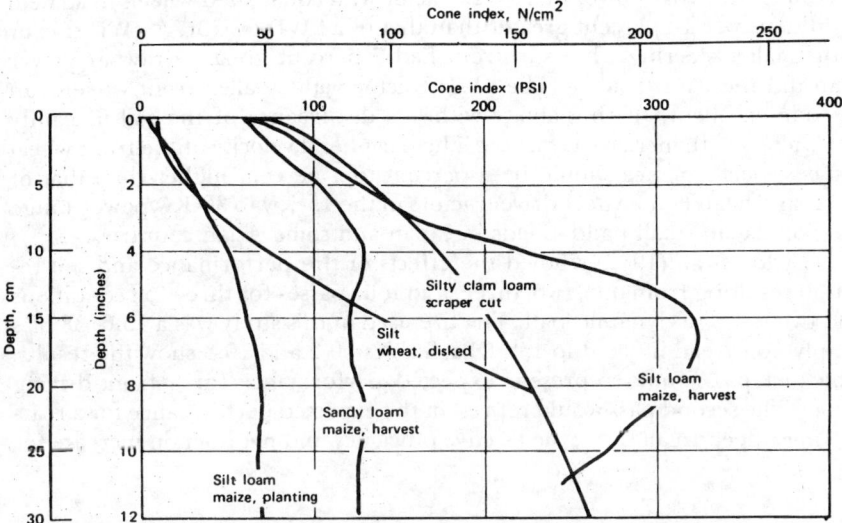

FIGURE 10-12 Typical cone index depth curves. (From Wismer and Luth 1974.)

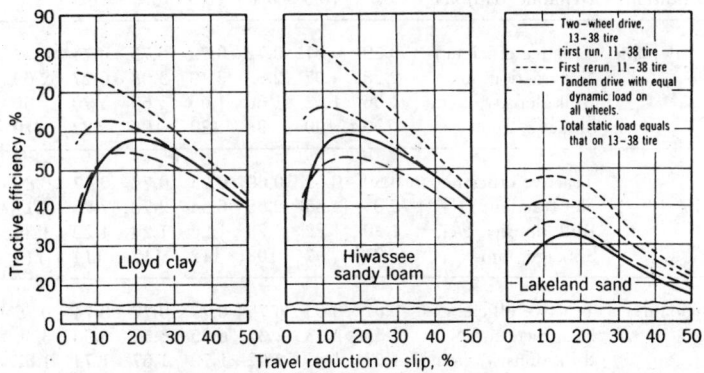

FIGURE 10-13 Power efficiency and travel reduction relationships for two-wheel-drive and tandem four-wheel-drive tractors. Data are for tractors of equal weight with the dynamic load equally distributed on all wheels for the tandem drive. (From I. F. Reed, A. W. Cooper, and C. A. Reeves 1959.)

maximum drawbar power of a 4WD tractor with equal-sized wheels in all field conditions was 14 percent greater than that of a 2WD tractor. A 4WD tractor with smaller steering wheels in front had 7 percent greater drawbar power than did the 2WD tractor. The 4WD tractor with smaller front wheels can provide excellent steering ability with less disturbance of the soil in paddy fields along with increased traction. The peripheral velocity of the front wheel for best performance should be 3 percent to 4 percent higher than that of the rear wheel. Four-wheel-drive tractors in the 15-Kw to 30-Kw power range are popular in small paddy fields in Japan and some Asian countries.

Taylor et al. (1982) studied the effects of tire performance and soil reaction resulting from one, two, three, and four passes for three soil conditions and two levels of dynamic load. The tire used in this study was a 13.6–38 bias six-ply, R-1 tread inflated to 152 KPa. Tables 10-2 and 10-3 show the results. The first pass would represent expected performance for a front-driving wheel. The second pass would represent the expected performance for a rear-driving wheel. In all cases, the tractive efficiency and net traction increase for

TABLE 10-2 Some Effects on Tire Performance and Soil Reaction Resulting from Four Passes of a Tire at 10 Percent TR in the Same Rut

Pass no.		1		2		3		4	
Initial soil conditions	Dynamic load, kN	8.1	16.3	8.1	16.3	8.1	16.3	8.1	16.3
Hiwassee sandy loam	Tractive efficiency	0.59	0.61	0.72	0.72	0.75	0.74	0.77	0.76
MC = 9.04%	Net traction, kN	2.28	4.97	2.83	6.49	3.08	6.27	2.69	6.95
BD = 1.34	Bulk density, g/cm³	1.56	1.62	1.60	1.66	1.63	1.70	1.65	1.71
	Sinkage, mm	100	100	98	120	106	122	110	124
Lloyd clay	Tractive efficiency	0.59	0.63	0.69	0.73	0.74	0.77	0.80	0.74
MC = 19.21%	Net traction, kN	2.03	4.85	2.58	6.54	2.78	7.26	3.11	6.79
BD = 1.04	Bulk density, g/cm³	1.20	1.22	1.23	1.26	1.24	1.29	1.26	1.29
	Sinkage, mm	91	97	108	112	112	114	113	116
Decatur silty loam	Tractive efficiency	0.61	0.63	0.75	0.75	0.77	0.74	0.76	0.74
MC = 16.62%	Net traction, kN	2.56	5.33	3.32	6.85	3.52	5.90	3.36	6.18
BD = 1.32	Bulk density, g/cm³	1.57	1.63	1.62	1.72	1.67	1.74	1.67	1.74
	Sinkage, mm	98	110	102	114	106	115	106	114

NOTE: TR = Travel reduction
MC = Moisture content
BD = Bulk density
Tire size = 13.6–38 bias six-ply, R-1 tread, inflated to 152 kPa

TABLE 10-3   Change in Bulk Density and Sinkage After the First Pass as a Percent of the Total Change After Four Passes of a Tire at 10 Percent TR in the Same Rut

Dynamic load, kN	Bulk density (g/cm³)		Sinkage (mm)	
	8.1	16.3	8.1	16.3
Hiwassee sandy loam	71	76	91	89
Lloyd clay	73	72	81	84
Decatur silty loam	71	74	92	96

NOTE: Tire size = 13.6–38 bias six-ply, R-1 tread, inflated to 152 kPa

a rear driving wheel (second pass). Compaction of the soil by a front-driving wheel provides a better traction surface for a rear-driving wheel.

Bailey and Burt (1981) studied the performance of tandem, dual, and single tires. Dual and single tires used were 13.6–38 bias six-ply, R-1 tread inflated to 152 KPa. Tandem tests were simulated by making two passes in the same lane with a single tire. Some results are shown in figure 10-14. It was concluded: (1) For a given travel reduction and dynamic load, the single and dual systems developed practically the same net traction. Duals have a greater load-carrying capacity than a single tire; therefore, they can develop greater net traction if the greater load-carrying capacity is utilized. (2) For a given travel reduction and at a reasonably high level of dynamic load, the tractive efficiency of the single tire system was usually slightly greater than that for the dual system. (3) For equal dynamic load on each system, tractive efficiency and net traction for a tandem system are at least equal to and, in most cases, higher than those for the single or dual systems. (4) On a per tire basis there was no consistent difference in tractive efficiency between the single and dual systems. Tractive efficiency for the tandem system on a per tire basis was consistently higher than for that for the single or dual system.

Tire Size, Load, and Air Pressure Relationship

The tire companies, through the Tire and Rim Association, have determined load and torque limits for each tire. This information (see Appendix) is published in the form of standards by SAE (J709d) and by ASAE (S295.2). Using this information, it is possible to select the minimum size tire for a given tractor load condition. Such a tire will be correctly sized for the vertical load and torque. However, it may be too small for the soil conditions. Ellis (1977) has simplified the problem by using the tractor power and operating speed

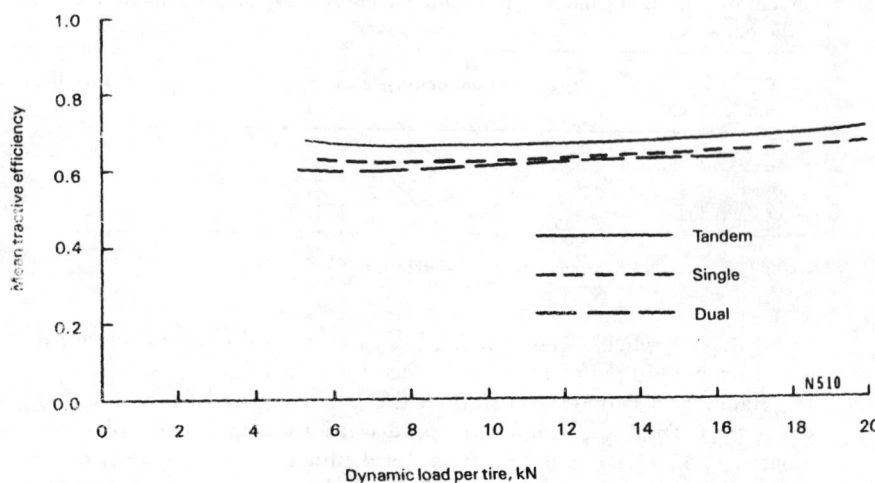

FIGURE 10-14 Tractive efficiency versus dynamic load per tire for the single, dual, and tandem systems. Norfolk sandy loam, loose surface, N5 test 10 percent travel reduction. (NTML Photo No. P10, 303b.) (From Bailey and Burt 1981.)

as a basis for selecting the proper tire. From this graph (see fig. 10-15), one can also select the proper combination of dual tires. Note that the use of dual tires does not double the power that can be transmitted by the tires.

Because a tractor may be used in a variety of soil conditions and loads, the manufacturer will have several sizes, treads, and ply ratings available for each tractor. One manufacturer of a popular 2WD tractor with 105 kW advertises ten different tires available plus five different dual arrangements for the rear-driving wheels. The same tractor also has six different front tire sizes available.

Radial-Ply Construction

The advantage of using radial-ply tractor tires as compared to bias-ply tires is not as pronounced for tractors as for highway vehicles, where the increase in life (mileage) and the decrease in fuel consumption justify the extra cost. Radial-ply tires have been used to a greater degree on tractors in Europe than in North America, possibly because European tractors are used more for highway transportation. In addition, the cost of fuel is much greater.

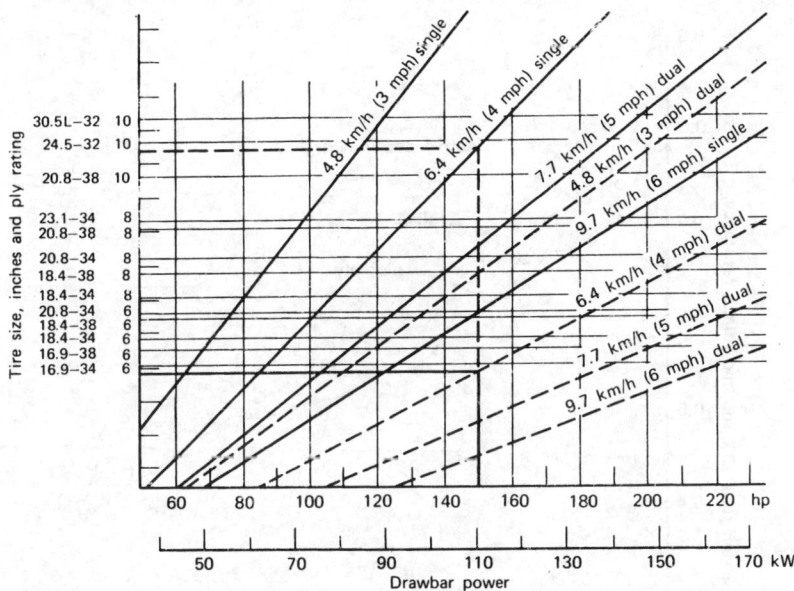

FIGURE 10-15   Tire selection chart based on drawbar power, speed, and option of single or dual rear tires. (From Ellis 1977. Courtesy Goodyear Tire and Rubber Co.)

The advantage of the radial-ply tractor tire in significantly improving the coefficient of traction under almost all conditions is shown in figure 10-16 from a study by Dwyer (1975). Taylor et al. (1976), at the National Tillage Machinery Laboratory, found a smaller advantage for the radial-ply tire. Taylor found that on five of seven soils, the coefficient of net traction at 15 percent slip increased 6 percent to 18 percent. However, the tractive efficiency of the radial-ply tire was only slightly higher. On soft soils, there was no advantage for the radial-ply tire.

Tread Design

The design of the tread on tires has been the subject of much debate because the performance of one tread compared to other treads is affected by the traction conditions and the performance criteria being used. Some of the available tire treads are illustrated in figures 10-17 and 10-18.

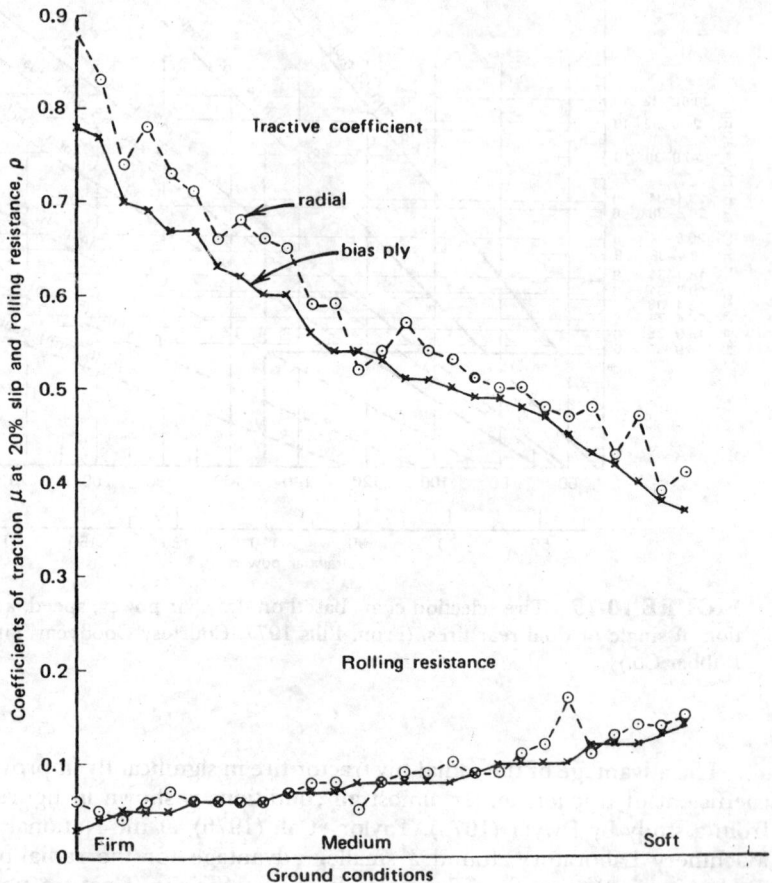

FIGURE 10-16 Comparison of coefficients of traction at 20 percent slip (*upper curves*) and rolling resistance (*lower curves*) for radial (*o*) and cross-ply (*x*) tires. (From Dwyer 1975.)

Some traction conditions that affect performance are:

1. Soil parameters (physical properties)
2. Presence of crop residues and cover crops
3. Direction of loading of tire (e.g., hillside use)
4. Load carried by tire
5. Tire pressure (deflection ratio*)

*Deflection ratio (deflection/tire section height)

FIGURE 10-17 Common North American types of tractor tire treads. (*a*) R-1 or general-purpose tread; (*b*) R-2 or deep lug cane and rice tread; (*c*) R-3 or flotation; and (*d*) R-4 or industrial lug. (Courtesy Goodyear Tire and Rubber Co.)

The decision as to which tire tread performs best is also dependent on the criteria being used. Some of the criteria are:

1. Tractive efficiency
2. Net tractive coefficient
3. Tire life
4. Soil compaction

(a)

(b)

FIGURE 10-18 (a) European-style tire with wider lugs for improved road wear. (Courtesy Goodyear Tire and Rubber Co.) (b) Japanese-style tire for rice soil conditions. (Courtesy Bridgestone Tire Co., Japan.)

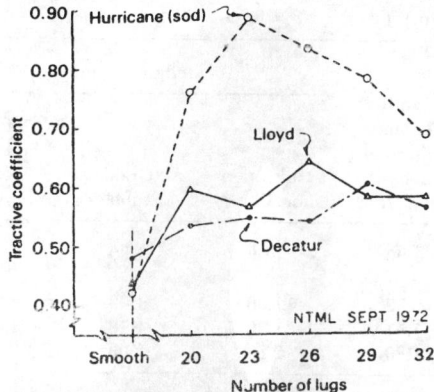

FIGURE 10-19 Effect of lug spacing on the dynamic traction ratio (pull number) of a 11.0–38 (279–965 mm) tractor tire.

Effect of Lug Spacing

A study was made by Taylor (1974) on the effect of lug spacing on 11.0–38 (279–965 mm) tires. The number of lugs per side on the five tires tested was 20, 23, 26, 29, and 32, which gave a pitch (in millimeters) of 238, 207, 178, 164, and 148. Taylor found (see fig. 10-19) that when the tires were tested on sod the maximum pull occurred when using the 23-lug tire. On the other soil conditions tested, lug spacing had little effect.

Traction Improvement

For certain soil conditions, traction aids are helpful. Table 10-4 shows the relative improvement of three traction aids. Strakes and halftracks are more commonly used in Europe. Table 10-5 shows the relative effect of adding weight and increasing the contact area (larger tires).

Tractors with both rubber tires and wheel extensions (strakes) are commonly used on weak surfaces such as rice paddy soils.

Rubber traction tires, as compared to steel traction wheels, have greatly improved the tractive efficiency, maneuverability, and comfort of farm tractors. Except on very firm soils, however, rubber tires have not increased the traction. In fact, under some conditions, such as when the surface of the soil is very wet and slick or when the soil is covered with a thick cover crop, the traction of rubber tractor tires is poor.

TABLE 10-4 Relative Improvement in Pull at 15 Percent Slip from Traction Aid on 13.6–28 (345–711 mm) Tires

| Traction Aid | Sandy Loam: Cultivated, Loose, and Dry | Medium Loam | | | Heavy Clay: Sparse Hay Aftermath |
		Cultivated and Damp	Wet Green Grain Stubble	Alfalfa Sod and Frozen Crust	
Air-filled tires only	1.00	1.00	1.00	1.00	1.00
Air-filled tires plus:					
Chains	.92	1.06	1.09	1.00	1.09
Strakes	1.51	1.56	1.78	1.76	2.86
Half-tracks	1.80	2.38	2.00	2.30	2.36

SOURCE: P. H. Southwell. "An Investigation of Traction and Traction Aids." *Trans of ASAE*, vol. 7, 1964.

TABLE 10-5 Relative Improvement in Pull at 15 Percent Slip from Various Tire Arrangements

| Tire and Arrangement | Sandy Loam: Cultivated, Loose, and Dry | Medium Loam | | | Heavy Clay: Sparse Hay Aftermath |
		Cultivated and Damp	Wet Green Grain Stubble	Alfalfa Sod and Frozen Crust	
Air-filled (13.6–28)	1.00	1.00	1.00	1.00	1.00
Air-filled dual (13.6–28)	1.05	—	1.24	1.35	1.41
Large air-filled (14.9–28)	.97	1.27	1.06	1.05	1.14
Solution-filled (13.6–28)	1.61	1.59	1.35	1.60	1.52
Solution-filled dual (13.6–28)	2.36	2.75	2.15	2.54	2.28

SOURCE: P. H. Southwell. "An Investigation of Traction and Traction Aids." *Trans of ASAE*, vol. 7, 1964.

When traction conditions are good, the largest improvement in traction can be made by simply adding more weight to the tractor drive wheel. However, when the surface soil is very wet, the internal friction, ϕ, approaches zero; therefore, an increase in the soil pressure will not increase traction significantly unless the traction device can "cut through" the low friction surface layer.

It was discussed earlier that traction performance in part is dependent on the amount of wheel slip. An electronic device is manufactured that provides the operator with "on the go" monitoring of percent of wheel slip and true ground speed. Such information can assist the operator in selecting the transmission gear setting and engine throttle setting for better tractive performance. The device is discussed in chapter 6.

Tracks

Tracks have been used for many years to reduce soil pressure and to increase traction on soft, loose soil surfaces that have low strength.

Tracks result in the best relative performance as compared with pneumatic tires when the tractor is operating at nearly maximum drawbar pull on soft, loose soil surfaces.

Tire Testing

The design of tires is determined almost entirely by the experimental method. Tire companies with the cooperation of tractor manufacturers must, of necessity, devote much time to testing tires in all conditions in which the tractor would normally be used. Several government agencies in the United States and other countries have facilities for testing tires and studying traction. The best-known facility in the United States is the National Soil Dynamics Laboratory (NSDL), which is part of the U.S. Department of Agriculture. Figure 10-20 shows the single-wheel dynomometer used at the NSDL. The dynamometer car, in this case, carries a computer on board which controls all aspects of the test and processes data on-line for many variables, some of which are pull, dynamic load, power, slip, inflation pressure, sidewall deformation, and stress of the tire at the tire-soil interface.

Traction Devices for Paddy Fields

The mechanization of rice production in some Asian countries has reached a high level of accomplishment (Tanaka 1984). In Japan approximately 95

FIGURE 10-20   Single-wheel dynamometer. (Courtesy U.S. Dept. of Agriculture, National Soil Dynamics Laboratory, Auburn, AL.)

percent of the operations are mechanized. The moisture content of paddy field soil is usually very high, and in many cases farm vehicles must operate in fields that are saturated or flooded. Trafficability of the surface soil layer is very poor, being extremely soft with low load-bearing capacity. Cone indices will typically be less than 50 N/cm. Performance of regular rubber-tired wheels is not acceptable because of high slippage and adhesion of sticky soil. Specially designed high-lug rubber tires as shown in figure 10-18(b) are commonly used in paddy fields.

Farm tractors and other farm vehicles that are operated in these conditions often require special devices used with tires or in place of tires. As an example of some typical devices, an open lug wheel and a float lug wheel are shown in figure 10-21. Figure 10-21(a) shows the open lug wheel for a two-wheeled power tiller. The average weight of a power tiller is in the 200- to 300-kg range, with average engine power of 7.0 Kw. The lug plates are welded to a circular rim. Many types and shapes of lug plates are in use for paddy field operations. Figure 10-21(b) shows a lug wheel used as an auxiliary device with a tractor tire. Float-type lugs are connected to the circular ring plate, and the lugs can be folded toward the center of the wheel when the tractor operates on a hard surface or paved road.

Development of traction and transport performance prediction equa-

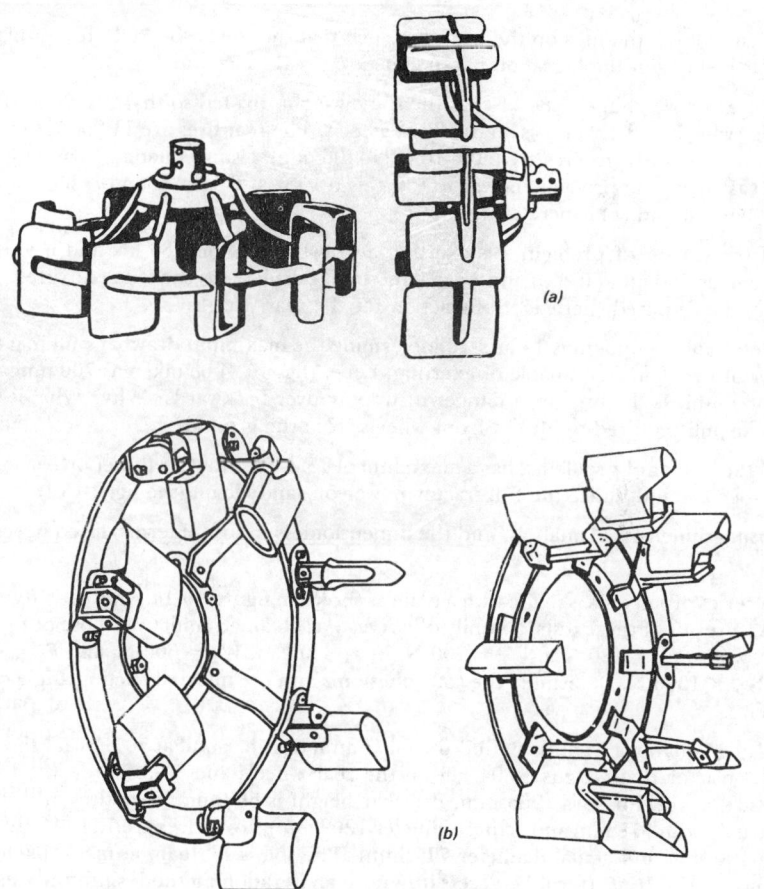

FIGURE 10-21 Lug wheels used in paddy fields. (*a*) Open lug wheel; (*b*) float-type lug wheel. (From Tanaka 1984.)

tions for paddy field conditions is extremely challenging. Tanaka (1984) provides a state-of-the-art overview of analytical work and an extensive bibliography.

PROBLEMS

1. Using equation 1, predict the maximum traction thrust of a track-type tractor with two tracks each 360 mm wide by 1680 mm long. The weight of the tractor is 31.75

kN. Assume that the lugs on the track are such that the soil is sheared off in a plane area at the ends of the lugs. Soil parameters are $c = 14$ kPa and $\phi = 30°$.

2. A tractor weighing 15.84 kN has the static weight divided so that 11.60 kN is on the rear wheels and 4.24 kN is on the front wheels. The rear tires are 11.25–36* (286–914 mm) and the front tires are 5.00–16 (127–406 mm). Using equation 14 and figure 10-12, calculate the probable power to tow this tractor at 6.5 km/h on (a) silt loam at maize harvest and (b) concrete.

3. If the tractor of problem 2 is exerting a drawbar pull of 7.80 kN and if y_f (fig. 10-4) equals 700 mm (also y_r) and $L_1 = 2160$ mm, calculate (a) the true rear-tire load, and (b) the required coefficient of traction for the rear wheels.

4. Referring to equations 14 and 16, determine the maximum drawbar pull that the tractor of problem 2 is capable of exerting. Let y_f (fig. 10-4) be taken as 700 mm and y_r as 460 mm. Is the tractor in danger of tipping over backwards? Why? What is the drawbar pull required to lift the front wheels from the ground?

5. If the tractor of problem 2 has a maximum of 13.4-kW drawbar power on concrete, what speed would utilize the full tractor power on sandy loam (see fig. 10-12)?

6. Using dimensional analysis, find the dimensionless ratios (pi terms) listed in equation 13.

7. A series of soil tests is made with a plate as shown in figure 10-1a, where $l = b = 10$ cm. As a result of these tests, a family of curves was obtained similar to those of figure 10-1b, where $F_1 = 180$ N at $W_1 = 200$ N, $F_2 = 340$ N at $W_2 = 600$ N, and $F_3 = 500$ N at $W_3 = 1000$ N. Determine the soil cohesion c and the internal friction angle ϕ of the soil.

8. Using equations 14 and 16, find the maximum drawbar pull at 20 percent slip for a 2WD tractor that weighs 13.94 kN on the rear wheels and 7.93 kN on the front wheels; the wheel base is 1963 mm; drawbar height is 508 mm; rear tires are 14.9–28 (actual width 378 mm and actual diameter 1209 mm); front tires are 7.50–16 (actual width 190 mm and actual diameter 711 mm). The soil is silt loam at maize planting (see fig. 10-12). *Hint:* It will be necessary to use an iteration method, since the weight distribution on the wheels while pulling will not be known until the drawbar pull is found. Assume wheel sinkage is such that CI can be evaluated through the zone from 0 to 15 cm.

9. A conventional 2WD tractor is to be converted to 4WD using a hydrostatic drive. The tractor specifications and performance are:
Max pto power = 60 kW
Drawbar height = 520 mm
Wheel base = 2500 mm
Rear tires are 18.4–34 ($d = 1661$ mm)
New front tires are 13.6–24 ($d = 1206$ mm)
Static front weight = 21,130 N

*Tire width and rim diameter in inches.

Static rear weight = 32,250 N

(a) What drawbar pull could be expected with 2WD on concrete at 7 percent slip?

(b) What drawbar pull could be expected with 4WD on concrete at 7 percent slip?

(c) What drawbar pull could be expected with 2WD on soft soil (CI = 45 psi) at 20 percent slip?

(d) What drawbar pull could be expected with 4WD on soft soil (CI = 45 psi) at 20 percent slip?

REFERENCES

Ashmore, C., J. L. Turner, and E. C. Burt. "Predicting Tractive Performance of Log-skidder Tires." ASAE Paper 85-1579, 1985.

Bailey, A. C., and E. C. Burt. "Performance of Tandem, Dual, and Single Tires." *Trans. of ASAE*, vol. 24, no. 5, 1981, p. 1103.

Burt, E. C., et al. "Combined Effects of Dynamic Load and Travel Reduction on Tire Performance." *Trans. of ASAE*, vol. 22, no. 1, 1979, p. 40.

Dwyer, M. J. *Some Aspects of Tyre Design and Their Effect on Agricultural Tractor Performance*. Institution of Mechanical Engineers, England, 1975.

Dwyer, M. J. "The Tractive Performance of Wheeled Vehicles." *J. Terramechanics*, vol. 21, no. 1, 1984, p. 19.

Dwyer, M. J., and G. Pearson, "A Field Comparison of the Tractive Performance of Two- and Four-Wheel Drive Tractors." *J. Agric. Eng'g Res.*, vol. 21, 1976, pp. 77–85.

Ellis, R. W. "Agricultural Tire Design Requirements and Selection Considerations." ASAE Distinguished Lecture Series (Tractor Design No. 3), Dec. 13, 1977.

Freitag, D. R. "A Dimensional Analysis of the Performance of Pneumatic Tires on Soft Soils." USAE Waterways Experiment Station, Technical Rep. No. 3-688, Aug. 1965.

Freitag, D. R. "Soil Dynamics as Related to Traction and Transport." Proceedings of International Conference on Soil Dynamics, vol. 4, Auburn, AL, June 1985.

Reed, I. F., A. W. Cooper, and C. A. Reeves. "Effects of Two-Wheel and Tandem Drives on Traction and Soil Compacting Stresses." *Trans. of ASAE*, vol. 2, no. 1, 1959, p. 22.

Tanaka, T. "Operation in Paddy Fields: State-of-the-Art Report." *J. Terramechanics*, vol. 21, no. 2, 1984, p. 153.

Taylor, J. H. "Lug Spacing Effect on Traction of Pneumatic Tractor Tires." *Trans. of ASAE*, vol. 17, no. 2, 1974, p. 195.

Taylor, J. H., E. C. Burt, and A. C. Bailey. "Radial Tire Performance in Firm and Soft Soils." *Trans. of ASAE*, vol. 19, no. 6, 1976, p. 1062.

Taylor, J. H., et al. "Multipass behavior of a pneumatic tire in tilled soils." *Trans. of ASAE*, vol. 25, no. 5, 1982, p. 1229.

Turnage, G. W. "Tire Selection and Performance Prediction for Off-Road Wheeled-Vehicle Operations." Proceedings of Fourth International Conference of the International Society for the Terrain Vehicle Systems, vol. 1, Stockholm, Sweden, Apr. 1972.

Wismer, R. D., and H. J. Luth, "Off-road Traction Prediction for Wheeled Vehicles." *Trans. of ASAE*, vol. 17, no. 1, 1974, p. 8.

SUGGESTED READINGS

Agricultural Engineers Yearbook, American Society of Agricultural Engineers, 1985.

Bailey, P. H. "The Comparative Performance of Some Traction Aids." *J. Agric. Eng'g Res.* (England), vol. 1, no. 1, 1956.

Barger, E. L., and J. Roberts. "Effect of Tire Wear on Tractor Performance." *Agric. Engr.*, vol. 20, May 1939, pp. 191–194.

Bekker, M. G. *Theory of Land Locomotion*. The University of Michigan Press, Ann Arbor, 1956.

Burt, E. C., R. L. Schafer, and J. H. Taylor, "Similitude of a Model Traction Device: Part I—Prediction of the Dynamic Traction Ratio, Part II—Prediction of Wheel Sinkage. *Trans. of ASAE*, vol. 17, no. 4, 1974, p. 662.

Caterpillar Performance Handbook, 3d ed. Caterpillar Tractor Company. Peoria, IL, Jan. 1973.

Davidson, J. B., E. V. Collins, and E. G. McKibben. "Tractive Efficiency of the Farm Tractor." *Iowa Agric. Expt. Sta. Res. Bull.*, vol. 189, Sept. 1935.

Deere & Co. *Fundamentals of Service—Tires and Tracks*. John Deere Service Publications, Moline, IL, 1970.

Domier, K. W., D. H. Friesen, and J. S. Townsend. "Traction Characteristics of Two-Wheel Drive, Four-Wheel Drive and Crawler Tractors." *Trans. of ASAE*, vol. 14, no. 3, 1971, p. 520.

Dwyer, M. J., D. W. Evernden, and M. McAllister. *Handbook of Agricultural Tyre Performance*, 2d ed. National Institute of Agricultural Engineering, Wrest Park, Silsoe, Bedford, England, Apr. 1975.

Firestone Tire and Rubber Company. *Agricultural Tire Engineering Data*, 1966.

Freitag, D., R. L. Schafer, and R. D. Wismer. "Similitude Studies of Soil Machine Systems." *Trans. of ASAE*, vol. 13, no. 2, 1970, p. 201.

Gee-Clough, D. "The Special Problem of Wetland Traction and Flotation." *J. Agric. Eng'g Res.*, vol. 32, 1985, pp. 279–288.

Gill, W. R., and G. E. Vandenberg. "Soil Dynamics in Tillage and Traction." *Agriculture Handbook No. 316*. Agricultural Research Service, U.S. Department of Agriculture, 1967.

Karafiath, L. L., and E. A. Nawatzki. *Soil Mechanics for Off-Road Vehicle Engineering*. TransTech Publications, Rockport, MA, 1978.

SUGGESTED READINGS   271

Lambe, T. W., and R. V. Whitman. *Soil Mechanics*. John Wiley & Sons, New York, 1978.

Langhaar, H. L. *Dimensional Analysis and Theory of Models*. John Wiley & Sons, New York, 1978.

McKibben, E. G., and J. B. Davidson. "Effort of Inflation Pressure on the Rolling Resistance of Pneumatic Implement Tires." *Agric. Engr.*, vol. 21, Jan. 1940, pp. 25–26.

McKibben, E. G., and J. B. Davidson. "Effect of Outside and Cross-sectional Diameters on the Rolling Resistance of Pneumatic Implement Tires." *Agric. Engr.*, vol. 21, Feb. 1940, pp. 57–58.

Murphy, G. *Similitude in Engineering*. Ronald Press, New York, 1950.

Rubber Manufacturers Association. *Care and Service of Farm Tires*. Rubber Manufacturers Association, Washington, DC, 1973.

Soehne, W. "Kraftubertranung Zwischen Schepperreiffen und Ackerboden (Stress Transmission Between Tractor Tires and Soils)." *Grundl. Landtech.*, vol. 3, 1952, pp. 75–78.

Southwell, P. H. "An Investigation of Traction and Traction Aids." *Trans. of ASAE*, vol. 7, no. 2, 1964, p. 190.

Taylor, James H. "Lug Angle Effect on Traction Performance of Pneumatic Tractor Tires." *Trans. of ASAE*, vol. 16, no. 1, 1973, p. 16

Taylor, James H. "Comparative Traction Performance of R-1, R-3, and R-4 Tractor Tires." *Trans. of ASAE*, vol. 19, no. 1, 1976, p. 14

Vandenberg, G. E., and W. R. Gill. "Pressure Distribution Between a Smooth Tire and the Soil." *Trans. of ASAE*, vol. 5, no. 2, 1962, p. 105.

11
MECHANICS OF THE TRACTOR CHASSIS

If the agricultural engineer who is specializing in farm power is to
make his best contribution to the agricultural engineering profession,
if as an agricultural engineer he is to earn and hold a recognized
place among other engineers, he must have more than the garage
mechanic's conception of the tractor. He must visualize the tractor as
a unit and have a clear conception of all forces acting upon it. He
must be informed concerning the fundamental laws of mathematics
and physics which govern its kinematic and dynamic response to
these forces. In no other way will he be able to make the tractor perform its maximum service.

E. G. McKibben (1927)

An understanding of the statics and dynamics of farm tractors is important
in the analysis of tractor performance, stability, ride, and handling. This
chapter provides an introduction to these individual areas while attempting
to emphasize their interrelationships.

Because of the introductory nature of this chapter, the two-dimensional
analysis of rear-wheel-driven tractors is emphasized. The suggested readings
at the end of the chapter provide additional information on methods of three-
as well as two-dimensional analysis.

Simplifying Assumptions

The following simplifying assumptions apply to the rear-wheel-driven tractor
shown in figures 11-1 and 11-2:

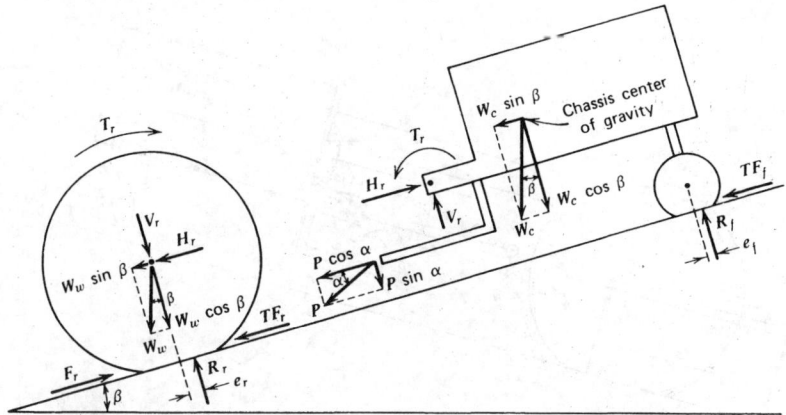

FIGURE 11-1 Free-body diagrams of the chassis and drive wheels of a rear-wheel-driven tractor.

1. The ground surface is planar and nondeformable.
2. The motion of the tractor can be analyzed as two dimensional.
3. Rotational motion of the front wheels is neglected.
4. Aerodynamic forces are neglected.

Equations of Motion

Figure 11-1 contains free-body diagrams of the two major components of the tractor: the chassis and the rear wheels. The traction mechanics developed in the section Traction Performance Equations in chapter 10 is used to describe the force systems acting on both the rear drive wheels and the front unpowered wheels. The weights of the rear wheels (W_w) and chassis (W_c) are decomposed into components acting parallel and perpendicular to the ground surface. Similarly, the external force P exerted at the chassis drawbar is broken down into a parallel component $P \cos \alpha$ (commonly called the drawbar pull) and a perpendicular component $P \sin \alpha$.

The kinematics used to describe the rear wheel and chassis motions are illustrated in figure 11-2. The translational motions of the rear wheels and chassis are referenced to the fixed or inertial XZ coordinate system. Such a coordinate system, in which the positive Z axis points downward, is consistent with the terminology widely used in other areas of vehicle dynamics. The angle of rotation of the drive wheels, ϕ_w, is measured relative to the tractor

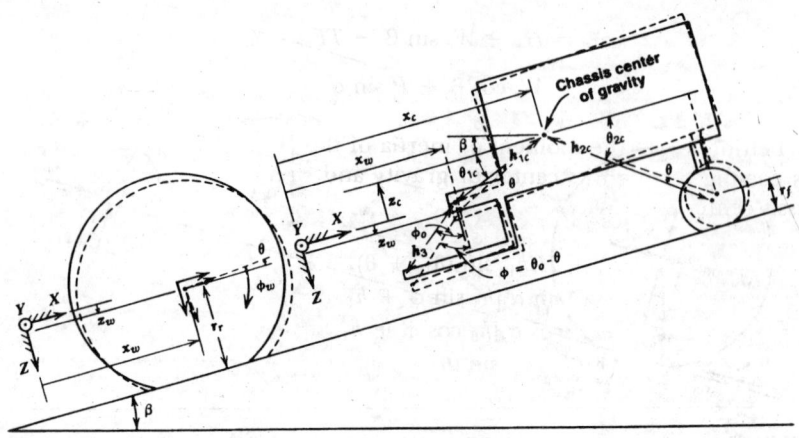

FIGURE 11-2 Kinematics associated with planar motion of the chassis and drive wheels. The dashed outline shows the position of the chassis and drive wheels after an angular displacement θ.

chassis so that the absolute angular motion of the drive wheels is given by the angle $\phi_w - \theta$, where θ is the pitch angle of the chassis.

Letting m_w be the mass of the rear wheels and summing forces on the rear wheels in the X and Z directions,

$$m_w \ddot{x}_w = F_r - TF_r - W_w \sin \beta - H_r \tag{1}$$

$$m_w \ddot{z}_w = V_r + W_w \cos \beta - R_r \tag{2}$$

Letting I_{yyw} be the moment of inertia of the wheels about the y or lateral axis passing through their center of gravity and summing moments about the rear axle center (assumed to be coincident with the center of gravity of the rear wheels),

$$I_{yyw} (\ddot{\phi}_w - \ddot{\theta}) = T_r - (F_r - TF_r)r_r - R_r e_r \tag{3}$$

However, in the traction mechanics of Chapter 10, e_r is considered to be equal to $(TF_r/R_r)r_r$, so that equation 3 may be written

$$I_{yyw} (\ddot{\phi}_w - \ddot{\theta}) = T_r - F_r r_r \tag{4}$$

Letting m_c be the mass of the chassis and summing forces on the chassis in the X and Z directions,

$$m_c\ddot{x}_c = H_r - W_c \sin \beta - TF_f - P \cos \alpha \qquad (5)$$

$$m_c\ddot{z}_c = W_c \cos \beta + P \sin \alpha - V_r - R_f \qquad (6)$$

Letting I_{yyc} be the moment of inertia of the chassis about the y or lateral axis passing through its center of gravity and summing moments about the chassis center of gravity,

$$\begin{aligned}
I_{yyc}\ddot{\theta} = {} & T_r + H_r h_{1c} \sin (\theta_{1c} + \theta) - V_r h_{1c} \cos (\theta_{1c} + \theta) \\
& + P \sin \alpha \, [h_3 \sin \phi + h_{1c} \cos (\theta_{1c} + \theta)] \\
& - P \cos \alpha \, [h_3 \cos \phi + h_{1c} \sin (\theta_{1c} + \theta)] \\
& - TF_f \, [h_{2c} \sin (\theta_{2c} - \theta) + r_f] \\
& + R_f \, [h_{2c} \cos (\theta_{2c} - \theta) + e_f]
\end{aligned} \qquad (7)$$

Equation 7 may be simplified, since the traction mechanics of the unpowered front wheels assumes that $e_f = (TF_f/R_f)r_f$.

Equations 1, 2, and 4 are the differential equations of motion governing the three degrees of freedom that the rear wheels possess for general planar motion. Equations 5, 6, and 7 are the corresponding equations for the chassis. However, the constraint that the rear wheels are attached to the chassis implies that the corresponding 6 degrees of freedom (x_w, z_w, ϕ_w, x_c, z_c, θ) are not independent.

Referring to figure 11-2, the following two equations express the constraint relation:

$$x_c = x_w + h_{1c} \cos (\theta_{1c} + \theta) \qquad (8)$$

$$z_c = z_w - h_{1c} \sin (\theta_{1c} + \theta) \qquad (9)$$

The two constraint equations reduce the number of degrees of freedom of the system from 6 to 4 and thus imply that the rear wheel–chassis system (the total tractor) can also be described by four independent differential equations. These four equations can be derived from equations 1, 2, 3, 5, 6, and 7 by eliminating the internal reactions H_r, V_r, and T_r, using constraint equations 8 and 9, and considering the relations between the locations of the centers of gravity of the rear wheels, chassis, and tractor. After some algebraic manipulation, the equations of motion for the tractor as a whole may be found. The free-body and kinematic diagrams associated with these equations are shown in figures 11-3 and 11-4.

$$m_t\ddot{x}_t = F_r - W_t \sin \beta - TF_f - TF_r - P \cos \alpha \qquad (10)$$

$$m_t\ddot{z}_t = W_t \cos \beta + P \sin \alpha - R_r - R_f \qquad (11)$$

$$
\begin{aligned}
I_{yyt}\ddot{\theta} = {} & I_{yyw}\ddot{\phi}_w + (F_r - TF_r)[h_{1t}\sin(\theta_{1t} + \theta) + r_r] \\
& - TF_f[h_{2t}\sin(\theta_{2t} - \theta) + r_f] \\
& + R_f[h_{2t}\cos(\theta_{2t} - \theta) + e_f] \\
& - R_r[h_{1t}\cos(\theta_{1t} + \theta) - e_r] \\
& + P\sin\alpha[h_3\sin\phi + h_{1t}\cos(\theta_{1t} + \theta)] \\
& - P\cos\alpha[h_3\cos\phi + h_{1t}\sin(\theta_{1t} + \theta)] \qquad (12)
\end{aligned}
$$

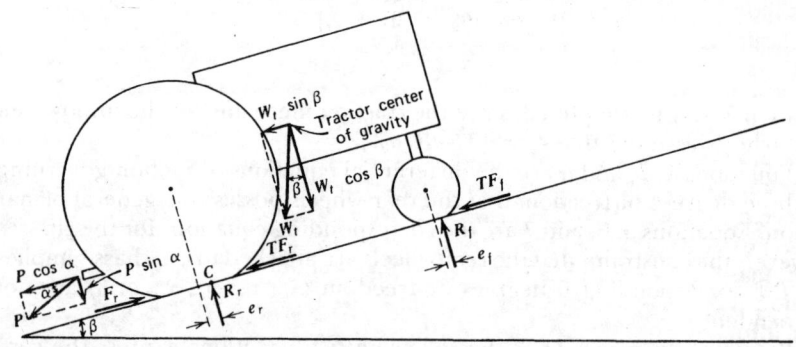

FIGURE 11-3   Free-body diagram of a rear-wheel-driven tractor.

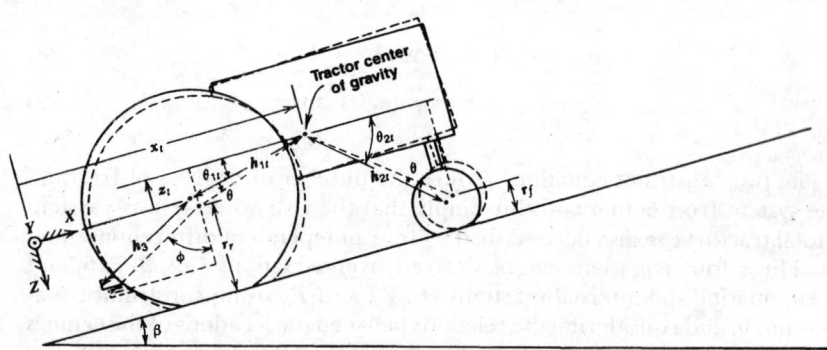

FIGURE 11-4   Kinematics associated with planar motion of the tractor. The dashed outline shows the position of the tractor after an angular displacement θ.

m_t is the mass of the tractor ($m_t = m_c + m_w$), and $\ddot{x}_t$ and $\ddot{z}_t$ are the translational accelerations of the tractor center of gravity in the X and Z directions. I_{yyt} is the moment of inertia of the entire tractor about the y or lateral axis passing through the center of gravity of that body.

The right-hand sides of equations 10 and 11 represent force summations in the X and Z directions, respectively, whereas the right-hand side of equation 12 is a summation of moments about the center of gravity of the tractor. Thus, equations 10 and 11 are simply the translational equations of motion that can be derived more directly from the free-body diagram of figure 11-3. Similarly, with the exception of the term $I_{yyw} \dot{\phi}_w$, equation 12 can be derived directly from a summation of moments about the center of gravity of the tractor. Equations 10, 11, and 12 in combination with equation 4 form the four independent differential equations of motion discussed previously.

Equation 10 governs the forward translational motion of the tractor. The gross tractive force F_r is seen to be the only force acting to propel the tractor forward. This force must exceed the sum of the rolling resistance forces acting at the wheels, the portion of the tractor weight acting parallel to the ground surface, and the drawbar pull in order for the tractor to accelerate forward. If the rear wheels are braked instead of driven, F_r changes direction and acts to decrease the forward speed of the tractor.

As discussed in the section Traction Performance Equations in chapter 10, F_r may be considered to be equal to the product of the coefficient of gross traction μ_g and the soil reaction R_r. The coefficient of gross traction is, in turn, a function of the slippage S of the rear wheels as well as of tire and soil surface parameters. Using equation 7 in chapter 10 and the kinematics illustrated in figure 11-2,

$$S = 1 - \dot{x}_w/(r_r \, \dot{\phi}_w) \qquad (13)$$

Equations 11 and 12 govern the vertical translation and rotation of the tractor. These equations are particularly important for describing tractor rearward overturning and vibration. Equation 12 may be simplified by using the relations $e_r = (TF_r/R_r)r_r$ and $e_f = (TF_f/R_f)r_f$.

If we consider that the power delivered to the rear axle is equal to the power output of the engine multiplied by an overall efficiency η, we can write

$$T_r\dot{\phi}_w = \eta T_e\dot{\phi}_e$$

where T_e is the engine torque produced at engine speed $\dot{\phi}_e$. Letting N be the ratio of engine speed to axle speed ($N = \dot{\phi}_e/\dot{\phi}_w$),

$$T_r = \eta N T_e$$

Thus, equation 4 can also be written

$$I_{yyw} (\ddot{\phi}_w - \ddot{\theta}) = \eta N T_e - F_r r_r \qquad (14)$$

For a given throttle position, T_e can be expressed as an empirical function of $\dot{\phi}_e$ and thus, given N, of $\dot{\phi}_w$. Thus, equation 14 describes how the engine delivers torque to the rear wheels and in turn is loaded by the gross tractive force acting on them.

Static Equilibrium Analysis—Force Analysis

Except for a few special cases, the inherent complexity and nonlinearity of equations 10, 11, 12, and 14 make their analytical solution difficult, if not impossible. Fortunately, with the aid of numerical integration methods and modern digital computers, it is possible to obtain solutions to these equations for quite general situations.

However, much useful information, particularly for evaluating tractor field performance, can be obtained from a static equilibrium analysis. For the static equilibrium situation, $\ddot{z}_t = 0$ and equation 11 becomes

$$W_t \cos \beta + P \sin \alpha - R_r - R_f = 0 \qquad (15)$$

Similarly, for static equilibrium, $\ddot{\theta} = \ddot{\phi}_w = 0$, and equation 12 could be used to represent the moment equilibrium of the tractor. However, since the sum of moments about any transverse axis must be zero for the tractor to be in static equilibrium, the location of the axis may be chosen to help simplify the resulting moment summation. A convenient axis to use passes through point C of figure 11-3. This axis has the advantage that the resulting moment arms of forces F_r, TF_r, R_r, and TF_f all become zero. Then, summing moments about point C, taking counterclockwise moments as positive, and assuming $\theta = 0$ when the tractor is in static equilibrium,

$$R_f (h_{1t} \cos \theta_{1t} + h_{2t} \cos \theta_{2t} + e_f - e_r)$$
$$+ W_t \sin \beta (r_r + h_{1t} \sin \theta_{1t}) - W_t \cos \beta (h_{1t} \cos \theta_{1t} - e_r)$$
$$+ P \cos \alpha (r_r - {}'h_3 \cos \phi) + P \sin \alpha (h_3 \sin \phi + e_r) = 0 \qquad (16)$$

Solving equation 16 for R_f,

$$R_f = \{W_t [(h_{1t} \cos \theta_{1t} - e_r) \cos \beta - (r_r + h_{1t} \sin \theta_{1t}) \sin \beta]$$
$$- P [(r_r - h_3 \cos \phi) \cos \alpha + (h_3 \sin \phi + e_r) \sin \alpha]\}$$
$$/\{h_{1t} \cos \theta_{1t} + h_{2t} \cos \theta_{2t} + e_f - e_r\} \qquad (17)$$

Although equation 17 is useful for computation, its complexity tends to

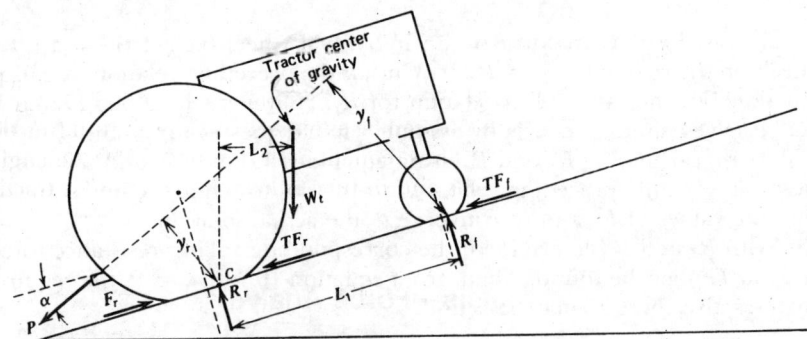

FIGURE 11-5   Free-body diagram illustrating the calculation of force R_f.

obscure its physical meaning. Using the notation of figure 11-5, equation 17 may be written

$$R_f = (W_t L_2 - P y_r)/L_1 \tag{18}$$

With R_f now known, equation 15 may be solved for R_r

$$
\begin{aligned}
R_r &= W_t \cos \beta + P \sin \alpha - R_f \\
&= W_t \cos \beta + P \sin \alpha - \frac{W_t L_2}{L_1} + \frac{P y_r}{L_1} \\
&= W_t \cos \beta - \frac{W_t L_2}{L_1} + \frac{P(y_r + L_1 \sin \alpha)}{L_1} \\
&= W_t \cos \beta - \frac{W_t L_2}{L_1} + \frac{P y_f}{L_1}
\end{aligned} \tag{19}
$$

The terms $P y_r/L_1$ and $P y_f/L_1$ express changes in the soil reactions R_f and R_r as a result of the drawbar force P. Although there is no actual shift of weight, the changes in the forces R_r and R_f are commonly known as weight transfer (see the section Traction Performance Equations in chapter 10).

The traction mechanics given in the section Traction Prediction from Dimensional Analysis in chapter 10 gives the following relation for e_f and e_r as functions of R_f and R_r, respectively.

$$e_i = \rho r_i = (1.2/C_{ni} + 0.04) r_i \qquad i = f, r$$

where $\qquad C_{ni} = CI_i b_i d_i/W_i \qquad i = f, r$

Here W_i is the normal force on an individual wheel (i.e., if there are two wheels on the rear axle, $W_r = R_r/2$). Thus, if the preceding relations are used, equation 17 is not an explicit relation for R_f. However, equations 17 and 19 can be solved simultaneously by assuming initial guesses for e_f and e_r (i.e., $e_i = 0.04\ r_i$), calculating R_f and R_r using equations 17 and 19, and then using these values in the preceding relation for e_f and e_r to refine the initial guesses until the values of R_f and R_r converge to the actual solution.

With R_f and R_r determined, the corresponding rolling resistance forces TF_f and TF_r may be found. Then, from equation 10 with $\ddot{x}_t = 0$, the required gross tractive force F_r may be found.

$$F_r = W_t \sin \beta + TF_f + TF_r + P \cos \alpha \tag{20}$$

Static Equilibrium Analysis—Maximum Achievable Drawbar Pull

The drawbar pull, $P \cos \alpha$, that can be continuously sustained by a tractor is, of course, an important factor in determining the tractor's productivity. The maximum drawbar pull that can be developed may be limited by one of three factors: stability, traction, or power. Although it is impossible to control the situations in which a tractor may be operated, the maximum drawbar pull should be limited by traction or power availability.

Stability
Through the weight transfer effect, the drawbar force P may be large enough to cause force R_f to become zero and thus endanger the stability of the tractor. The drawbar force P_s required for this situation to occur can be found from equation 18 by letting $R_f = 0$.

$$P_s = W_t L_2/y_r \tag{21}$$

Thus P_s may be increased by increasing the weight of the tractor, W_t, by moving forward the center of gravity of the tractor (increasing L_2), or by lowering and/or moving forward the drawbar hitch point (decreasing y_r).

The generation of sufficient lateral forces for steering the tractor also depends on the value of R_f. Thus from the standpoint of steering control, the maximum value of P should be somewhat less than P_s.

Traction
The maximum achievable drawbar force P may also be limited by the tractive conditions of the surface on which the tractor is operating. As mentioned, the gross tractive coefficient is a function of both drive wheel slippage and

tire and soil parameters. As discussed in the section Traction Prediction from Dimensional Analysis in chapter 10, the gross tractive coefficient μ_g versus slippage S relation has the following functional form

$$\mu_g = F/R = \mu_{max} (1 - e^{-0.3C_nS}) \tag{22}$$

where the dimensionless parameter C_n incorporates tire and soil parameters as well as the vertical load on the drive wheel. The exponential nature of equation 22 indicates that the maximum value of μ_g obtainable is μ_{max}.

For a given drawbar force P, equations 18 and 19 may be used to calculate R_f and R_r from which the rolling resistance forces TF_f and TF_r may then be determined. Equation 20 then gives the gross tractive force F_r required. Dividing F_r by R_r determines the required gross tractive coefficient μ_g. If μ_g is greater than μ_{max}, it is obvious that the required tractive coefficient cannot be obtained under the given operating conditions.

If the required tractive coefficient is less than but close to the value of μ_{max} and if sufficient engine power is available, the given drawbar force may be developed, but the slippage of the drive wheels may be so great that operation under such conditions would be impractical.

Power

Assuming that sufficient traction is available, a check should be made to determine if sufficient engine torque is available for developing the given drawbar force. Starting with the required gross tractive force F_r, the engine torque T_e necessary may be determined from equation 14, where $\ddot{\theta} = \ddot{\phi}_w = 0$.

$$T_e = (F_r r_r)/(\eta N) \tag{23}$$

If T_e is less than the maximum torque that can be produced by the engine, the engine has sufficient power for the tractor to pull the given load.

Operating Conditions

To complete the analysis of the tractor operating conditions for a given drawbar pull, the engine's torque-speed relation may be used to estimate the steady-state engine speed $\dot{\phi}_e$ at which torque T_e may be produced. The resulting rear wheel speed, $\dot{\phi}_w$, is then

$$\dot{\phi}_w = \dot{\phi}_e/N \tag{24}$$

Since the required gross tractive coefficient μ_g has already been determined, the tractive coefficient (μ_g)–slippage (S) relation may be used to find the slippage of the drive wheels. For example, using equation 22,

$$e^{-0.3C_nS} = 1 - \mu_g/\mu_{max}$$
$$S = - [\ln (1 - \mu_g/\mu_{max})]/(0.3C_n) \tag{25}$$

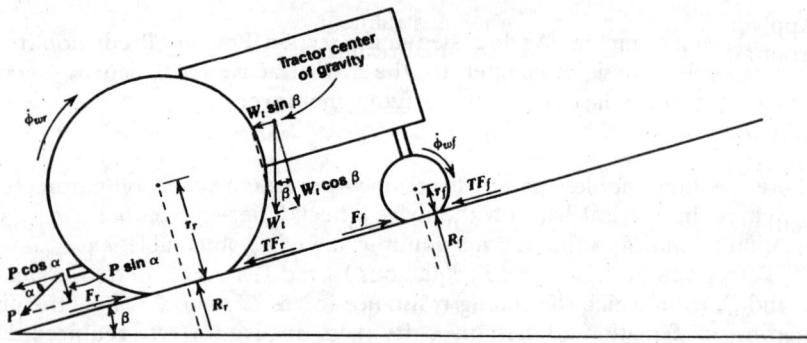

FIGURE 11-6   Free-body diagram of a four-wheel-drive tractor.

Given the slippage of the drive wheels, equation 13 may be used to calculate the forward velocity of the drive wheels, $\dot{x}_w$, which is equal to the forward velocity of the tractor.

$$\dot{x}_w = r_r \, \dot{\phi}_w (1 - S) \tag{26}$$

Static Equilibrium Analysis—Four-Wheel-Drive

Figure 11-6 is a free-body diagram of a four-wheel-drive tractor. The only difference from the free-body diagram of the rear-wheel-driven tractor of figure 11-3 is the addition of a gross tractive force F_f acting on the front wheels, which are turning at a rotational speed $\dot{\phi}_{wf}$.

Equations 17 and 19 are still applicable to computing forces R_f and R_r, which again can be used to compute TF_f and TF_r. However, equation 20 becomes

$$F_r + F_f = W_t \sin \beta + TF_f + TF_r + P \cos \alpha \tag{27}$$

Equation 27 has two unknowns, F_r and F_f, and the problem is to determine how the two axles divide the total tractive force requirement.

By applying equation 22 to both wheels, equation 27 can be written

$$\mu_{max} (1 - e^{-0.3C_{nr}S_r}) R_r + \mu_{max} (1 - e^{-0.3C_{nf}S_f}) R_f =$$
$$W_t \sin \beta + TF_f + TF_r + P \cos \alpha \tag{28}$$

where the subscripts r and f added to the variables C_n and S denote rear and front, respectively. Equation 28 is now in terms of the unknowns S_r and S_f.

Applying equation 26 with the constraint that the forward velocities of the front and rear axle centers are equal results in

$$\dot{x}_w = r_f \dot{\phi}_{wf} (1 - S_f) = r_r \dot{\phi}_{wr} (1 - S_r)$$

so that

$$S_f = 1 - [(r_r \dot{\phi}_{wr})/(r_f \dot{\phi}_{wf})] (1 - S_r)$$

Letting $N_f = \dot{\phi}_e/\dot{\phi}_{wf}$ and $N_r = \dot{\phi}_e/\dot{\phi}_{wr}$,

$$S_f = 1 - [(r_r N_f)/(r_f N_r)] (1 - S_r) \tag{29}$$

Equation 29 can now be used in equation 28 to provide a relationship that can be solved iteratively for S_r. With S_r determined, S_f and, hence, F_r and F_f can be calculated. The engine torque required becomes

$$T_e = \frac{1}{\eta} \left(\frac{F_r r_r}{N_r} + \frac{F_f r_f}{N_f} \right)$$

The engine torque-speed relation can again be used to determine the engine speed $\dot{\phi}_e$, and, hence, the wheel speeds $\dot{\phi}_{wf}$ and $\dot{\phi}_{wr}$ and the forward speed $\dot{x}_w$.

Equation 28 indicates that for the same drawbar loading, powering the front wheels of a rear-wheel-driven tractor will reduce the slippage of the rear wheels, resulting in the tractor's ability to pull the given loading at a higher forward travel speed, thus increasing its productivity. Since the rear wheels of a four-wheel-drive tractor usually follow in the tracks left by the front wheels, the rear wheels will usually encounter higher-strength soil. The higher-strength soil is reflected in equation 28 by an increase in the wheel numeric C_{nr}, which also has the effect of lowering the required rear drive wheel slippage S_r.

Longitudinal Stability

There are several situations in which a tractor may be in danger of rearward overturning. The equations associated with the free-body diagrams of figure 11-1 are particularly useful for the analysis of these situations. In such situations in which the front wheels have left the ground, forces R_f and TF_f will be zero. With use of D'Alembert's principle, the tractor chassis may be considered to be in static equilibrium, allowing moments to be summed about any point. D'Alembert's principle is applied by adding two fictitious forces, often called inertial forces, acting at the chassis center of gravity: $m_c \ddot{x}_c$ in the $-X$ direction and $m_c \ddot{z}_c$ in the $-Z$ direction. In addition, a fictitious clockwise moment $I_{yyc} \ddot{\theta}$ is applied to the chassis. Now, with the tractor in the assumed static equilibrium state, moments may be summed about the rear axle, thus eliminating the presence of the internal reaction forces V_r and H_r from the resulting equation. Using equations 8 and 9 to express $\ddot{x}_c$ and $\ddot{z}_c$ in terms of $\ddot{x}_w$, $\ddot{z}_w$, $\ddot{\theta}$, and $\dot{\theta}^2$, the resulting moment equation may be written

$$
\begin{aligned}
(I_{yyc} + m_c h_{1c}^2)\ddot{\theta} = \ & T_r - W_c h_{1c} \cos(\theta_{1c} + \beta + \theta) \\
& - P \cos\alpha\, h_3 \cos(\phi_0 - \theta) \\
& + P \sin\alpha\, h_3 \sin(\phi_0 - \theta) \\
& + m_c \ddot{z}_w h_{1c} \cos(\theta_{1c} + \theta) \\
& + m_c \ddot{x}_w h_{1c} \sin(\theta_{1c} + \theta)
\end{aligned} \tag{30}
$$

Equation 30 indicates that the parallel component, $P \cos\alpha$, of the drawbar force exerts a moment tending to resist positive rotation of the chassis while an overturning moment is exerted by the perpendicular component, $P \sin\alpha$. One situation that has led to many rearward overturns is the application of the drawbar force P at a point near or above the rear axle. In such a situation, the angle of inclination α of the drawbar force with the ground surface may also be fairly large. As a result, the moment arm $h_3 \cos(\phi_0 - \theta)$ of the parallel component may be small or even negative, while the perpendicular component $P \sin\alpha$ is increased. As equation 30 indicates, such a situation makes the development of a positive angular acceleration $\ddot{\theta}$ more probable, and thus the chances for a rearward overturn are increased.

Even though sufficient traction and power are available to develop a drawbar force in excess of P_s, the drawbar force required to cause the front wheels to just lose contact with the ground, the tractor may not be in danger of overturning if the drawbar load is properly hitched to the tractor. Equation 30 and figure 11-2 indicate that the moment arm, $h_3 \cos(\phi_0 - \theta)$, of the parallel drawbar force component, $P \cos\alpha$, tends to increase as the tractor rotates counterclockwise through a positive angle θ. Simultaneously, the moment arm, $h_3 \sin(\phi_0 - \theta)$, of the perpendicular component, $P \sin\alpha$, tends to decrease as θ increases. Thus, for a properly hitched load, rotation of the tractor chassis may tend to stabilize the tractor, thus providing an explanation for situations in which a tractor may be in equilibrium with the front wheels of the tractor off the ground.

Another situation in which a rearward overturn may take place occurs when a post or log is chained to the rear drive wheels in an effort to free a tractor that has become immobilized in soft soil. If the post or log prevents the rear wheels from turning when the tractor clutch is released, sufficient rear axle torque T_r may be developed to overturn the tractor.

Equation 30 indicates that the moment arm, $h_{1c} \cos(\theta_{1c} + \beta + \theta)$, of the chassis weight decreases as the rotation angle θ increases. In addition, this moment arm is also decreased by operation on a slope and is influenced by the height of the chassis center of gravity. Thus the longitudinal stability of the tractor may be increased by adding weight to the tractor chassis in such a manner as to move forward and lower the chassis center of gravity. Such an action would also probably have the favorable effect of increasing the moment of inertia of the chassis.

When no drawbar force P is applied, a tractor becomes statically unstable when the angle $(\theta_{1c} + \beta + \theta)$ reaches 90°. However, in a dynamic sense, the tractor may become unstable at a considerably smaller angle. At this angle, the angular velocity $\dot{\theta}$ may be sufficient to allow the tractor to become statically unstable even though the rear axle torque T_r and drawbar force P may be reduced to zero. In such a situation where T_r, P, $\ddot{x}_w$, and $\ddot{z}_w$ are assumed to be zero, equation 30 becomes

$$(I_{yyc} + m_c h_{1c}^2)\ddot{\theta} = -W_c h_{1c} \cos{(\theta_{1c} + \beta + \theta)} \tag{31}$$

Equation 31 can be integrated using the identities $\ddot{\theta}\,d\theta = \dot{\theta}\,d\dot{\theta}$ and $d\theta = d(\theta_{1c} + \beta + \theta)$. Thus, by multiplying both sides of equation 31 by $d\theta$,

$$(I_{yyc} + m_c h_{1c}^2)\ddot{\theta}\,d\theta = -W_c h_{1c} \cos{(\theta_{1c} + \beta + \theta)}\,d\theta$$
$$(I_{yyc} + m_c h_{1c}^2)\dot{\theta}\,d\dot{\theta} = -W_c h_{1c} \cos{(\theta_{1c} + \beta + \theta)}\,d(\theta_{1c} + \beta + \theta) \tag{32}$$

Both sides of equation 32 may now be integrated once appropriate limits are chosen. Assume that for a given angle of rotation θ_s, we desire to find the angular velocity $\dot{\theta}_s$ just sufficient to cause the chassis to become statically unstable. Then the appropriate limits are $\theta_{1c} + \beta + \theta = \theta_{1c} + \beta + \theta_s$ when $\dot{\theta} = \dot{\theta}_s$, and $\theta_{1c} + \beta + \theta = \pi/2$ when $\dot{\theta} = 0$ (the tractor becomes statically unstable just as the angular velocity $\dot{\theta}$ becomes zero).

Integrating equation 32 with the above limits results in

$$(I_{yyc} + m_c h_{1c}^2)\dot{\theta}_s^2/2 = W_c h_{1c}(1 - \sin{[\theta_{1c} + \beta + \theta_s]}) \tag{33}$$

Equation 33 may be used to estimate the angular velocity $\dot{\theta}_s$ required for rotation of the tractor chassis to the point of static instability starting from a given angle of rotation θ_s. Equation 33 may also be derived from the work-energy theorem of mechanics.

The Tractor as a 2-Degree-of-Freedom Vibratory System

The ride motions of tractors have historically been associated with their bounce (vertical translation) and pitch motions. Considerable insight into these motions can be obtained from linearization of equations 11 and 12. Assume that the tractor is traveling at a constant forward speed ($\ddot{x}_t = \dot{\phi}_w = 0$) on level

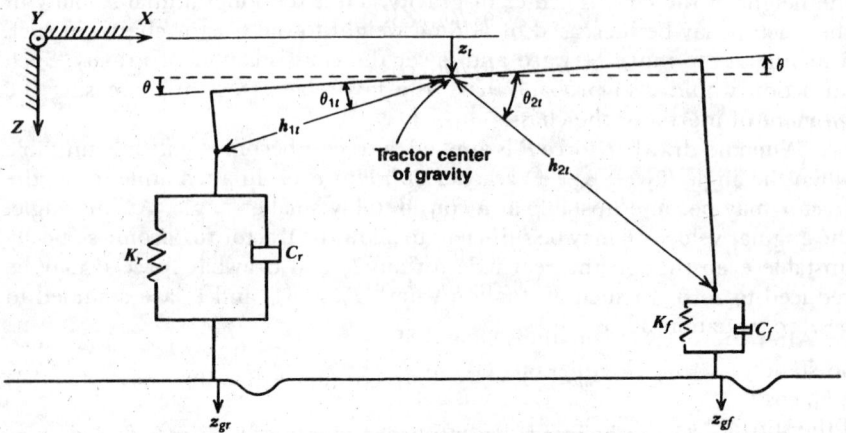

FIGURE 11-7 Representation of the tractor as a 2-degree-of-freedom spring-mass-damper system.

ground $(\beta = 0)$, under no drawbar load $(P = 0)$ and that the forces R_r and R_f pass through the rear and front wheel centers respectively $(e_r = e_f = 0)$. Note in equation 12 that the moment arms of the forces F_r, TF_f, and TF_r are equal to each other. Further from equation 10, $F_r - TF_f - TF_r = 0$. Under these conditions, equations 11 and 12 become

$$m_t\ddot{z}_t = W_t - R_r - R_f \tag{34}$$

$$I_{yyt}\ddot{\theta} = R_f h_{2t} \cos (\theta_{2t} - \theta) - R_r h_{1t} \cos (\theta_{1t} + \theta) \tag{35}$$

The tires, which represent the only suspension element besides a seat suspension on conventional tractors, can be idealized as parallel combinations of a linear spring and damper as shown in figure 11-7. The spring and damping rates are the sums of the corresponding rates for each individual tire. For example, K_r is the sum of the spring rates of the rear tires.

z_{gr} and z_{gf} represent the time histories of the ground displacements encountered by the rear and front tires, respectively. z_{gr} and z_{gf} would normally be thought of as functions of the forward travel of the tractor. However, through use of the constant forward velocity of the tractor, z_{gr} and z_{gf} may be converted into functions of time.

Note that the change in the definition of z_t caused by the vertical translation of the inertial coordinate system shown in figure 11-7 has no effect on the equations of motion.

When the tractor is in static equilibrium $(\theta = z_t = 0)$, forces R_r and R_f take the values R_{rs} and R_{fs}, respectively, where

$$R_{rs} = W_t h_{2t} \cos \theta_{2t}/(h_{1t} \cos \theta_{1t} + h_{2t} \cos \theta_{2t}) \qquad (36)$$

$$R_{fs} = W_t h_{1t} \cos \theta_{1t}/(h_{1t} \cos \theta_{1t} + h_{2t} \cos \theta_{2t}) \qquad (37)$$

For the displacements shown in figure 11-7,

$$R_r = R_{rs} + K_r[z_t + h_{1t} \sin (\theta_{1t} + \theta) - h_{1t} \sin \theta_{1t} - z_{gr}] \qquad (38)$$
$$+ C_r[\dot{z}_t + h_{1t} \dot{\theta} \cos (\theta_{1t} + \theta) - \dot{z}_{gr}]$$

$$R_f = R_{fs} + K_f[z_t + h_{2t} \sin (\theta_{2t} - \theta) - h_{2t} \sin \theta_{2t} - z_{gf}] \qquad (39)$$
$$+ C_f[\dot{z}_t - h_{2t} \dot{\theta} \cos (\theta_{2t} - \theta) - \dot{z}_{gf}]$$

After substituting equations 38 and 39 into equations 34 and 35, utilizing the static equilibrium equations ($W_t - R_{rs} - R_{fs} = 0$ and $R_{fs}h_{2t} \cos \theta_{2t} - R_{rs}h_{1t} \cos \theta_{1t} = 0$), using the trigonometric identities for the sine and cosine of the sum and difference of two angles, and assuming that since θ is considered to be small, $\sin \theta \approx \theta$, $\cos \theta \approx 1$, and that second-order terms such as $\theta \dot{\theta}$ may be neglected, equations 34 and 35 may be linearized, yielding

$$\ddot{z}_t + K_1\dot{z}_t + K_2\dot{\theta} + K_3 z_t + K_4\theta = (K_r z_{gr} + K_f z_{gf} + C_r\dot{z}_{gr} + C_f\dot{z}_{gf})/m_t \qquad (40)$$

$$\ddot{\theta} + K_5\dot{z}_t + K_6\dot{\theta} + K_7 z_t + K_8\theta = [(K_r z_{gr} + C_r\dot{z}_{gr})(h_{1t} \cos \theta_{1t}) \qquad (451)$$
$$- (K_f z_{gf} + C_f\dot{z}_{gf})(h_{2t} \cos \theta_{2t})]/I_{yyt}$$

where
$$K_1 = (C_f + C_r)/m_t$$
$$K_2 = (C_r h_{1t} \cos \theta_{1t} - C_f h_{2t} \cos \theta_{2t})/m_t$$
$$K_3 = (K_f + K_r)/m_t$$
$$K_4 = (K_r h_{1t} \cos \theta_{1t} - K_f h_{2t} \cos \theta_{2t})/m_t$$
$$K_5 = (C_r h_{1t} \cos \theta_{1t} - C_f h_{2t} \cos \theta_{2t})/I_{yyt}$$
$$K_6 = (C_f[h_{2t} \cos \theta_{2t}]^2 + C_r[h_{1t} \cos \theta_{1t}]^2)/I_{yyt}$$
$$K_7 = (K_r h_{1t} \cos \theta_{1t} - K_f h_{2t} \cos \theta_{2t})/I_{yyt}$$
$$K_8 = (K_f[h_{2t} \cos \theta_{2t}]^2 + K_r[h_{1t} \cos \theta_{1t}]^2$$
$$- [R_{fs} - K_f z_{gf} - C_f\dot{z}_{gf}][h_{2t} \sin \theta_{2t}]$$
$$- [R_{rs} - K_r z_{gr} - C_r\dot{z}_{gr}][h_{1t} \sin \theta_{1t}])/I_{yyt}$$

Equations 40 and 41 define the response of the linearized system to terrain excitation. If the terrain excitation is random in nature, a frequency analysis of the time histories of the system response will usually show prominent peaks at the natural frequencies of the system, since the damping provided by the tires is usually relatively small. The natural frequencies of the system are important in seat suspension design, since the natural frequency of the seat suspension must be somewhat less than the major frequency input to the seat if the suspension is to attenuate the input. For free vibration ($z_{gf} = z_{gr} = 0$), the natural frequencies may be determined by setting the damping rates C_f and C_r to zero. Equations 40 and 41 then become

$$\ddot{z}_t + K_3 z_t + K_4 \theta = 0 \tag{42}$$

$$\ddot{\theta} + K_7 z_t + K_8 \theta = 0 \tag{43}$$

Assume that a periodic solution exists for equations 42 and 43 of the form

$$z_t = Z \sin \omega t \tag{44}$$

$$\theta = \Theta \sin \omega t \tag{45}$$

Substituting equations 44 and 45 into equations 42 and 43 and simplifying

$$(K_3 - \omega^2)Z + K_4 \Theta = 0 \tag{46}$$

$$K_7 Z + (K_8 - \omega^2)\Theta = 0 \tag{47}$$

Equations 46 and 47 have a nontrivial solution only if the determinant of the coefficient matrix formed from these equations is zero. Thus

$$(K_3 - \omega^2)(K_8 - \omega^2) - K_4 K_7 = 0$$

or

$$\omega^4 - (K_3 + K_8)\omega^2 + (K_3 K_8 - K_4 K_7) = 0$$

Using the quadratic formula to solve for ω^2

$$\omega^2 = \frac{(K_3 + K_8) \pm \sqrt{(K_3 + K_8)^2 - 4(K_3 K_8 - K_4 K_7)}}{2} \tag{48}$$

The two values of ω^2 given by equation 48 lead to the two natural frequencies of the system.

Each natural frequency is associated with a corresponding mode of vibration defined by the ratio Z/Θ obtainable from either equation 46 or 47.

$$\frac{Z}{\Theta} = \frac{K_4}{\omega^2 - K_3} = \frac{\omega^2 - K_8}{K_7} \tag{49}$$

The mode of vibration may be illustrated by assuming an arbitrary value for the amplitude Θ. The amplitude Z may then be calculated from equation 49. The two amplitudes may then be combined with a scaled drawing of the tractor to indicate the mode of vibration as shown in figure 11-8. In figure 11-8, the mode of vibration is represented by what might be described as a

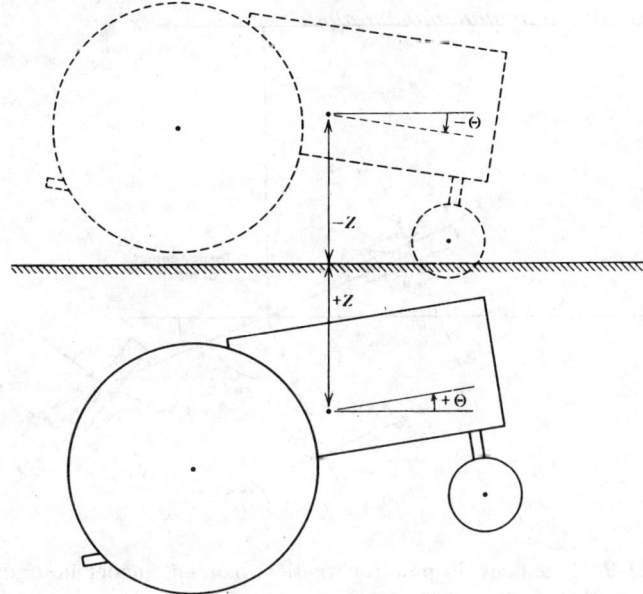

FIGURE 11-8 Method of illustrating a mode shape.

double-exposure photograph. The solid line drawing of the tractor represents one extreme position of the mode of vibration (sin $\omega t = 1$), and the dashed line drawing illustrates the other extreme (sin $\omega t = -1$).

Of course, the amplitude of the motion indicated by a mode shape drawing such as figure 11-8 has no relation to the amplitude of motion that would result if that mode were excited experimentally or in the field. However, the mode shape drawing does provide a valuable insight into the type of motion associated with each of the natural frequencies of the system.

Transient and Steady-State Handling

Figure 11-9 illustrates a simplified model that can be used to describe tractor transient handling. The wheels on the front and rear axles have been replaced by an equivalent single wheel at each axle having lateral force properties equivalent to the total of the wheels on that axle.

The model has 2 degrees of freedom, the lateral translational velocity, v, of the center of gravity and the yaw angular velocity, r, of the tractor about its center of gravity. The forward velocity, u, of the center of gravity is assumed

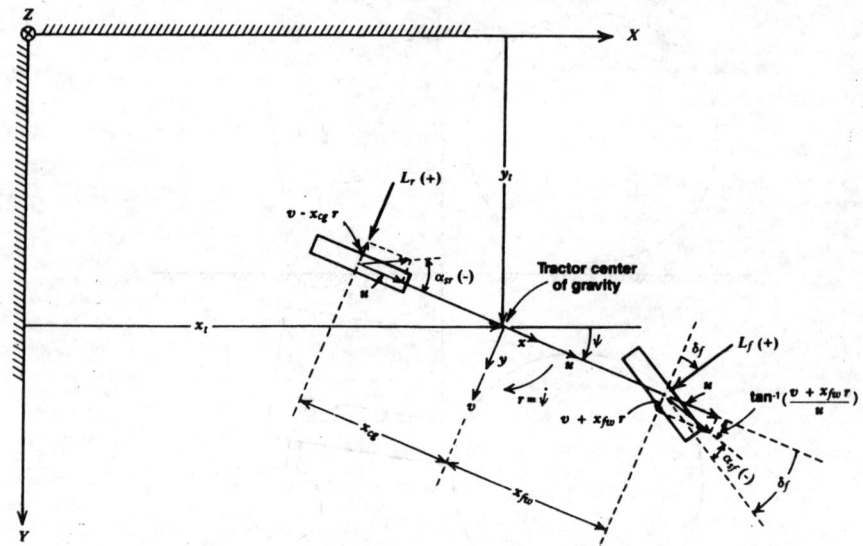

FIGURE 11-9   Free-body diagram for transient handling model illustrating front and rear slip angles.

to be constant, whereas the steer angle, δ_f, of the front wheels is assumed to be a known function of time.

A pneumatic tire can be considered to develop a lateral force whenever the direction in which the tire is headed differs from the direction of the plane of the wheel itself. This difference in directions, called the tire slip angle, is expressed in terms of the velocities of the wheel in the vehicle coordinate system. Figure 11-9 illustrates the slip angles of the front, α_{sf}, and rear, α_{sr}, wheels. From that figure,

$$\alpha_{sf} = \tan^{-1}\left((v + x_{fw}r)/u\right) - \delta_f$$
$$\alpha_{sr} = \tan^{-1}\left((v - x_{cg}r)/u\right)$$

The lateral force developed by a tire at a given vertical load takes the form shown in figure 11-10. The lateral force varies nearly linearly with slip angle at small slip angles and reaches a maximum value asymptotically. The slope of the curve at the origin is termed the cornering stiffness, C_α. Thus, for small slip angles, $L = -C_\alpha \alpha_s$; i.e., a positive slip angle produces a negative lateral force. If the lateral force versus slip angle relation is nondimensionalized by dividing the lateral force by the normal force on the tire, the resulting ratio, μ_l, is termed the cornering coefficient.

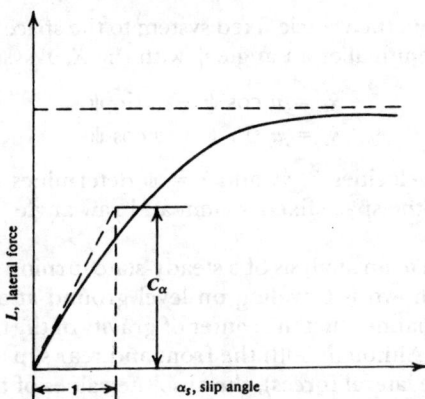

FIGURE 11-10 Lateral force–slip angle functional relationship.

The equations of motion for the tractor in figure 11-9 are most easily expressed in terms of the vehicle coordinate system. In such a system, the lateral acceleration of the center of gravity is $\dot{v} + ur$. Thus,

$$m_t (\dot{v} + ur) = L_f \cos \delta_f + L_r \tag{50}$$

Letting I_{zzt} be the moment of inertia of the tractor about the vertical (or z) axis through its center of gravity,

$$I_{zzt} \dot{r} = (L_f \cos \delta_f) x_{fw} - L_r x_{cg} \tag{51}$$

In transport situations, the steer and slip angles will be small so that the preceding equations can be linearized yielding

$$m_t (\dot{v} + ur) = -C_{\alpha f} \left[\frac{v + x_{fw} r}{u} - \delta_f \right] - C_{\alpha r} \left[\frac{v - x_{cg} r}{u} \right] \tag{52}$$

$$I_{zzt} \dot{r} = -C_{\alpha f} \left[\frac{v + x_{fw} r}{u} - \delta_f \right] x_{fw} + C_{\alpha r} \left[\frac{v - x_{cg} r}{u} \right] x_{cg} \tag{53}$$

Note that $C_{\alpha f}$ and $C_{\alpha r}$ represent the total cornering stiffness at the front and rear axles, respectively.

Equations 52 and 53 may be integrated once to find the lateral velocity, v, and yaw velocity, r, resulting from a given time history of the front wheel steer angle, δ_f. To find the trajectory of the center of gravity of the tractor in the X, Y space fixed system, one must first transform the velocity of the center of gravity from the vehicle fixed system to the space fixed system. Since the vehicle x, y system makes an angle ψ with the X, Y system,

$$\dot{x}_t = u \cos \psi - v \sin \psi$$
$$\dot{y}_t = u \sin \psi + v \cos \psi$$

Integration of the velocities $\dot{x}_t$, $\dot{y}_t$, and $r = \dot{\psi}$ determines the location of the center of gravity in the space fixed system and yaw angle, ψ, of the tractor as functions of time.

Let us now turn to an analysis of a steady-state turning situation. In figure 11-11, the tractor shown is traveling on level ground at a constant forward velocity u in such a manner that the center of gravity of the tractor is traversing a circle of radius R. Although both the front and rear slip angles are negative (resulting in positive lateral forces), the absolute values of the angles are used in the geometric analysis.

From figure 11-11, $L/R = \alpha_{sr} + \delta_f - \alpha_{sf}$

or

$$\delta_f = L/R + \alpha_{sf} - \alpha_{sr}$$

But, for this steady-state situation, $\dot{v} = \dot{r} = 0$ and $r = u/R$. Thus, with δ_f small, equations 50 and 51 become

$$m_t (\dot{v} + ur) = m_t u^2/R = L_f + L_r \tag{54}$$

$$I_{zzt} \dot{r} = 0 = L_f x_{fw} - L_r x_{cg} \tag{55}$$

From equation 55, $L_r = L_f (x_{fw}/x_{cg})$. Substituting in equation 54,

$$m_t u^2/R = L_f (1 + x_{fw}/x_{cg}) = L_f \frac{(x_{cg} + x_{fw})}{x_{cg}} = L_f \frac{L}{x_{cg}}$$

Thus,

$$L_f = \frac{x_{cg}}{L} m_t \frac{u^2}{R} = \frac{x_{cg}}{L} \frac{W_t}{g} \frac{u^2}{R}$$

But $x_{cg} W_t/L = W_f$ is the portion of the tractor weight, W_t, statically supported at the front axle so that $L_f = (W_f/g)(u^2/R)$. Similarly, $L_r = (W_r/g)(u^2/R)$ where W_r is the portion supported at the rear axle. Then

$$L_f = \frac{W_f}{g} \frac{u^2}{R} = C_{\alpha f} \alpha_{sf}$$

$$L_r = \frac{W_r}{g} \frac{u^2}{R} = C_{\alpha r} \alpha_{sr}$$

or

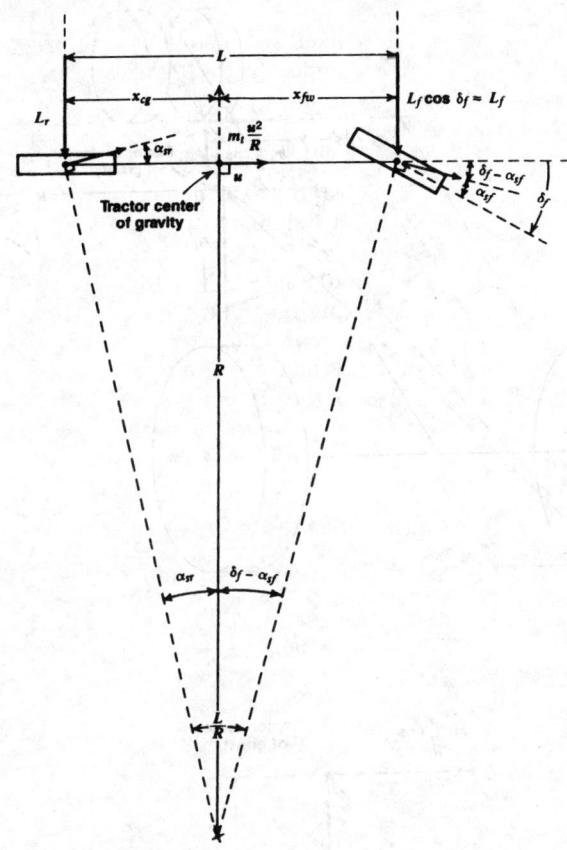

FIGURE 11-11 Tractor making a steady-state turn of radius R at constant forward speed u.

$$\alpha_{sf} = \frac{1}{C_{\alpha f}} \frac{W_f}{g} \frac{u^2}{R}$$

$$\alpha_{sr} = \frac{1}{C_{\alpha r}} \frac{W_r}{g} \frac{u^2}{R}$$

Thus,

$$\delta_f = \frac{L}{R} + \frac{u^2}{gR} \left(\frac{W_f}{C_{\alpha f}} - \frac{W_r}{C_{\alpha r}} \right) \tag{56}$$

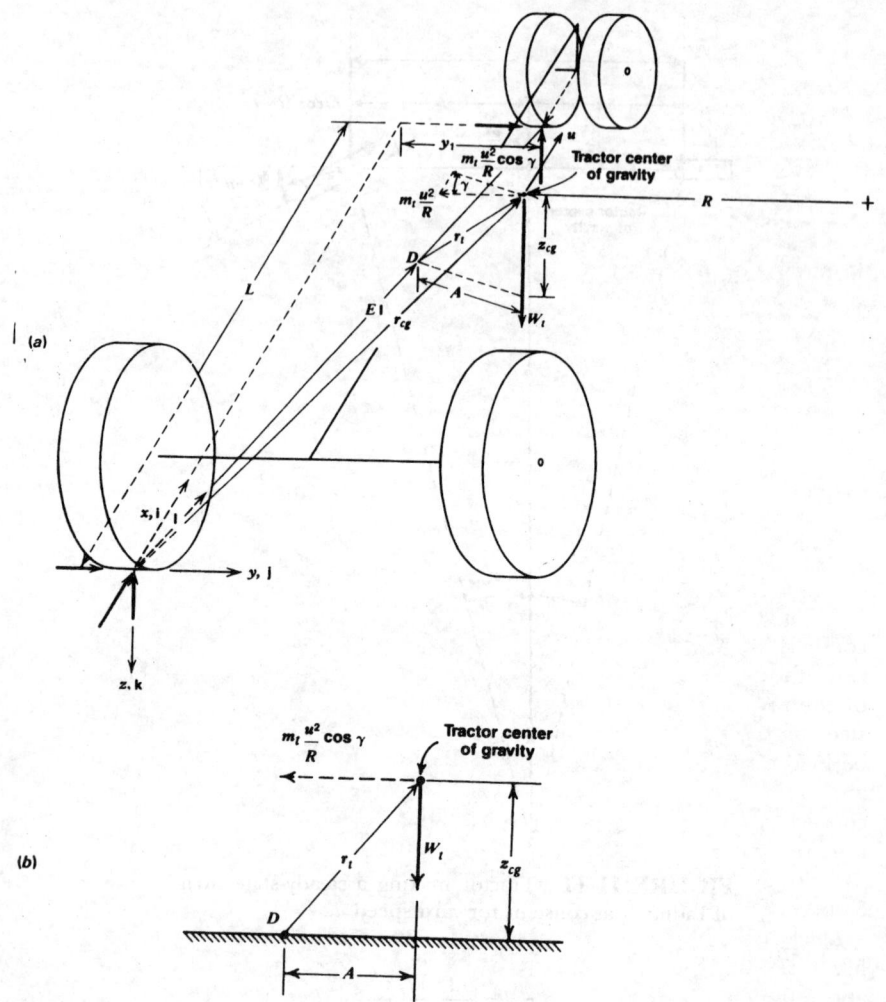

FIGURE 11-12 (*a*) Free-body diagram of a tractor about to overturn laterally **as a** result of making a steady-state turn. (*b*) Plane containing the tipping motion of the tractor. (*c*) Top view of tractor.

Equation 56 is the fundamental equation describing the steady-state handling behavior of a road vehicle. The quantity L/R is termed the Ackerman steer angle and is the front wheel steer angle required for the vehicle to move in a turn of radius R at low speed.

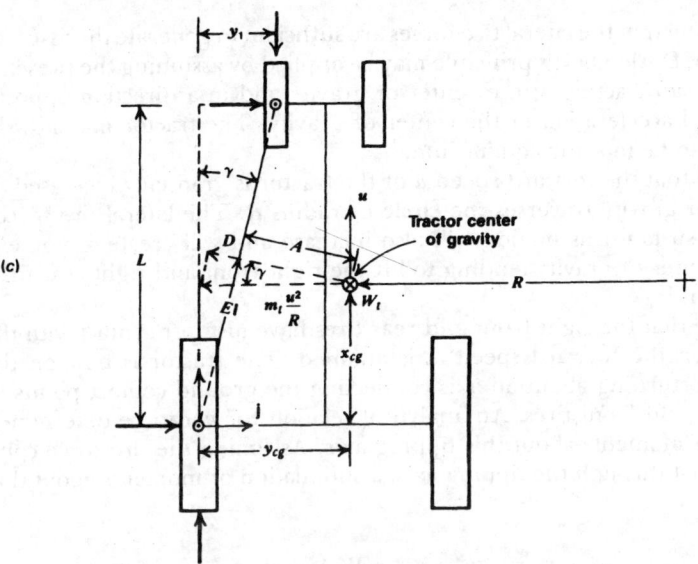

For higher speeds, the term $(W_f/C_{\alpha f} - W_r/C_{\alpha r})$, called the understeer coefficient, determines the steering behavior of the vehicle. If this term is zero, the steer angle required to negotiate a turn of radius R is independent of the forward speed u. Such a vehicle would be termed as having neutral steering. If the understeer coefficient is positive, the wheel steer angle must be increased with increasing speed. This condition is termed understeering and is the situation desired on most road vehicles. Finally, if the understeer coefficient is negative, the wheel steer angle δ_f must be decreased as the speed u increases. Such a condition is called oversteering.

For an oversteering vehicle, the speed at which the required steer angle δ_f becomes zero is termed the critical speed, u_c. From equation 56, $u_c^2 = -Lg/(W_f/C_{\alpha f} - W_r/C_{\alpha r})$. By examining the eigenvalues of equations 52 and 53, it can also be shown that an oversteering vehicle is directionally unstable above the critical speed.

Lateral Stability in a Steady-State Turn

The geometric configuration of tricycle tractors, combined with their ability, aided by individual brakes on the drive wheels, to make sharp turns at moderately high travel speeds, can result in a potential lateral overturning situation. Figure 11-12 illustrates a tricycle tractor in the steady-state circular turn analyzed in the previous section.

Assuming that the lateral tire forces are sufficient to generate the assumed acceleration, D'Alembert's principle may be applied by assuming the presence of a force $m_t u^2/R$ acting at the center of gravity and in a direction opposite to the lateral acceleration of the center of gravity. The tractor may now be considered to be in static equilibrium.

Assume that the forward speed u of the tractor is gradually increased as the center of gravity traverses the circle of radius R. The lateral tire forces required to sustain this motion will also increase and will create a moment about the center of gravity tending to lift the right front and right rear tires off the ground.

Assume that the right front and rear tires have just lost contact with the ground when the forward speed u_s is attained. The tractor is now on the verge of overturning about an axis connecting the ground contact points of the left rear and front tires. An analytical relation for u_s can be determined by summing moments about this tipping axis. Assuming the tire force components all act through the tipping axis, a summation of moments about that axis produces

$$m_t \frac{u_s^2}{R} \cos \gamma\, z_{cg} - W_t A = 0$$

Thus

$$u_s = \sqrt{\frac{gAR}{z_{cg} \cos \gamma}} \tag{57}$$

where g is the acceleration of gravity. γ, the angle between the assumed force $m_t u^2/R$ and the tipping plane, is given by $\tan^{-1}(y_1/L)$. Equation 57 indicates u_s is decreased as the radius of the turn or the moment arm A of the tractor weight is decreased or as the height z_{cg} of the center of gravity is increased.

To use equation 57, a relation for A must be determined from the tractor geometry. Let the xyz coordinate system and associated unit vector system $\mathbf{i}, \mathbf{j}$, and $\mathbf{k}$ shown in figure 11-12 have its origin at the point on the ground surface directly beneath the center of the left rear wheel. Relative to this coordinate system, the tipping axis is defined by the unit vector $\mathbf{l}$.

$$\mathbf{l} = (L/B)\mathbf{i} + (y_1/B)\mathbf{j} \tag{58}$$

where L and y_1 locate the point directly beneath the left front wheel center where the corresponding tire forces are assumed to act and $B = \sqrt{L^2 + y_1^2}$.

The plane containing the motion of the center of gravity as it rotates about the tipping axis intersects the tipping axis at point D. The vector from the origin of the xyz system to point D is denoted $E\mathbf{l}$. Given the vector $\mathbf{r}_{cg} = x_{cg}\mathbf{i} + y_{cg}\mathbf{j} - z_{cg}\mathbf{k}$ locating the center of gravity of the tractor ($x_{cg} = h_{1t} \cos \theta_{1t}$, $z_{cg} = r_r + h_{1t} \sin \theta_{1t}$), it follows that E may be found by finding the component of $\mathbf{r}_{cg}$ in the $\mathbf{l}$ direction. This can be done by taking the dot or scalar product of the $\mathbf{r}_{cg}$ and $\mathbf{l}$ vectors.

$$E = \mathbf{r}_{cg} \cdot \mathbf{l} = (x_{cg}L + y_{cg}y_1)/B \tag{59}$$

The vector $\mathbf{r}_t$ from point D to the center of gravity of the tractor can now be found using the relation $\mathbf{r}_t = \mathbf{r}_{cg} - E\mathbf{l}$.

$$\mathbf{r}_t = (x_{cg} - EL/B)\mathbf{i} + (y_{cg} - Ey_1/B)\mathbf{j} - z_{cg}\mathbf{k}$$

The value of A is just the component of $\mathbf{r}_t$ in the direction lying in the ground surface and perpendicular to the tipping axis. Thus

$$A = \sqrt{(x_{cg} - EL/B)^2 + (y_{cg} - Ey_1/B)^2} \tag{60}$$

The addition of a front end loader to a tricycle tractor may considerably reduce the lateral stability of the tractor in a turning situation. If a load is transported in the bucket of the loader with the bucket raised, the center of gravity of the tractor-loader combination will probably be both raised (increasing z_{cg}) and moved forward (decreasing A) as compared to the center of gravity location of the tractor alone. Thus the value of u_s for the loader equipped tractor may be considerably less than for the tractor alone.

A tractor with a pivoting front axle assembly may tip sideways about two axes. The initial tipping motion can be assumed to be about an axis connecting the point on the ground surface directly beneath the center of the rear wheel remaining on the ground to the front axle pivot. The entire tractor except the front axle assembly may rotate about this axis until the tipping part of the tractor strikes stops on the front axle assembly. At this point, the entire tractor may continue to tip about a second tipping axis. This axis may be assumed to connect the ground contact points of the front and rear tires remaining in contact with the ground during the overturning motion.

Three-Dimensional Static Analysis

The analyses presented up to this point have been limited to particular two-dimensional cases of more general three-dimensional motions. This section provides an introduction to three-dimensional analysis through the static force and steady-state motion analysis of the rear-wheel-driven, single-front-wheeled tricycle tractor shown in figure 11-13.

The tractor is assumed moving with constant longitudinal and lateral velocities u and v and at a fixed yaw or heading angle ψ relative to the X, Y, Z space fixed coordinate system. Note that $\psi = 0°$ corresponds to operation up the slope β, and $\psi = \pm 90°$ corresponds to operating on a side slope.

To simplify the force analysis, the center of gravity and drawbar location are assumed to lie in the vertical plane containing the longitudinal centerline

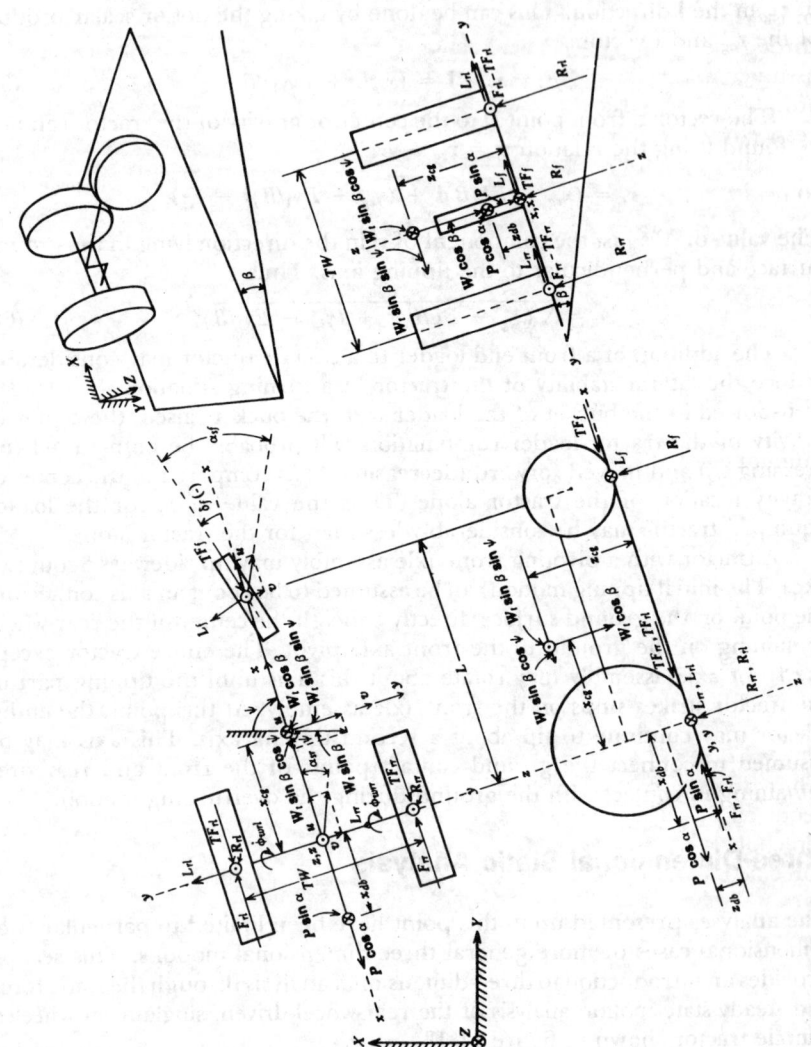

FIGURE 11-13 Free-body diagram of tricycle tractor.

of the tractor. In addition, the normal forces on the tires are assumed to act through the respective wheel centers, and the front wheel steer angle δ_f is assumed small so that the rolling resistance and lateral force acting on the front wheel can be assumed to act in the x and y vehicle fixed directions respectively.

The classical approach to solving for the forces acting on the tractor is to begin by summing forces in the x, y, and z vehicle fixed directions.

$$\Sigma F_x = 0 = F_{rr} + F_{rl} - W_t \sin \beta \cos \psi$$
$$- TF_f - TF_{rr} - TF_{rl} - P \cos \alpha \quad (61)$$

$$\Sigma F_y = 0 = W_t \sin \beta \sin \psi - L_f - L_{rr} - L_{rl} \quad (62)$$

$$\Sigma F_z = 0 = W_t \cos \beta + P \sin \alpha - R_{rr} - R_{rl} - R_f \quad (63)$$

Summing moments about the x-x, y-y, and z-z axes shown in figure 11-13,

$$\Sigma M_{xx} = 0 = W_t \sin \beta \sin \psi \, z_{cg} + R_{rl} \, TW/2 - R_{rr} \, TW/2 \quad (64)$$

$$\Sigma M_{yy} = 0 = R_f L - W_t \cos \beta \, x_{cg} + W_t \sin \beta \cos \psi \, z_{cg} +$$
$$P \sin \alpha \, x_{db} + P \cos \alpha \, z_{db} \quad (65)$$

$$\Sigma M_{zz} = 0 = -L_f L + W_t \sin \beta \sin \psi \, x_{cg} + (F_{rl} - TF_{rl}) \, TW/2$$
$$- (F_{rr} - TF_{rr}) \, TW/2 \quad (66)$$

We now have six equations in the 11 unknown forces acting on the tractor. Thus, to allow a solution, five additional relations between the forces are required. Three of these relations come from the relation of the section Traction Prediction from Dimensional Analysis in chapter 10 for the towed force acting on a wheel.

$$TF_i/R_i = 1.2/C_{ni} + 0.04 \quad i = f, rr, rl \quad (67),(68),(69)$$

Let us assume the tractor is equipped with a differential that delivers equal torques to each rear wheel. With the use of the traction mechanics of the section Traction Performance Equations in chapter 10, this constraint is equivalent to assuming that the gross tractive forces on the rear wheels are equal.

$$F_{rr} = F_{rl} \quad (70)$$

The final relation comes from the fact that since the rear wheels operate with equal slip angles, the cornering coefficient, μ_l (the ratio of lateral to normal force), is the same for both rear wheels. Thus,

$$L_{rr}/R_{rr} = L_{rl}/R_{rl} \tag{71}$$

With 11 equations relating the 11 unknown forces, a unique solution can now be found. In a more general situation, the corresponding equations might be nonlinear and require a simultaneous, iterative solution using a numerical technique. However, under the assumptions made previously, it is possible to solve the preceding set of equations analytically.

From equation 65,

$$R_f = (W_t \cos \beta \, x_{cg} - W_t \sin \beta \cos \psi \, z_{cg} - P \sin \alpha \, x_{db} - P \cos \alpha \, z_{db})/L$$

This relation is analogous to equation 17.

From equations 63 and 64,

$$R_{rr} + R_{rl} = W_t \cos \beta + P \sin \alpha - R_f \tag{72}$$

$$R_{rr} - R_{rl} = (2/TW) \, W_t \sin \beta \sin \psi \, z_{cg} \tag{73}$$

Adding equations 72 and 73 and then subtracting equation 73 from equation 72,

$$R_{rr} = (W_t \cos \beta + P \sin \alpha - R_f)/2 + W_t \sin \beta \sin \psi \, z_{cg}/TW$$

$$R_{rl} = (W_t \cos \beta + P \sin \alpha - R_f)/2 - W_t \sin \beta \sin \psi \, z_{cg}/TW$$

Note the downhill (right for ψ positive) wheel carries a larger normal force than the uphill (left) wheel.

With R_f, R_{rr}, and R_{rl} determined, equations 67, 68, and 69 allow TF_f, TF_{rr}, and TF_{rl} to be calculated. Then from equations 61 and 70,

$$F_{rr} + F_{rl} = 2F_{rr} = 2F_{rl} = W_t \sin \beta \cos \psi + TF_f + TF_{rr} + TF_{rl} + P \cos \alpha \tag{74}$$

From equation 66,

$$L_f = (W_t \sin \beta \sin \psi \, x_{cg} - (F_{rr} - TF_{rr}) \, TW/2 + (F_{rl} - TF_{rl}) \, TW/2)/L \tag{75}$$

From equations 62 and 71,

$$L_{rr} + L_{rl} = W_t \sin \beta \sin \psi - L_f$$

$$L_{rl} = (R_{rl}/R_{rr}) \, L_{rr}$$

so that

$$L_{rr} = (W_t \sin \beta \sin \psi - L_f)/(1 + R_{rl}/R_{rr}) \qquad (76)$$

With the forces determined, we can now turn to finding the velocity components u and v. The torque required of the engine is

$$T_e = (F_{rr} + F_{rl}) \, r_r/(\eta N)$$

From the engine's torque-speed relation, the required engine speed, $\dot{\phi}_e$, can be determined. The corresponding average rear wheel speed, $\dot{\phi}_{wa}$, is then

$$\dot{\phi}_{wa} = \dot{\phi}_e/N = (\dot{\phi}_{wrr} + \dot{\phi}_{wrl})/2 \qquad (77)$$

where the latter relation expresses the constraint imposed by the differential on the wheel speeds.

The slippage required of each rear wheel to develop the required gross tractive force can be determined by a generalization of equation 25,

$$S_{rr} = -[\ln (1 - F_{rr}/(R_{rr} \, \mu_{max})]/(0.3 \, C_{nrr})$$

$$S_{rl} = -[\ln (1 - F_{rl}/(R_{rl} \, \mu_{max})]/(0.3 \, C_{nrl})$$

Generalizing equation 26 and applying it to both rear wheels,

$$u = r_r\dot{\phi}_{wrr} (1 - S_{rr}) = r_r\dot{\phi}_{wrl} (1 - S_{rl})$$

Using equation 77,

$$\dot{\phi}_{wrr} = \dot{\phi}_{wrl} \frac{(1 - S_{rl})}{(1 - S_{rr})} = 2\dot{\phi}_{wa} - \dot{\phi}_{wrl}$$

Solving for $\dot{\phi}_{wrl}$,

$$\dot{\phi}_{wrl} = \frac{2\dot{\phi}_{wa}}{1 + \dfrac{(1 - S_{rl})}{(1 - S_{rr})}}$$

Then

$$u = r_r \, \dot{\phi}_{wrl} (1 - S_{rl})$$

Determining the required cornering coefficient $\mu_l = L_{rr}/R_{rr} = L_{rl}/R_{rl}$ and knowing the relationship for μ_l as a function of slip angle, the slip angle, α_{sr}, of the rear wheels can be determined. Since $\alpha_{sr} = \tan^{-1}(v/u)$, v can now be determined to complete the velocity analysis.

A driven or braked wheel, operating near its tractive limit, becomes limited in the amount of lateral force it can develop. One often used method of expressing this limitation is termed the friction ellipse concept, which is illustrated in figure 11-14.

Suppose the wheel is operating at point A on its tractive coefficient, μ_g, versus slip relation (fig. 11-14[a]). This operating point corresponds to the points B and C on the ellipse of figure 11-14(b). In turn, the cornering coef-

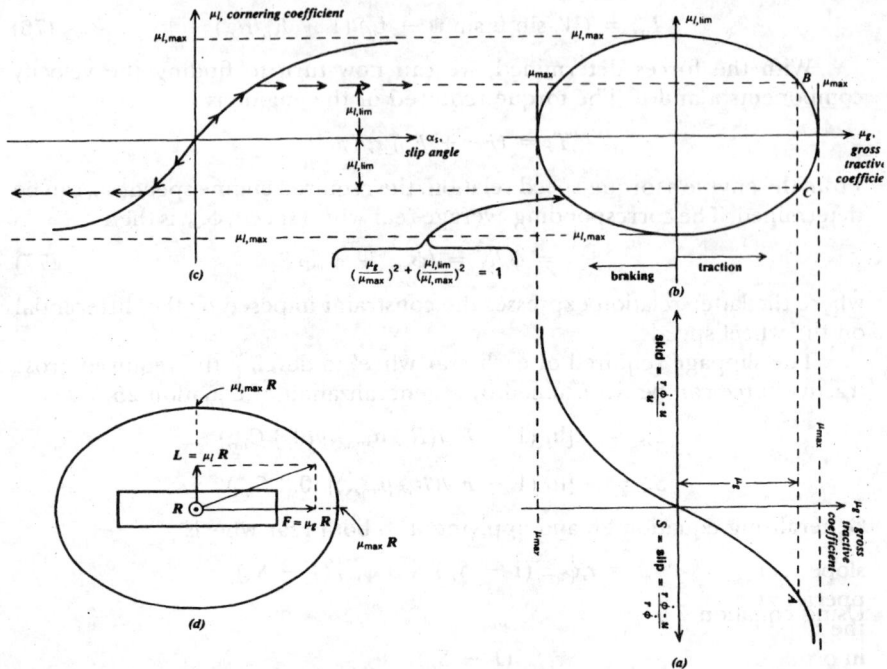

FIGURE 11-14 The friction ellipse concept. (*a*) Tractive coefficient-slip relation, (*b*) the friction ellipse, (*c*) cornering coefficient-slip angle relation, and (*d*) tractive and lateral forces on wheel.

ficient, μ_l, is limited, as shown by the arrows of figure 11-14(*c*). Thus, the wheel can operate at any point inside the ellipse (fig. 11-14[*d*]). When on the ellipse itself, the wheel is using all the available force-generating capacity of the surface upon which it is operating. If the force analysis indicates operation at a point outside the ellipse, the wheel will have exceeded the capacity of the surface to generate the required forces and a spinout ($u = 0$), locked wheel ($\dot{\phi} = 0$), or sliding condition will exist.

Returning to the analysis and assuming the lateral force capacity of the rear tires has not been exceeded, the cornering coefficient required of the front wheel is $\mu_l = L_f/R_f$. Inverting the cornering coefficient–slip angle relationship for the front wheel gives the slip angle, α_{sf}, required. Since $\alpha_{sf} = \tan^{-1}(u/v) - \delta_f$, the front wheel steer angle δ_f can be determined, which completes the analysis.

The presence of the lateral velocity v indicates that the tractor is moving forward along a direction that is not parallel to its own longitudinal centerline. This situation is most obvious when a tractor is operating on a substantial side

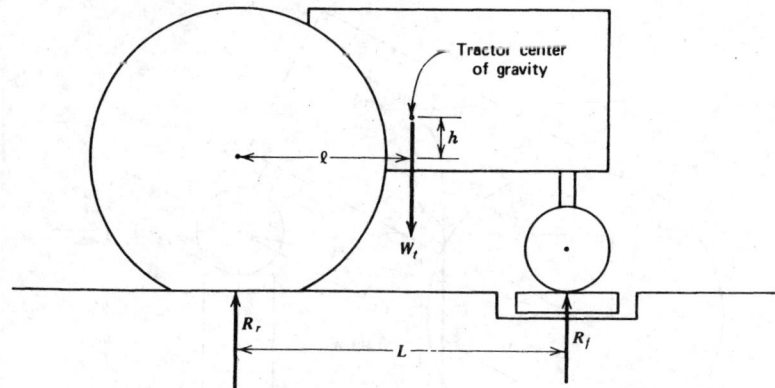

FIGURE 11-15 Determination of the longitudinal location of the center of gravity using the weighing method.

slope such as illustrated in figure 11-13. In such a situation, the tractor may operate at a heading angle ψ such that the rear axle is operating farther down the slope than the front axle with the front wheel(s) steered up the slope just in order to develop the lateral forces to be able to move parallel to the slope. If the tractor is also operating under a drawbar loading, the friction ellipse concept explains why it may be difficult to maintain a desired heading. In such a case, the rear wheels may not be able to develop the lateral forces required to keep the tractor on the desired course, resulting in the rear axle of the tractor sliding down the slope.

Center of Gravity Determination

In the preceding analysis, the location of the center of gravity of the tractor was assumed to be known. Since most tractors are composed of many comparatively irregularly shaped parts, it is difficult to analytically find the center of gravity of a tractor still in the design stage. However, several methods exist for experimentally determining the center of gravity location of an assembled tractor. The locations determined experimentally are then often useful in estimating the center of gravity location of a new tractor design.

Only the weighing method for center of gravity location will be discussed in this section. Several other methods are discussed in the suggested readings at the end of the chapter.

Given the tractor weight W_t and wheelbase, $L = h_{1t} \cos \theta_{1t} + h_{2t} \cos \theta_{2t}$, the longitudinal location, $\ell = x_{cg} = h_{1t} \cos \theta_{1t}$, of the center of gravity may be

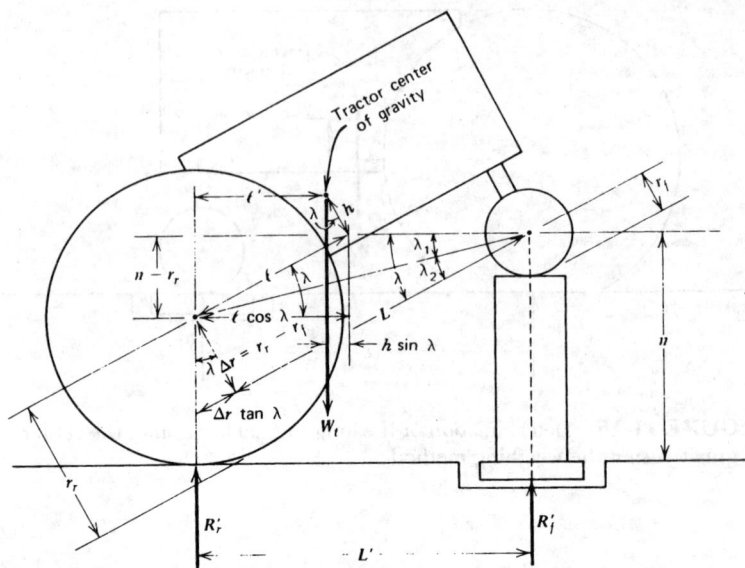

FIGURE 11-16   Determination of the vertical location of the center of
gravity using the weighing method.

found by placing the front axle of the tractor on a scale to determine force
R_f (fig. 11-15). Then

$$\ell = R_f L / W_t \qquad (78)$$

As a check, the reaction R_r may be found and the distance $L - \ell$ determined.

Since most tractors are approximately symmetrical about the vertical plane
perpendicular to the axles and passing midway between the wheels, the lateral
location of the center of gravity will normally be quite close to this plane.
Given the wheel tread setting, the lateral location may be found by weighing
one of the sides of the tractor. Again weighing the other side can serve as a
check.

The measurements required for finding the longitudinal and lateral locations of the center of gravity are straightforward. However, the determination of the height, $h = h_{1t} \sin \theta_{1t}$, is considerably more difficult. Again the
weighing method can be used with either the front or rear axle elevated.
Figure 11-16 illustrates the geometry involved in the following derivation.

Summing moments about the rear axle,

$$\ell' = R_f' L' / W_t$$

But from the geometry of figure 11-16,

$$\ell' = \ell \cos \lambda - h \sin \lambda$$

Thus

$$R_f' L'/W_t = \ell \cos \lambda - h \sin \lambda \tag{79}$$

But

$$L' = (L + \Delta r \tan \lambda) \cos \lambda \tag{80}$$

Substituting equation 80 into equation 79, dividing by $\cos \lambda$, and solving for h,

$$h = \frac{W_t \ell - R_f' L)}{W_t \tan \lambda} - \frac{R_f' \Delta r}{W_t} \tag{81}$$

The angle λ is the only quantity in equation 81 that cannot be directly measured. However, $\lambda = \lambda_1 + \lambda_2$, where

$$\tan \lambda_1 = (n - r_r)/L'$$

$$\tan \lambda_2 = \Delta r/L$$

To eliminate the need for measuring L' for use in calculating $\tan \lambda_1$,

$$L' = \sqrt{L^2 + (\Delta r)^2 - (n - r_r)^2}$$

Two assumptions implicitly made in the preceding analysis should be pointed out: (1) the tires are assumed rigid, and (2) fluid shifts occurring in the fuel, coolant, and oil compartments when the tractor is tilted are ignored. For accurate measurement of h, the front wheels should be elevated by an amount sufficient to obtain a significant difference between the reactions R_f and R_f'.

The Nebraska test report for a tractor contains the center of gravity location of the unballasted tractor with full fuel tank and serviced for operation.

Moment of Inertia Determination

In the preceding analyses, the moments of inertia of the chassis, rear wheels, and complete tractor were assumed to be known about transverse axes passing through the center of gravity of each of these bodies. As was the case for the center of gravity location, the many, often irregularly shaped parts composing a tractor make it difficult to analytically calculate the moments of inertia of a tractor about the longitudinal, transverse, or vertical axes passing through its center of gravity.

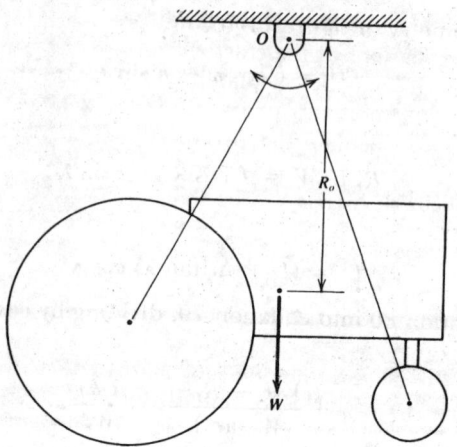

FIGURE 11-17   Determination of the pitch
moment of inertia using the pendulum method.

The moments of inertia about the transverse (pitch) and longitudinal
(roll) axes passing through the center of gravity of the tractor may be measured
using the setup illustrated in figure 11-17. The tractor and the supporting
sling form a compound pendulum that will oscillate with a period T given by

$$T = 2\pi \sqrt{\frac{I_o}{WR_o}} \tag{82}$$

where I_o =   moment of inertia of tractor plus sling about pivot O
            W =   weight of tractor plus sling
            R_o =   distance between pivot O and the center of gravity of
                     the tractor plus sling

Measurements of the distance R_o and the period of oscillation allow the calculation of I_o.

Often the weight and moment of inertia of the sling can be neglected
with respect to the corresponding quantities for the tractor. If so, the moment
of inertia of the tractor about its center of gravity, I_t, can be computed using
the parallel axis theorem.

$$I_t = I_o - m_t R_o^2 \tag{83}$$

For accurate measurement of I_t, the distance R_o should be made as small as
practical.

The yaw moment of inertia may be measured using a trifilar pendulum.
For such a measurement, the tractor is supported either directly or on a

platform by three vertical cables. If the length of the cables is known and the period for small oscillations of the tractor about the vertical is measured, the yaw moment of inertia may be calculated.

By measuring the moments of inertia with the tractor tilted instead of level, it is possible to calculate the products of inertia. In many cases, the tractor will be very nearly symmetric about the xz plane. In such a case, the only nonzero product of inertia will be that associated with the x and z axes.

PROBLEMS

The following problems illustrate how the subject matter of the chapter may be applied to the analysis of a given tractor. Since results from one problem may be needed as input data to the other problems, answers are given at the end of each problem.

1. The weighing method is used to determine the center of gravity location of the unballasted tractor. Because of tractor symmetry, the center of gravity is known to lie in the vertical plane perpendicular to the axles and passing midway between the wheels. (a) With the tractor level, the weights supported by the rear (R_r) and front (R_f) axles are measured. The following values are recorded: $R_r = 35.954$ kN, $R_f = 14.533$ kN. The wheelbase of the tractor, L, is 2708.9 mm. Find the longitudinal location, ℓ, of the center of gravity of the tractor. ($\ell = 779.8$ mm) (b) With the center of the front wheels raised to $n = 1600$ mm above the level surface used in (a), the weight supported by the rear axle, R'_r, is measured as 38.307 kN. Values of 483.1 and 875.0 mm are determined for the radii of the front (r_f) and rear (r_r) wheels respectively. Find the height h of the center of gravity of the tractor above the rear axle center. ($h = 194.5$ mm)

2. To determine the pitch moment of inertia of the tractor about its center of gravity, the tractor is supported as shown in figure 11-17. The lengths of the supporting cables connecting the axles to the pivot are adjusted so that the tractor is level in the equilibrium position. Because the distance R_o between the pivot and the tractor center of gravity is difficult to measure directly, the vertical distance between the pivot and the rear axle center is measured and found to be 2200 mm. The tractor is set into oscillation and the elapsed time for several cycles recorded. The average period is determined as 3.247 s. Assuming the weight and moment of inertia of the supporting cables are negligible, find the pitch moment of inertia I_{yyt} of the tractor about its center of gravity ($I_{yyt} = 6335$ kg m^2)

3. One of the two rear wheels of the tractor is removed and found to weigh 5.648 kN. Because of its symmetry, the center of gravity of the wheel is assumed to be coincident with the center of the wheel. The moment of inertia of the wheel about a transverse axis through its center of gravity is measured using the pendulum method and found to be 136.1 kg m^2. (a) Show analytically that the center of gravity of the chassis lies on the extension of the line connecting the center of the rear axle and the center of gravity of the tractor. That is, show $\theta_{1c} = \theta_{1t}$.

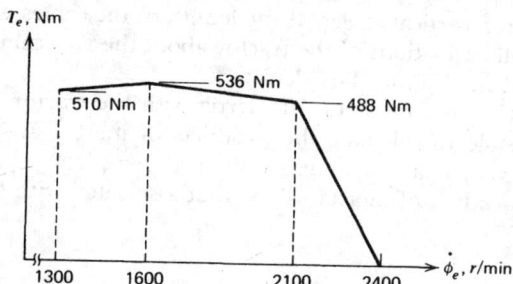

FIGURE 11-18 Linearized torque-speed relation for the tractor engine.

Hint: Recall that the center of gravity of a body made of several parts can be found if the center of gravity locations and weights of the individual parts are known. The tractor is such a body composed of the chassis and rear wheels.

(*b*) Find the weight of the chassis (W_c), the location of the center of gravity of the chassis ($h_{1c} \cos \theta_{1c}$ and $h_{1c} \sin \theta_{1c}$), and the pitch moment of inertia (I_{yyc}) of the chassis about its center of gravity.

Hint: Recall that, by using the parallel axis theorem, the moment of inertia of a body made of several parts can be found if the moments of inertia of the parts about their centers of gravity are known in addition to the locations of the individual part centers of gravity relative to the center of gravity of the complete body.

$$(W_c = 39.191 \text{ kN}; \ h_{1c} \cos \theta_{1c} = 1004.5 \text{ mm}; \ h_{1c} \sin \theta_{1c} = 250.6 \text{ mm};$$
$$I_{yyc} = 5104 \text{ kg m}^2)$$

4. (*a*) Determine the steady-state drawbar pull, $P_s \cos \alpha$, required to just lift the front wheels of the tractor off the ground ($R_f = 0$) when the tractor is operating on level ground. The drawbar force P applied to the tractor is inclined downward from the horizontal at an angle α of 15°. Assume $TF_r/R_r = 0.04$ in the relation $e_r = (TF_r/R_r)r_r$. The drawbar force is applied 906.8 mm behind and 386.1 mm below the rear axle center. ($P_s \cos \alpha = 50.73$ kN) (*b*) If the gross tractive coefficient μ_g has the form $\mu_g = 0.75(1 - e^{-0.3C_nS})$, show the tractor cannot develop the tractive force required for developing this drawbar pull.

5. Suppose the drawbar hitch point used in problem 4 is raised by 250 mm above the standard location. (*a*) What drawbar pull is now required to just lift the front wheels off the ground when the tractor is operating on level ground? The inclination of the drawbar force with the horizontal is maintained at 15°. ($P_s \cos \alpha = 37.935$ kN) (*b*) Show that sufficient traction is available to pull this load and calculate the resulting slippage of the drive wheels. Each drive wheel has a diameter of 1.75 m and a width of .467 m. The tractor is operating on a firm soil for which the cone index CI is 180 N/cm². Remember that the equation $\mu_g = 0.75(1 - e^{-0.3C_nS})$ applies to a single wheel. (slip = .150 = 15.0 percent) (*c*) The linearized full throttle torque (T_e)-speed ($\dot\phi_e$) relation for the tractor engine is shown in figure 11-18.

The tractor has eight forward speeds for which the gear ratios, N, are (the subscript indicates the gear)

$$N_1 = 205 \quad N_3 = 95 \quad N_5 = 60 \quad N_7 = 35$$
$$N_2 = 130 \quad N_4 = 75 \quad N_6 = 45. \quad N_8 = 20$$

The efficiency of power transmission, η, between the engine and the rear axle is assumed to be independent of the gear used and equal to 0.85. Show that sufficient engine torque is available to pull the load in third gear and calculate the resulting engine and axle speeds ($\dot\phi_e = 2131$ rpm, $\dot\phi_w = 2.35$ rad/s) (d) Using the drive wheel slip calculated in (b), what is the forward speed of the tractor? Assume the rolling radius r_r for the rear wheels is equal to half the wheel diameter. ($\dot x_w = 1.75$ m/s)

6. (a) Assume the tractor is exerting a drawbar pull $P \cos \alpha$ of 3ι operating in third gear on a level surface. The drawbar force P is inclined at an aug.ι α of 15° to the horizontal and is applied at the standard drawbar location. The tractor is operating on a firm soil with a cone index of 180 N/cm². Assume $TF_r/R_r = 0.04 = TF_f/R_f$ in the relations $e_r = (TF_r/R_r)r_r$ and $e_f = (TF_f/R_f)r_f$. Determine the resulting forward speed of the tractor ($\dot x_w = 1.89$ m/s) (b) If the tractor is operating under the same conditions as in (a) but on a slope of 10°, what is the resulting forward speed of the tractor? ($\dot x_w = 1.53$ m/s)

7. Assume the front wheels of the tractor are powered and that the now four-wheel-drive tractor is operating under the same conditions as those of problem 6(a). Assume the cone index for the soil encountered by the rear wheels is the same as for the soil encountered by the front wheels and that the gear ratio N_f for the front wheels in third gear is 50. Each front tire has a section width of 0.279 m. What is the resulting forward speed of the tractor? ($\dot x_w = 1.95$ m/s)

8. The tractor has become immobilized while traveling under no drawbar load on soft ground. The rear wheels have dug into the ground to such an extent that the tractor is on an effective slope β of 15°. In an effort to free the tractor, the operator chains a post in front of the rear wheels and releases the clutch with the transmission in first gear. With the rear wheels restrained from moving ($\ddot x_w = \ddot z_w = \dot\phi_w = 0$), the tractor begins to tip backward. When the angle of rotation θ reaches 20°, the operator depresses the clutch. At this time, the angular velocity $\dot\theta$ is 1.6 rad/s. Is the tractor in danger of overturning? (Since the angular velocity, θ_s, required to bring the tractor to the point of static instability is 1.46 rad/s, the tractor is in danger of overturning.)

9. Estimate the natural frequencies and illustrate the modes of vibration of the tractor considering it as a linear two-degree-of-freedom system. By linearizing the appropriate load-deflection curves, the spring rates for a single front and a single rear tire are estimated to be 287.4 N/mm and 337.7 N/mm respectively. (Natural frequencies are 2.21 and 3.38 Hz. The corresponding mode shapes are shown in fig. 11-19.)

10. (a) If the cornering stiffnesses of a single front and a single rear tire are 35.8 and 141.1 kN/radian, respectively, determine the understeer coefficient. (.076 radians) (b) Will the tractor tend to understeer or oversteer? (understeer)

11. The tractor is put into a steady-state turn such that the center of gravity of the

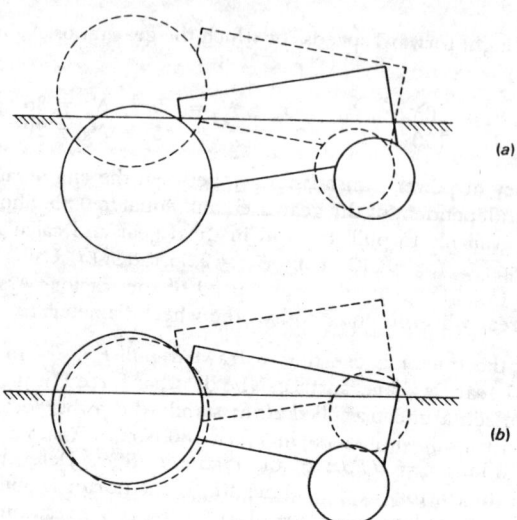

FIGURE 11-19   Modes of vibration corresponding to the (a) 2.21-Hz and (b) 3.38-Hz natural frequencies of the tractor.

tractor traverses a circle of radius 7 m. The tractor is of the tricycle type and has a front wheel tread setting of 0.5 m and a rear wheel tread setting of 2 m. If the forward speed of the tractor is 6.4 m/s, is the tractor in danger of turning over sideways? Assume the tractor is traveling on a firm surface on which the tires can develop the lateral forces required by the specific turn (No. $u_s = 7.09$ m/s) (b) A front-end loader weighing 4000 N is attached to the tractor. A 5000 N load is to be carried in the loader bucket while the bucket is raised to its maximum height. In this position, the center of gravity of the loader plus the bucket load is estimated to be 2 m ahead of and 1.75 m above the rear axle. Is the loader-equipped tractor in danger of overturning in the turn described in (a)? (Yes, $u_s = 6.21$ m/s)

12.   Assume the tractor is of the single front wheel tricycle type analyzed in the section on Three-Dimensional Static Analysis. The tractor is in steady-state motion traveling with a heading angle ψ of 85° on a slope β of 10°. The tractor is operating on soil with a cone index of 150 N/cm² and under a drawbar pull $P \cos \alpha$ of 15,000 N. The drawbar force P is inclined downward at an angle α of 12° to the horizontal and is applied at the standard drawbar location. The rear tread width TW is 2 m. (a) Using the analysis in the section Three-Dimensional Static Analysis, determine the forces acting on the tractor. Use the towed force relation $TF_i/R_i = 1.2/C_{ni} + 0.04$ to calculate the rolling resistance on each wheel. ($R_f = 10.235$ kN, $R_{rr} = 26.006$ kN, $R_{rl} = 16.667$ kN, $TF_f = .720$ kN, $TF_{rr} = 1.702$ kN, $TF_{rl} = .939$ kN, $F_{rr} = F_{rl} = 9.564$ kN, $L_f = 2.795$ kN, $L_{rr} = 3.618$ kN, $L_{rl} = 2.319$ kN) (b) Use the friction ellipse concept to verify that the surface can support the tractive and lateral forces required on the rear wheels. Assume the cornering coefficient (μ_l) – slip angle (α_s) relation is the same for the front and rear wheels and has the form $\mu_l - 0.50 (1 - e^{-\alpha_s/3})$, where α_s is expressed in degrees. (c) Find the velocity components u and v and the required steer angle, δ_f, for the front wheel. ($u = 2.58$ m/s, $v = 0.044$ m/s, $\delta_f = -1.39°$)

SUGGESTED READINGS

Bekker, M. G. *Introduction to Terrain-Vehicle Systems*. The University of Michigan Press, Ann Arbor, 1969.

Berenyi, T., R. L. Pershing, and B. E. Romig. "Vehicle Mission Simulation." SAE Paper 730693, 1973.

Captain, K. M., A. B. Boghani, and D. N. Wormley. "Analytical Tire Models for Dynamic Vehicle Simulation." *Vehicle System Dynamics*, vol. 8, 1979, pp. 1–32.

Chisholm, C. J. "A Mathematical Model of Tractor Overturning and Impact Behavior." *J. Agric. Eng'g Res.*, vol. 24, 1979, pp. 375–394.

Chu, M. L., and G. R. Doyle. "Nondeterministic Analysis of a Four-Wheeled Model Vehicle Traversing a Simulated Random Terrain." SAE Paper 780789, 1978.

Claar, P. W. II, W. F. Buchele, and P. N. Sheth. "Off Road Vehicle Ride: Review of Concepts and Design Evaluation with Computer Simulation." SAE Paper 801023, 1980.

Clark, S. J., "Lagrangian Methods Applied to Off-the-Road Vehicle Dynamics." ASAE Paper 72-554, 1972.

Crolla, D. A. "Off-Road Vehicle Dynamics." *Vehicle System Dynamics*, vol. 10, 1981, pp. 253–266.

Crolla, D. A., and H. B. Spencer. "Tractor Handling During Control Loss on Sloping Ground." *Vehicle System Dynamics*, vol. 13, 1984, pp. 1–17.

Davis, D. C., and G. E. Rehkugler, "Agricultural Wheel-Tractor Overturns—Part I: Mathematical Model—Part II: Mathematical Model Verification by Scale-Model Study." *Trans. of ASAE*, vol. 17, 1974, pp. 477–488, 492.

Dwyer, M. J. "The Braking Performance of Tractor-Trailer Combinations." *J. Agric. Eng'g Res.*, vol. 15, 1970, pp. 148–162.

Ellis, J. R. *Vehicle Dynamics*. Business Books Ltd., London, 1969.

Gibson, H. G., K. C. Elliott, and S. P. E. Persson. "Side Slope Stability of Articulated-Frame Logging Tractors." *J. Terramechanics*, vol. 8, 1971, pp. 65–79.

Goering, C. E., and W. F. Buchele. "Computer Simulation of an Unsprung Vehicle." *Trans. of ASAE*, vol. 10, 1967, pp. 272–280.

Goran, M. B., and G. W. Hurlong, Jr. "Determining Vehicle Inertial Properties for Simulation Studies." *Bendix Technical J.*, vol. 6, 1973, pp. 53–57.

Grace, L. S., C. J. Biller, and K. H. Means. "A Survey of Tractor Stability Analysis." ASAE Paper 83-1618, 1983.

Grecenko, A. "Operation on Steep Slopes: State-of-the-Art Report." *J. Terramechanics*, vol. 21, 1984, pp. 181–194.

Horton, D. N. L., and D. A. Crolla. "The Handling Behavior of Off-Road Vehicles." *Int. J. Vehicle Design*, vol. 5, 1984, pp. 197–218.

Koch, J. A., W. F. Buchele, and S. J. Marley. "Verification of a Mathematical Model to Predict Tractor Tipping Behavior." *Trans. of ASAE*, vol. 13, 1970, pp. 67–72, 76.

Larson, D. L., D. W. Smith, and J. B. Liljedahl. "The Dynamics of Three-Dimensional Tractor Motion." *Trans. of ASAE*, vol. 19, 1976, pp. 195–200.

Matthews, J., and J. D. C. Talamo. "Ride Comfort for Tractor Operators III. Investigation of Tractor Dynamics by Analogue Computer Simulation." *J. Agric. Eng'g Res.*, vol. 12, 1965, pp. 93–108.

McCormick, E. "Some Engineering Implications of High Speed Farming." *Agric. Engr.*, vol. 22, May 1941, pp. 165–167.

McKibben, E. G. "The Kinematics and Dynamics of the Wheel Type Farm Tractor." *Agric. Engr.*, vol. 8, Jan.–July 1927, pp. 15–16, 39–40, 43, 58–60, 90–93, 119–122, 155–160, 187–189.

Mitchell, B. W., G. L. Zachariah, and J. B. Liljedahl. "Prediction and Control of Tractor Stability to Prevent Rearward Overturning." *Trans. of ASAE*, vol. 15, 1972, pp. 838–844.

Pershing, R. L. "Simulating Tractive Performance." SAE Paper 710525, 1971.

Pershing, R. L., and R. R. Yoerger. "Steady-State Behavior of Tractors on Roadside Slopes." SAE Paper 903B, 1964.

Pershing, R. L., and R. R. Yoerger. "Simulation of Tractors for Transient Response." *Trans. of ASAE*, vol. 12, 1969, pp. 715–719.

Plackett, C. W. "A Review of Force Prediction Methods for Off-Road Wheels." *J. Agric. Eng'g Res.*, vol. 31, 1985, pp. 1–29.

Raney, J. P., J. B. Liljedahl, and R. Cohen. "The Dynamic Behavior of Farm Tractors." *Trans. of ASAE*, vol. 4, 1961, pp. 215–218, 221.

Rehkugler, G. E., V. Kumar, and D. C. Davis. "Simulation of Tractor Accidents and Overturns." *Trans. of ASAE*, vol. 19, 1976, pp. 602–609, 613.

Rehkugler, G. E. "Simulation of Articulated Steer Four-Wheel Drive Agricultural Tractor Motion and Overturns." *Trans. of ASAE*, vol. 23, 1980, pp. 2–8, 13.

Rehkugler, G. E. "Tractor Steering Dynamics—Simulated and Measured." *Trans. of ASAE*, vol. 25, 1982, pp. 1512–1515.

Sack, H. W. "Longitudinal Stability of Tractors." *Agric. Engr.*, vol. 37, May 1956, pp. 328–333.

SAE Recommended Practice J670e. Vehicle Dynamics Terminology.

SAE Standard J874 OCT85. Center of Gravity Test Code.

Smith, D. W. "Computer Simulation of Tractor Ride for Design Evaluation." SAE Paper 770704, 1977.

Smith, D. W. "The Influence of Drawbar Position on Tractor Rearward Stability." ASAE Paper 84-1560, 1984.

Smith, D. W., and J. B. Liljedahl. "Simulation of Rearward Overturning of Farm Tractors." *Trans. of ASAE*, vol. 15, 1972, pp. 818–821.

Smith, D. W., J. V. Perumpral, and J. B. Liljedahl. "The Kinematics of Tractor Sideways Overturning." *Trans. of ASAE*, vol. 17, 1974, pp. 1–3.

Song, A., B. K. Huang, and H. D. Bowen. "Simulating a Powered Model Wheel-Tractor on Soft Ground." ASAE Paper 85-1054, 1985.

Spencer, H. B. "Stability and Control of Two-Wheel Drive Tractors and Machinery on Sloping Ground." *J. Agric. Eng'g Res.*, vol. 23, 1978, pp. 169–188.

Stayner, R. M., T. S. Collins, and J. A. Lines. "Tractor Ride Vibration Simulation as an Aid to Design." *J. Agric. Eng'g Res.*, vol. 29, 1984, pp. 345–355.

Steinbruegge, G. W. "Improved Methods of Locating Centers of Gravity." *Trans. of ASAE*, vol. 12, 1969, pp. 681–684, 689.

Taborek, J. J. *Mechanics of Vehicles*. The Penton Publishing Company, Cleveland, 1957.

Ulrich, A., and H. Gohlich. "Driving Dynamics of Tractors with and without Mounted Machines at Higher Velocities." *Grund. Landtechnik*, vol. 33, 1983, pp. 108–115.

Unruh, D. H. "Determination of Wheel Loader Static and Dynamic Stability." SAE Paper 710526, 1971.

Winkler, C. B. "Measurement of Inertial Properties and Suspension Parameters of Heavy Highway Vehicles." SAE Paper 730182, 1973.

Wismer, R. D., and H. J. Luth. "Off-Road Traction Prediction for Wheeled Vehicles." *Trans. of ASAE*, vol. 17, 1974, pp. 8–10, 14.

Wolken, L. P., and R. R. Yoerger. "Dynamic Response of a Prime Mover to Random Inputs." *Trans. of ASAE*, vol. 17, 1974, pp. 468–473.

Wong, J. Y. *Theory of Ground Vehicles*. John Wiley & Sons, New York, 1978.

Worthington, W. H. "Evaluation of Factors Affecting the Operating Stability of Wheel Tractors." *Agric. Engr.*, vol. 30, March–April, 1949, pp. 119–123, 179–183.

Xie, L., and P. W. Claar II. "Simulation of Agricultural Tractor-Trailer System Stability." SAE Paper 851530, 1985.

Zoz, F. M. "Predicting Tractor Field Performance." *Trans. of ASAE*, vol. 15, 1972, pp. 249–255.

12

HYDRAULIC SYSTEMS AND CONTROLS

This chapter discusses hydraulic systems, their control systems, and some of the principles of tractor hitches.

Hydraulic Component Symbols

Hydraulic systems have become so complicated that it is much easier to use symbols to describe them. This language, *Graphic Symbols for Fluid Power Diagrams,* must necessarily be learned before proceeding to a discussion of hydraulic systems.

The symbols are shown in Appendix B along with some examples of their use. Some advantages of using symbols are:

1. Their use simplifies communication and saves drawing time.
2. Symbols can be used to convey the functional requirements of a component, or an assembly of components. Thus, a designer's concept of the component will not be biased by seeing a drawing or model of the component being designed or redesigned.

Symbolic representation of hydraulic circuits does not take the place of drawings for manufacturing purposes.

Hydraulic Components

A hydraulic system consists of part or all of the following components:

1. Pump
2. Motor

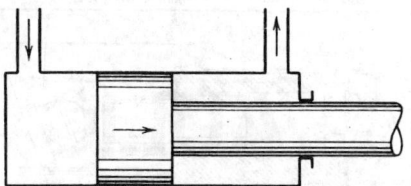

FIGURE 12-1   Schematic diagram of
double-acting cylinder.

3. Valves
4. Lines and connections
5. Heat exchanger
6. Sump (supply tank)
7. Accumulator (stored energy)
8. Controls (manual or automatic)
9. Fluid
10. Actuators
11. Filters

   Pumps and motors are often quite similar and can sometimes be inter-
changed in their purpose. The simplest type of pump or motor is a hydraulic
cylinder (fig. 12-1). When hydraulic cylinders are arranged axially, as shown
in figure 12-2, the rate of flow through the pump can be regulated by controlling the angle between the piston block and the swash plate, a common
method of control on a hydrostatic transmission.
   Radial piston pumps (fig. 12-3) can also be used as motors. The displacement of a radial piston pump can be controlled by allowing the pressure to
lift the pistons off the eccentric. By this method the pump unloads and does
not do any work except when the pressure drops sufficiently to force the
pistons back onto the eccentric.
   A spur-gear pump is shown in figure 12-4. It is normally used on tractor
hydraulic systems of lower pressure.
   The spur-gear, the internal-gear pump (fig. 12-5), the gerotor-gear pump
(fig. 12-6), and the vane-type pump (fig. 12-7) are all used on tractor hydraulic
systems where lower pressures are used.

Motor Performance*

Because efficiencies are often high, very accurate instrumentation is required
if the motor is externally loaded because a small error in measurement of

*This section also applies to pumps. Equations 1 and 2 should be inverted when used for pumps.

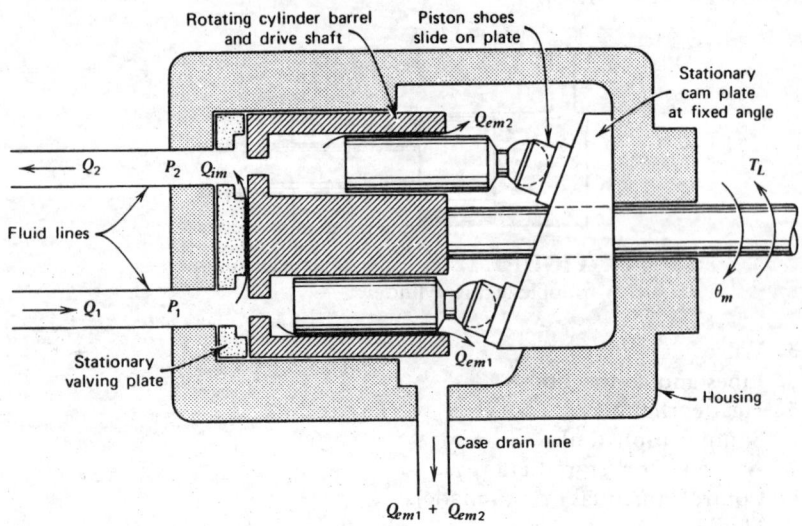

FIGURE 12-2 Schematic diagram of a fixed-displacement, axial-piston motor.

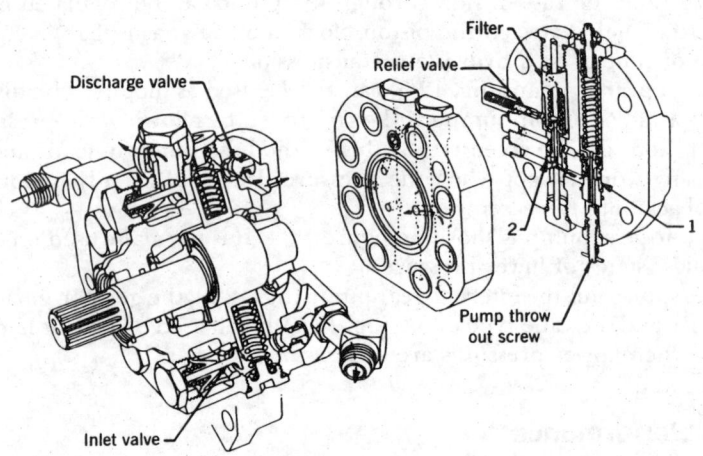

FIGURE 12-3 Radial piston pump, with variable displacement. (Courtesy Deere & Company.)

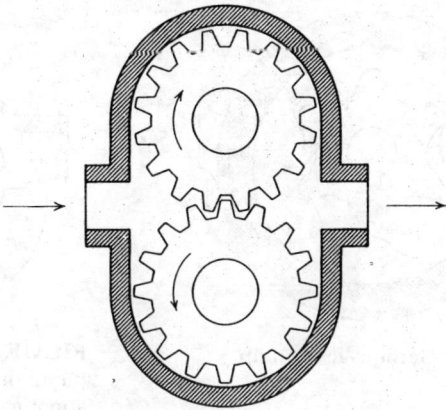

FIGURE 12-4 Spur-gear pump.

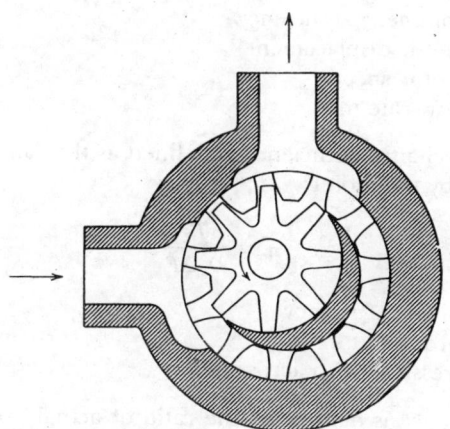

FIGURE 12-5 Internal-gear pump.

input or output power may be larger than the losses involved. Therefore, it is better to measure the losses and then compute motor performance with external loads.

The volumetric efficiency is defined as the ratio of flow that results in motor speed (the ideal flow) to the flow supplied to the motor. By definition:

$$\eta_v \equiv \frac{D_m \dot{\theta}_m}{Q_1} \tag{1}$$

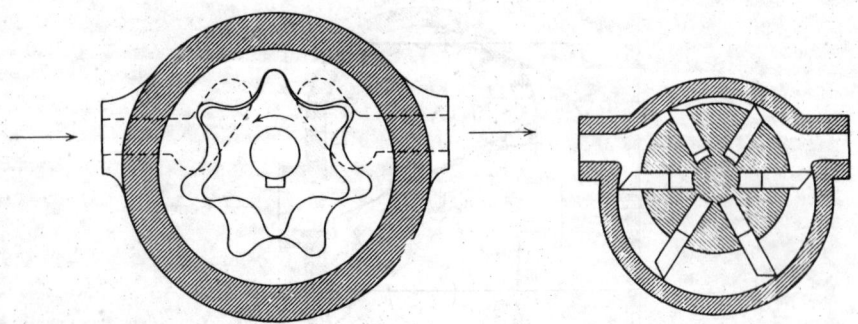

FIGURE 12-6 Gerotor-gear pump.   **FIGURE 12-7** Vane-type
                                              pump that can also be used as
                                              a motor.

where η_v = volumetric efficiency
         D_m = motor displacement
         $\dot{\theta}_m$ = motor speed
         Q_1 = flow rate in

The torque or mechanical efficiency is defined as the ratio of actual to ideal torque delivered by the motor:

$$\eta_t \equiv \frac{T_L}{D_m \Delta P} \qquad (2)$$

where η_t = torque efficiency
         T_L = torque out
         ΔP = pressure drop across motor

The overall efficiency is defined as the ratio of actual power output to the hydraulic power supplied.

$$\eta_{oa} \equiv \frac{P_{\text{out}}}{P_{\text{in}}} = \frac{T_L \dot{\theta}_m}{Q_1 \Delta P} = \frac{T_L}{D_m \Delta P} \times \frac{D_m \dot{\theta}_m}{Q_1} = \eta_v \eta_t \qquad (3)$$

If the efficiencies (volumetric, torque, and overall) are plotted as a function of the dimensionless term, $\mu \dot{\theta}_m/P_1$, where μ is the absolute viscosity, a typical set of curves, as shown in figure 12-8, will be generated.

From figure 12-8 it is clear that a motor has an optimum combination of pressure and speed. Fluid viscosity normally will not be a variable. As pressure increases, or the shaft speed decreases, the motor approaches a stall condition at which efficiency approaches zero. As pressure becomes very small and/or

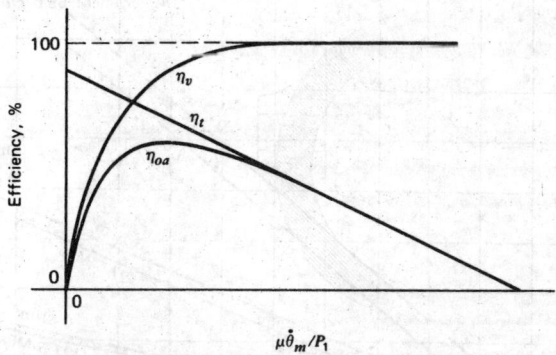

FIGURE 12-8   Typical efficiency curves for a motor.

the shaft velocity becomes very high, the motor approaches a condition just overcoming friction, and again the motor becomes very inefficient.

Performance curves for a typical radial piston pump are shown in figure 12-9. Note that the flow rate is affected by the pressure. The difference between each pair of curves is the internal leakage in the pump. It is left for the student to calculate and compare the ideal flow with the flow at zero pressure.

Accumulators

An accumulator, figure 12-10, is the only convenient device for storing energy in a hydraulic system. The one shown is a diaphragm type. The "spring" is a charge of inert nitrogen. It should be apparent that the use of a gas such as air for the "spring" would be very dangerous because of the possibility of mixing hydraulic oil with hot oxygen, which has the potential for combustion similar to that with a diesel engine.

An accumulator in hydraulic systems on agricultural equipment is used to supply energy when the demand of the motors and cylinders is greater than the capacity of the pump. An accumulator is also intended for standby or emergency use in case of a pump or engine failure for braking and steering of off-highway trucks and vehicles.

Accumulators are available in several sizes and pressure and temperature ratings and have several bladder materials so as to be compatible with most fluids. An accumulator also serves as a shock absorber to reduce maximum stresses if the system is subjected to unusual dynamic loads.

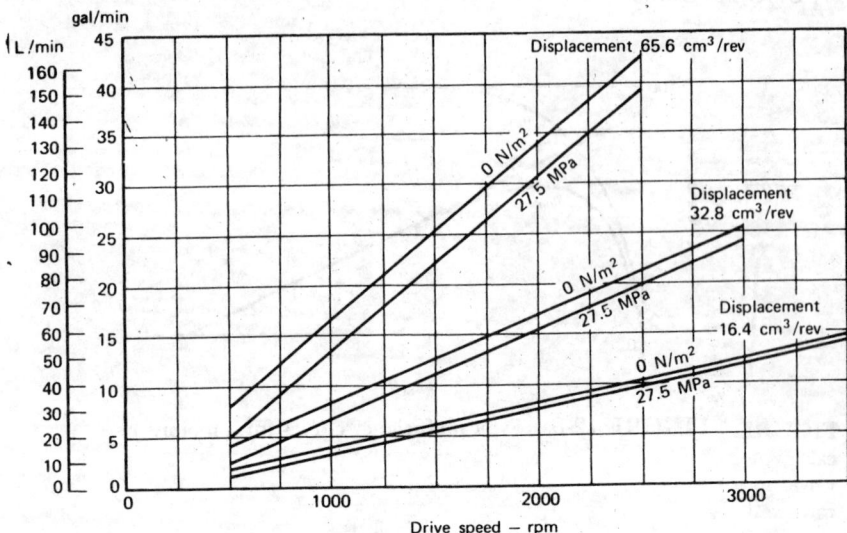

FIGURE 12-9 Typical performance curves for radial piston pump. (Courtesy Robert Bosch GmbH.)

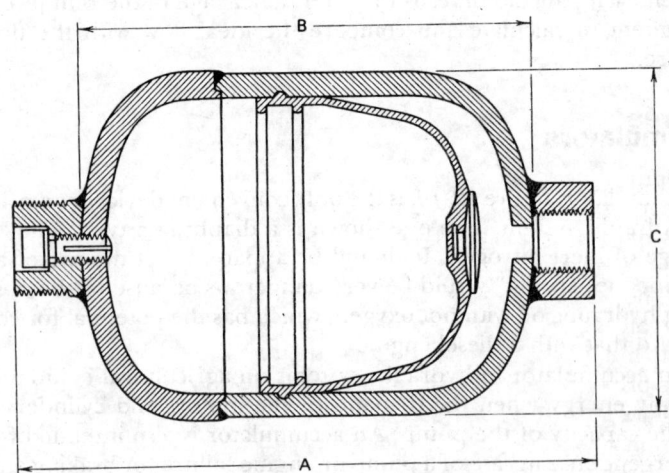

FIGURE 12-10 Bladder-type accumulator. (Courtesy Robert Bosch GmbH.)

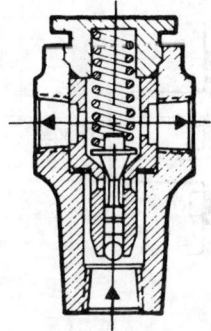

FIGURE 12-11 Typical pilot-operated relief valve, nonadjustable. Flow rates and pressure can be altered by changing internal parts. (Courtesy Robert Bosch GmbH.)

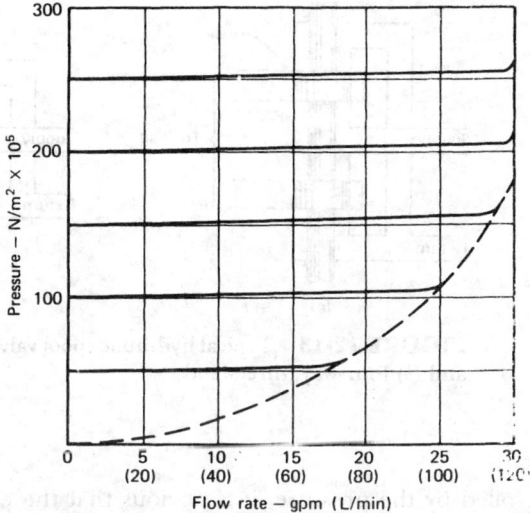

FIGURE 12-12 Control characteristics and operating limits of the pilot-operated relief valve shown in figure 12-11. (Courtesy Robert Bosch GmbH.)

Valves

The purposes of valves in a hydraulic system are many. The type and design are dependent largely upon the valve's purpose. Some of the more common types of valves are (1) control, (2) pressure reducing, (3) priority, and (4) relief.

The relief valve (fig. 12-11) shown is the nonadjustable type; however, such valves are available with an adjusting screw on the spring. Pumps commonly have a relief valve built into them. Flow characteristics and operating limits of the relief valve shown in figure 12-11 are shown in figure 12-12.

The most common type of control valve is the spool valve. A schematic diagram of such a valve is shown in figure 12-13. A spool valve will have at least three ports and two lands. If the width of the land is less than the width of the port in the valve sleeve, the valve is said to be *open center*. A *closed-center* valve has land widths greater than the port width when the spool is in neutral. The major advantage of the closed-center valve is realized when it is used with a variable displacement pump. If the displacement of the pump is con-

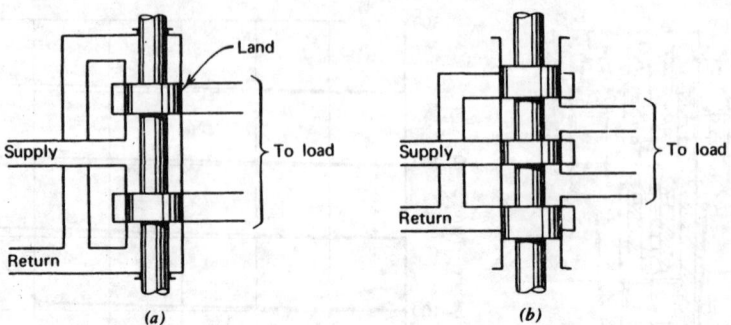

FIGURE 12-13   Typical hydraulic spool valves. (*a*) Four-way, two-land and (*b*) four-way, three-land.

trolled by the pressure, it is obvious that the efficiency of a system using a closed-center valve would be greater than that of an open-center system.

There are many types of valves, but only two more will be mentioned. A priority valve (see fig. 12-14) is used to give priority to one specific function, such as steering or brakes, which would be more critical from the standpoint of safety.

A pressure-reducing valve is shown in figures 12-22 and 12-23. Its function is to lower the nominal hydraulic pressure to a desired level for an operation such as controlling the differential lock, the pto clutch, and brake.

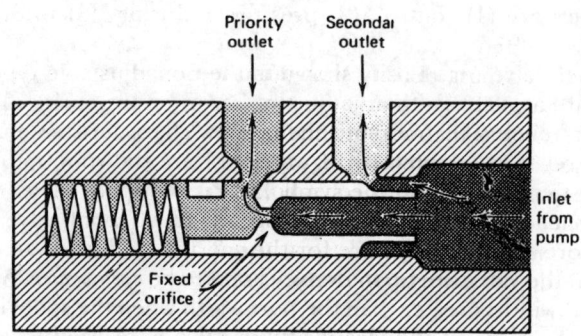

FIGURE 12-14   Priority flow-divider valve.

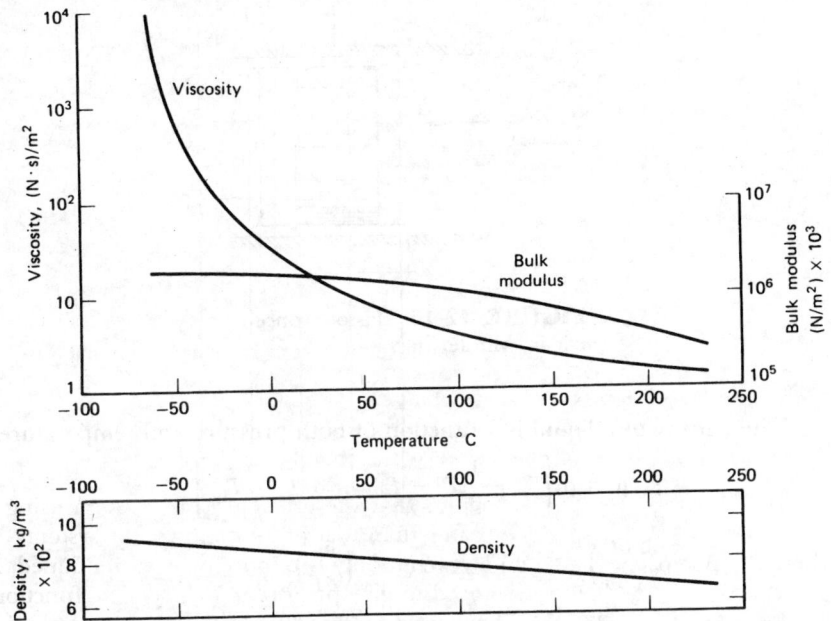

FIGURE 12-15 Absolute viscosity, bulk modulus, and density for a MIL H-5606B hydraulic fluid.

Hydraulic Fluids

Mass density is defined as mass per unit of volume. The symbol ρ is used to designate mass density with units of kilograms per liter (kg/L) or kg/m^3.

Specific gravity is the ratio of the mass density of a substance at a certain temperature to the mass density of water at the same temperature. Specific gravity is dimensionless, and the symbol SG is often used. The temperature must be specified. The petroleum industry in the United States has selected 15.1°C as a standard temperature for the specific gravity of hydraulic fluids. The effect of temperature upon density and bulk modulus is shown in figure 12-15.

The specific gravity of hydraulic fluids ranges from about 0.8 for petroleum-base fluids to as high as 1.5 for the chlorinated hydrocarbons.

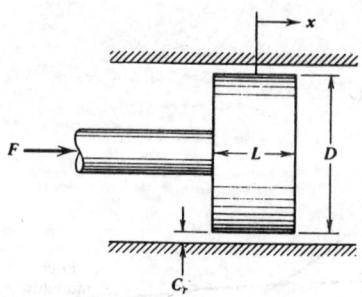

FIGURE 12-16 Piston concentric in cylinder.

The density of a liquid is a function of both pressure and temperature.

$$\rho = \rho_0\left[1 + \frac{1}{\beta}(P - P_0) - \alpha(T - T_0)\right] \tag{4}$$

where $\quad \beta \equiv \rho_0\left(\dfrac{\delta P}{\delta \rho}\right)_T \quad$ and $\quad \alpha \equiv \dfrac{1}{\rho_0}\left(\dfrac{\delta \rho}{\delta T}\right)_P$

Equation 4 is the linearized equation of state for a liquid. The mass density increases as pressure is increased and decreases with temperature increase. Because mass density is mass divided by volume, equivalent expressions for β and α are

$$\beta = -V_0\left(\frac{\delta P}{\delta V}\right)_T \tag{5}$$

$$\alpha = \frac{1}{V_0}\left(\frac{\delta V}{\delta T}\right)_P \tag{6}$$

where V is the total volume and V_0 is the initial total volume of the liquid.

The quantity β is the change in pressure divided by the fractional change in volume at a constant temperature and is called the *bulk modulus* of the liquid. The bulk modulus is always a positive quantity, for $(\delta P/\delta V)_T$ is always negative and has a value of about 1.5×10^6 kPa for petroleum fluids. However, values this large are rarely achieved in practice because the bulk modulus decreases sharply with small amounts of air entrained in the liquid. The bulk modulus is the most important fluid property in determining the dynamic performance of hydraulic systems because it relates to the "stiffness" of the liquid. The reciprocal of β is often designated c and called the compressibility of the liquid. The bulk modulus for a MIL-H-5606B hydraulic fluid is shown in figure 12-15.

The quantity α is the fractional change in volume resulting from a change in temperature and is called the *cubical expansion coefficient*. The cubical expansion coefficient for petroleum-base fluids is about $0.9 \times 10^{-3}/°C$.

Viscosity is an important property of any fluid. It is necessary for hydrodynamic lubrication, and a suitable value is required for many other purposes. Close-fitting surfaces in relative motion occur in most hydraulic components. If the viscosity of the fluid is too low, leakage flows increase; if the viscosity is too large, component efficiencies decrease because of additional power loss in fluid friction. Viscosity is of such significance that it is common practice to designate the fluid by its viscosity at a certain temperature; for example, oil with 150 SUS at 54°C might be such a fluid designation.

Referring to the piston and cylinder of figure 12-16 in which the radial clearance C_r is filled with a fluid, Newton observed that a force was necessary to cause relative motion. This force is a measure of the internal friction of the fluid or its resistance to shear; it is proportional to the area in contact and to the velocity and is inversely proportional to the film thickness. Therefore,

$$F = \mu A \frac{\dot{x}}{C_r}$$

The constant of proportionality μ is known as the absolute viscosity (the terms "dynamic viscosity" and "coefficient of viscosity" are also used) of the fluid. For this case, since $A = \pi DL$, we obtain

$$F = \frac{\pi DL\mu}{C_r}\dot{x} \tag{7}$$

In SI units, absolute viscosity has units of $N \cdot s/m^2$.

The ratio of absolute viscosity to mass density occurs in many equations (e.g., Navier-Stokes, Reynolds number) and is measured by many viscometers. This ratio is, by definition, the kinematic viscosity v of the fluid; that is,

$$v = \frac{\mu}{\rho} \tag{8}$$

In SI units, kinematic viscosity has units of m^2s.

Kinematic viscosity is easily measured using many instruments. The best known of these in the United States is the Saybolt Universal Viscometer. This instrument measures the time for 60 cm^3 of a sample to flow through a tube 1.76 mm in diameter and 12.25 mm long at a constant temperature. The resulting time in seconds is called Saybolt Universal Seconds (SUS or SSU). SUS is commonly used to designate liquid viscosity in the petroleum industry. However, SUS does not have the appropriate units to be of use in engineering computations and equations and must be converted to other measures. The equivalent kinematic viscosity in centistokes v is closely approximated by

$$v = 0.216 \text{ SUS} - \frac{166}{\text{SUS}} \tag{9}$$

for SUS greater than 32. The tables in ASTM D446-53 should be consulted if very accurate conversion is desired.

It is unfortunate that the name "viscosity" is attached to the quantity defined by equation 8, for it has confused the basic definition of viscosity and made conversion between absolute and kinematic, with the many measures of each, a trying experience. A nomogram, figure 12-17, has been prepared to facilitate conversion of measures of viscosity that are used in the United States.

The viscosity of liquids decreases markedly as the temperature rises and increases slightly with a pressure increase.

Two thermal properties of liquids, specific heat and thermal conductivity, are of importance, especially in the design of hydraulic power supplies. The *specific heat* is the amount of heat required to raise the temperature of a unit mass by 1°. The symbol C_v is used to designate specific heat, and a typical value for petroleum base fluids is $C_v = 2.09$ kJ/(kg $\cdot$ K).

Thermal conductivity is the rate of heat flow through an area for a temperature gradient. For petroleum base oils, the thermal conductivity is about 1.5 J/(m^2 $\cdot$ s $\cdot$ °C $\cdot$ m).

The effective or total bulk modulus β_e is

$$\frac{1}{\beta_e} = \frac{\Delta V_t}{V_t \Delta P} \tag{10}$$

The bulk modulus of a liquid is

$$\beta_l = -\frac{V_t \Delta P}{\Delta V_1} \tag{11}$$

The bulk modulus of a gas is

$$\beta_g = -\frac{V_g \Delta P}{\Delta V_g} \tag{12}$$

The negative signs in equations 11 and 12 indicate a decrease in volume with pressure increase. The quantity

$$\beta_c = \frac{V_t \Delta P}{\Delta V_c} \tag{13}$$

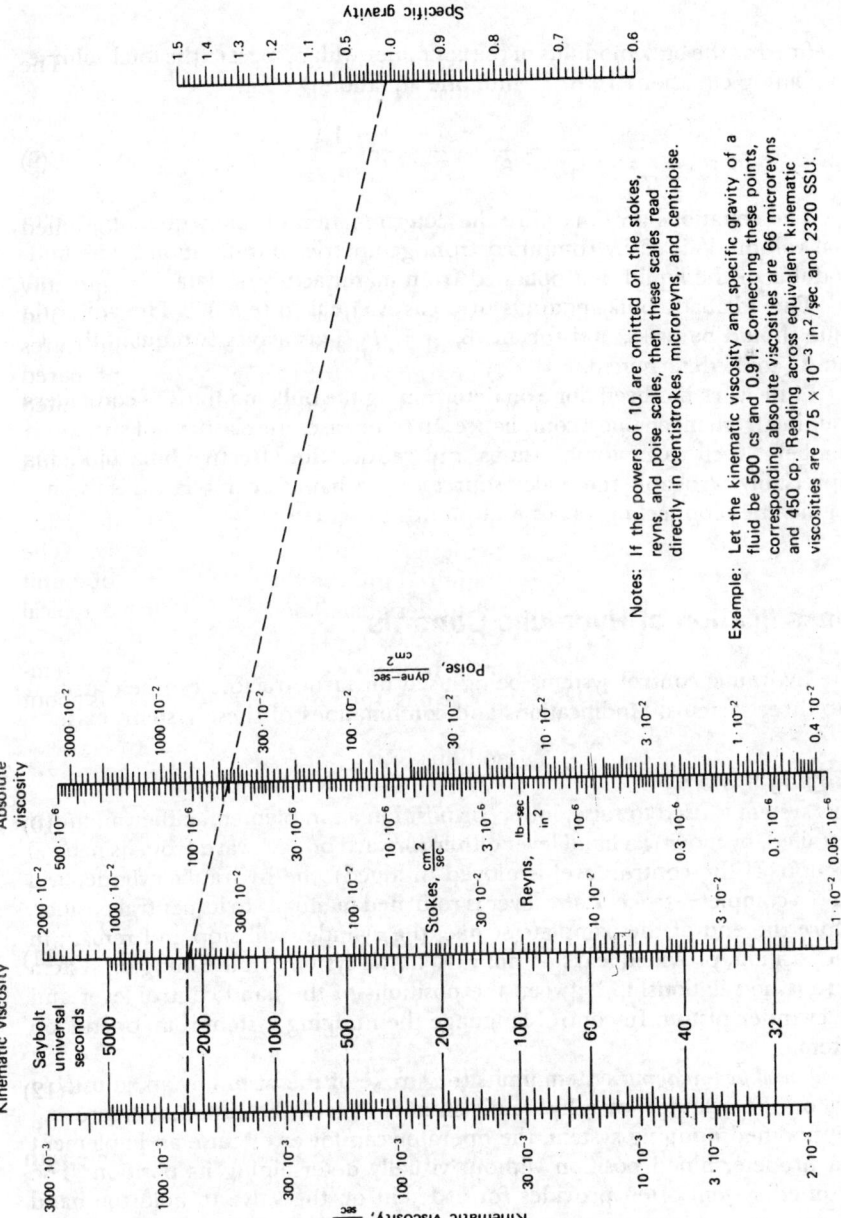

Notes: If the powers of 10 are omitted on the stokes, reyns, and poise scales, then these scales read directly in centistrokes, microreyns, and centipoise.

Example: Let the kinematic viscosity and specific gravity of a fluid be 500 cs and 0.91. Connecting these points, corresponding absolute viscosities are 66 microreyns and 450 cp. Reading across equivalent kinematic viscosities are 775×10^{-3} in²/sec and 2320 SSU.

FIGURE 12-17 Viscosity conversion chart.

is defined as the bulk modulus of the container with respect to the total volume. Combining equations 10 to 13 into one equation, we have

$$\frac{1}{\beta_e} = \frac{1}{\beta_c} + \frac{1}{\beta_l} + \frac{V_g}{V_t}\left(\frac{1}{\beta_g}\right) \tag{14}$$

The equations given require the determination of many quantities. The total volume V_t is easily computed from geometric considerations. The bulk modulus of the liquid β_l is obtained from manufacturers' data.

The adiabatic bulk modulus of a gas is equal to $(c_p/c_v)P$. The adiabatic value should be used; and for air, $\beta_g = 1.4P$. This leaves two quantities, V_g and β_c, to be determined.

Little work has been done on determining the bulk modulus of containers resulting from mechanical compliance. In some cases the elasticity of structural members, such as motor housings, can reduce the effective bulk modulus appreciably. Probably the major source of mechanical compliance is the hydraulic lines connecting valves and pumps to actuators.

Classification of Hydraulic Controls

The hydraulic control systems being used on farm tractors can be classified into three systems. Modifications and combinations of these systems exist.

Nudging System

This system is used to raise, lower, or position an implement, either mounted or trailed, by moving a hand lever either forward or backward from its neutral position. If the control level is moved (nudged), the hydraulic cylinder will move a complete stroke. If the lever is returned manually to its neutral position before the end of the complete stroke, the cylinder will stop and remain in that position, provided leaks do not exist in the system. In the nudging system there is no relationship between the positions of the hand control lever and the cylinder piston. In control language the nudging system is an open-loop system.

A *modified nudging* system limits the stroke of the piston by an adjustable stop on the cylinder, or a hydraulic shutoff to lock the piston. When using the modified nudging system, the operator can lower or raise an implement to a predetermined position without visually determining its position. The modified system often provides for a detent on the valve to hold the hand lever unaided until the piston reaches the end of its stroke or an adjustable stop, at which time the valve is returned to neutral.

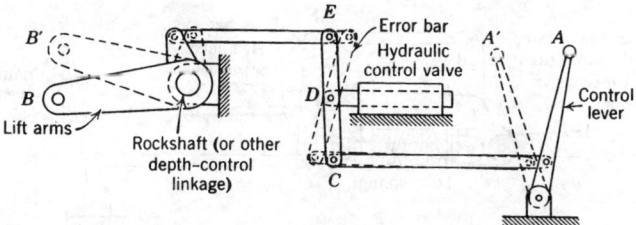

FIGURE 12-18 Elementary hydraulic position control. (From W. S. Hockey, The Institution of Mechanical Engineers, Nov. 1961.)

Automatic Position-Control System

This system provides automatic control of an attached implement and allows the operator to preselect and to position the implement as determined by the position of the hand control lever. The relative positions of the hand lever and of the hydraulic cylinder are always identical. Within the limit of the relief valve controlling the maximum pressure, the hydraulic cylinder will automatically move the implement to its predetermined position and maintain it there, regardless of any leakage in the system.

The position-control system is normally associated with a three-point hitch system. It is not used with a remote cylinder. The most elementary form of a position control system is shown in figure 12-18.

Automatic Draft-Control System

These systems will automatically raise or lower an implement as the draft or resistance of the attached implement increases or decreases. The sensing device, which tells the hydraulic system to lower or raise the hitch system, is located on either the lower links or the upper link, depending on the size of the tractor. The position of the hand-control lever, in effect, establishes the draft to be maintained. For example, a draft-control system on a tractor pulling a plow will raise and lower the plow to maintain a constant force on the sensing device. A simple form of a draft-control system is shown in figure 12-19.

Draft Sensing

When the draft-control system was first developed by the late Harry Ferguson, the draft-sensing device was located on the upper link and responded to a compressive force. For a close-coupled implement—for example, a two-bot-

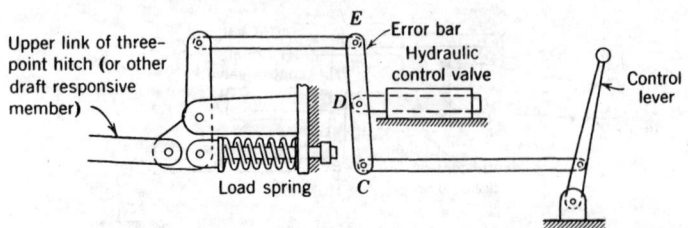

FIGURE 12-19   Elementary hydraulic draft control. (From W. S. Hockey, The Institution of Mechanical Engineers, Nov. 1961.)

tom plow—the upper link will normally be in compression, and as the draft increases, the compressive force will increase. It can be shown that the compressive force in the upper link becomes smaller as the size of an integrally mounted plow increases, and will often be a tension force for mounted plows of four or five bottoms and larger. Thus it becomes more difficult to use the force on the upper link to sense the draft, or change in draft, of the plow.

Automatic Control

There is little doubt that automatic control of the tractor will increase in the future. This is partly because such control of some of the functions will allow the tractor to be operated with more precision. Also, automatic control will, of course, relieve the operator of operating so many controls.

A brief introduction to automatic control theory will be made; however, readers are encouraged to study automatic control theory by reading some of the references listed at the end of the chapter, especially if they should become involved in the design of complex hydraulic systems.

Most hydraulic control systems, including those that are automatic, are a combination of hydraulic (oil) and mechanical (e.g., linkage, masses, springs) processes. In the future, electronic components likely will be added to the systems on tractors.

An automatic control, or feedback, system compares the output signal to the input signal and uses the difference to change the output. Some background in automatic control theory is necessary for analysis of hydraulic systems. The use of computers has made such analyses much easier.

A hydraulic control system is analyzed for two purposes. First, an analysis must be made to properly match or size the components of a system after the function and constraints have been specified. Second, the more difficult analysis is of the dynamic behavior of the system. Since a hydraulic control system has feedback, there is a good possibility that the system will be unstable; that

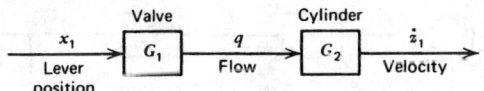

FIGURE 12-20   Open-loop control of a hydraulic cylinder.

is, it will vibrate. This latter problem requires the use of automatic control theory so as to eliminate, or reduce to an acceptable level, the instability or vibration in the system.

A simple example of control block mathematics follows. If a valve, G_1, is used to direct fluid to a cylinder, G_2, as shown in figure 12-20, the system is said to be *open loop* or there is no automatic control. However, such a system may have a feedback loop (not shown). The cylinder may have a stroke-control device that returns the fluid to the sump when the cylinder reaches the end of the stroke. Or perhaps the pressure will build up at the end of the stroke, causing the valve to return to neutral. In most cases the operator becomes part of the control by sensing (seeing) when the cylinder has reached the desired position, at which point the operator returns the valve control lever to neutral.

The operator is quite capable of being part of the control system. However, it is likely that he or she also has additional tasks to perform. The object of an automatic control system is to control the system without the aid of a human operator.

The speed of engines on tractors has been controlled for many years by a mechanical device called a governor. Although the engine speed is affected somewhat by the load on the engine, the "control" by the governor is acceptable to most operators. Hydraulic devices have also been used as governors on engines.

Let us return to figure 12-20. The transfer function of the valve, G_1, is the ratio of the output q to the input x_1 (q/x_1), and the transfer function of G_2 is $\dot{z}_1/q$. These can be combined by simply multiplying $G_1 \times G_2$. The transfer function is

$$\left(\frac{q}{x_1}\right) \cdot \left(\frac{\dot{z}_1}{q}\right) = \frac{\dot{z}_1}{x_1}$$

If we place G_1 and G_2 into a closed-loop system as in figure 12-21, the operator can forget about the control if the system was designed correctly. F is now in the feedback loop. The velocity, $\dot{z}_1$, of the cylinder can now be "fed back" to the valve by means of the signaling device, F_1, through the summing junction so as to correct the position of the valve.

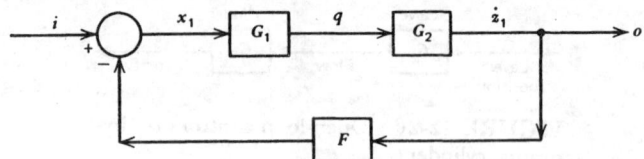

FIGURE 12-21   Closed-loop control system.

The overall transfer function, o/i, is

$$o/i = \frac{G_1 G_2}{1 + G_1 G_2 F} \qquad (15)$$

Until now we have not discussed the physical meaning of G_1, G_2, and F (the blocks). G_1 can be a valve whose output is the flow rate (measured in volume/time) and the input is the motion of the spool. Thus, G_1 is the output/input ratio (or flow gain) with the units of (L/s)/mm.

G_2 includes the dynamic behavior of the system (the cylinder plus the hydraulic oil plus the masses attached to the cylinder).

F is the feedback loop, which might be either a mechanical linkage attached to an "error bar" (see fig. 12-18) or a pressure line to sense the load on, or the position of, the output of the cylinder, as in hydrostatic power steering.

Further information on automatic control theory can be obtained from the references at the end of the chapter. A concise and yet complete treatise on automatic hydraulic, or fluid power, control systems is provided by Merritt (1967).

The method of automatically controlling a three-point hitch can be visually grasped by a thorough study of figure 12-22. This illustration shows the hydraulic system controlling the three-point hitch on a series of large tractors by a major manufacturer.

Complete Hydraulic System

The complete hydraulic system used on one tractor is shown schematically in figure 12-23 and is illustrated using symbols in figure 12-24. This system includes a circuit for a remote cylinder, a circuit for the three-point hitch servo valve, a circuit for steering, a circuit for brakes, a circuit for the pto clutch, a circuit for the differential lock, and several circuits for lubrication since most of the power train is lubricated by hydraulic fluid.

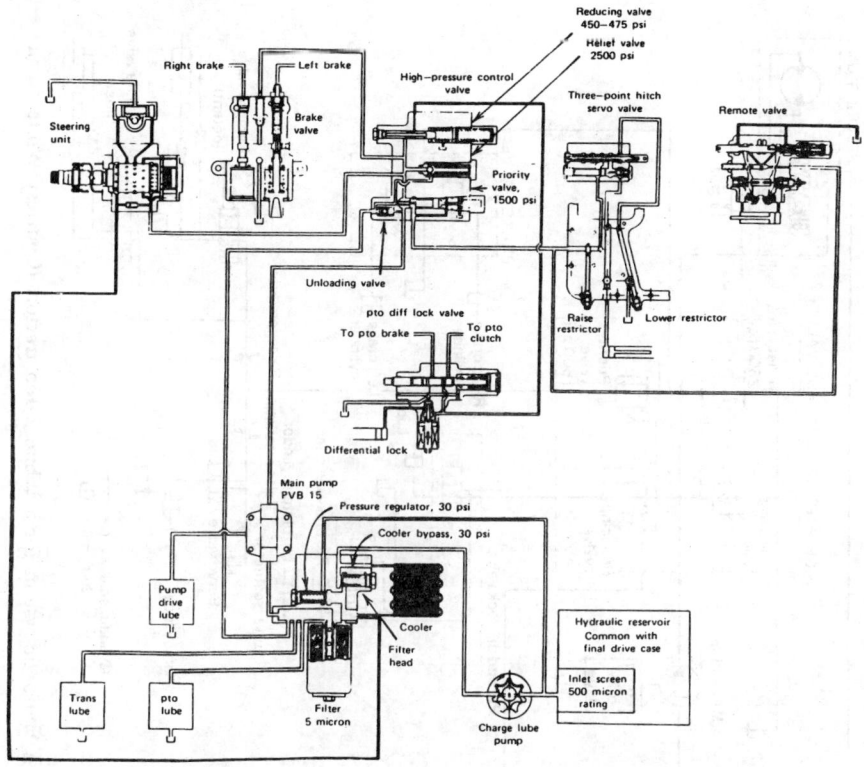

FIGURE 12-22 Hydraulic system for a White 2-155 tractor. (Courtesy White Farm Equipment Co.)

Power Steering*

Tractors having 30 kW or more power generally will have power steering. A quarter-century of power-steering development on farm tractors has focused almost completely on two basic fluid power types.

1. The hydromechanical systems
2. The full hydraulic systems, commonly called hydrostatic power steering

*This section is abstracted from the ASAE Distinguished Lecture Series, No. 1, by R. A. Wittren entitled "Power Steering for Agricultural Tractors," presented at the winter meeting of ASAE, Chicago, Ill., Dec. 17, 1975.

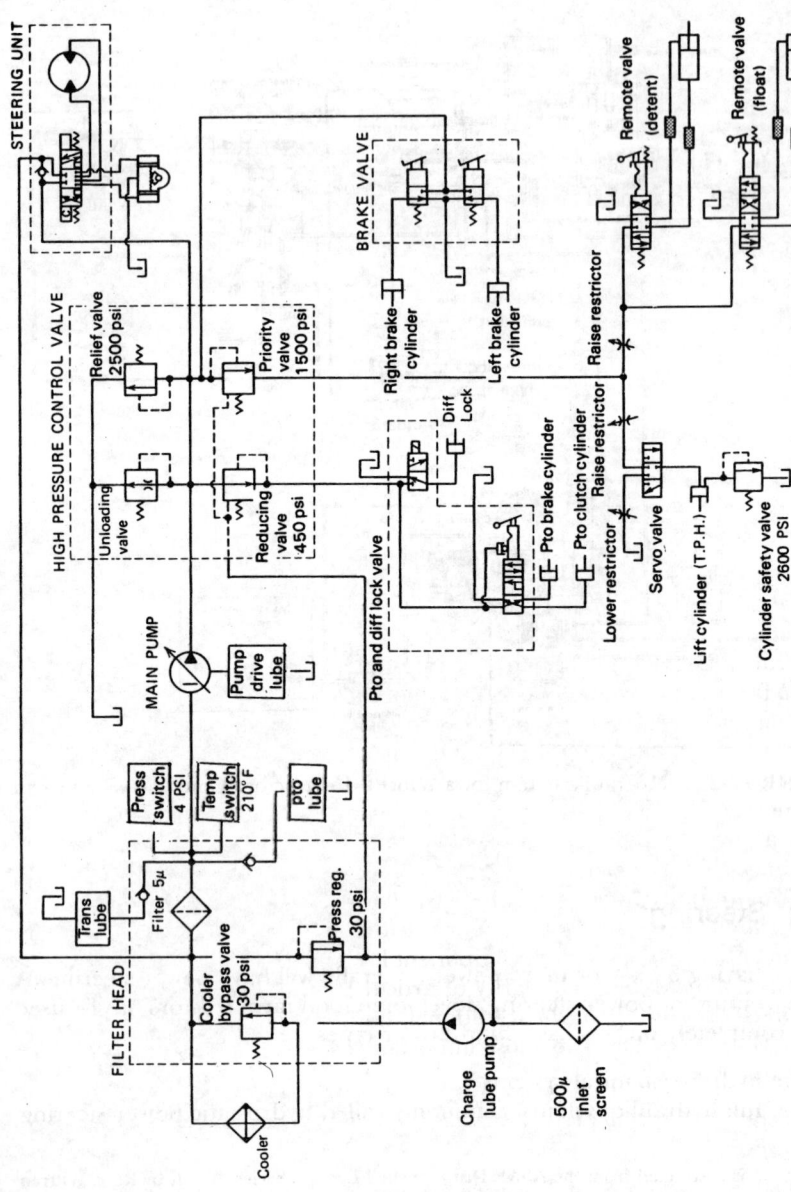

FIGURE 12-23 Schematic diagram of a hydraulic system for a White 2-155 tractor. (Courtesy White Farm Equipment Co.)

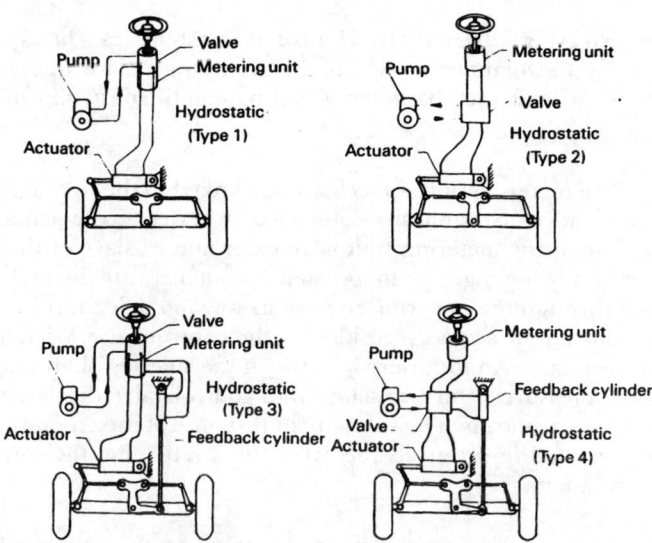

FIGURE 12-24 Basic component arrangement for a hydrostatic power steering. (From Wittren 1975.)

Hydrostatic Power Steering

The term "hydrostatic," although technically applicable to all power steering used on farm tractors today, has by common usage come to mean those systems requiring no mechanical linkage between the driver's steering wheel and the steered wheels. In these systems, fluid under pressure is used not only to power the load but also to provide hydraulic feedback from the load and to transmit manual effort to the load when the power source is unavailable. The lack of a positive, fixed-position relationship or index between the steering wheel and the ground wheels is considered a shortcoming by some when comparing the hydrostatic systems to the hydromechanical types.

The modern trend toward larger tractors featuring innovative improvements in chassis configuration, component location, and operator placement, as well as the increasing popularity of articulated, four-wheel-drive vehicles, often precludes the use of the traditional, familiar, mechanical steering linkages of yesterday. The two most important advantages of the hydrostatic system are:

1. Flexibility of installation
2. Lower cost

The most distinguishing feature common to hydrostatic steering systems is the use of a positive displacement flow metering or measuring device coupled to the steering wheel shaft. Hydrostatic systems can be conveniently categorized by the manner in which this metering device operates in the control loop. At least four basic types can be identified, as shown in figure 12-24.

Type 1   The metering unit is mechanically linked to the steering shaft and control valve and is hydraulically connected in series to the actuator. With this arrangement, the metering unit is an extension or slave of the actuator, the two always moving together in lockstep fashion because the high-pressure oil is routed through the metering unit on its way to the load. It provides the remote monitoring of actuator position at the control valve location, known as position feedback. An input error between the steering shaft and the metering unit is measured and translated into control valve displacement by a suitable mechanical means. The subsequent response of the actuator-metering unit to the directed flow cancels the error, thus returning the control valve to the null position.

Type 2   The metering unit is rigidly coupled to the steering shaft and hydraulically connected in series with the actuator, but in parallel with the control valve pilot chambers. Here the metering unit functions as a transducer to convert steering wheel rotation and input torque into flow and pressure to displace the control valve. The resulting flow of high-pressure oil again passes through the metering unit before entering the actuator. The incremental actuator motion, which continues after the steering wheel stops, hydraulically recenters the control valve.

Type 3   The metering unit is mechanically linked to the steering shaft and control valve as in type 1 but is hydraulically connected to a separate feedback displacement device, which in turn is linked to the output motion. This permits the control circuit to be isolated from the power circuit, with advantages to be discussed later.

Type 4   The metering unit is rigidly attached to the steering shaft and hydraulically connected to the control valve as in type 2, but it is hydraulically coupled to a separate feedback device as in type 3. This is a simpler, lower-cost arrangement than type 3 but introduces an operational factor also discussed later.

Parameters that influence power requirements include the following:

1. Tire loading
2. Road surface and soil conditions

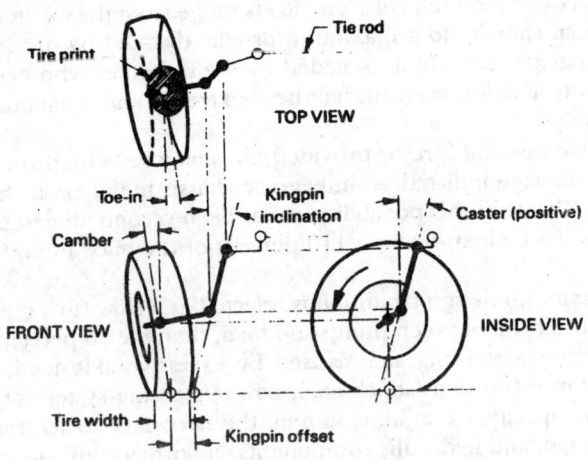

FIGURE 12-25 Tractor steering geometry. (From Wittren 1975.)

3. Tire inflation pressures
4. Tire sizes and tread patterns
5. Kingpin inclination
6. Caster angle
7. Camber angle
8. Kingpin offset or scrub radius
9. Toe-in and toe-out
10. Tread setting
11. Travel speeds
12. Steering rates
13. System efficiency
14. Front-end type (tricycle, single wheel, standard)
15. Tractive and braking forces
16. Chassis type

Many references are available on the mechanics of steering geometry and the analysis of linkage forces. Most of this literature deals with highway vehicles; however, the design techniques for the determination of steering forces with a stationary vehicle are valid for agricultural tractors using similar linkage geometry. Since the heaviest steering loads with Ackerman-type steering (fig. 12-25) usually occur with a stationary tractor on dry, clean concrete, this condition provides a convenient standard for calculating maximum power requirements.

On Ackerman-steer-type tractors, tire loading is the most significant variable affecting power requirements. Tire loads range from the minimum needed for longitudinal stability to a maximum usually dictated by tire load rating. Rated tire capacities are often exceeded by the customer who expects some degree of controllability even under these extreme and sometimes abusive conditions.

If excessive steering force is provided, tires may be twisted from the rims or structural damage inflicted on linkage or chassis members if the tires become lodged. To avoid this possibility, it has been recommended that power steering forces be limited to about 110 percent of the maximum design condition.

If the maximum design condition is selected as a slow turn creating a 10 percent pressure drop between pump and load, then the 10 percent overload will occur during a steering stall caused by excessive axle loads or wheel obstructions. Since transient shock loads to the steering system of twice the rated steering capacity are quite common, the necessary structural strength of wheels, linkage, and hydraulic components will provide an adequate safety factor with respect to maximum pump pressure. Often a power-steering system will display excellent performance under average to maximum load conditions, only to exhibit instability, erratic behavior, or an undesirable change in responsiveness when lighter load conditions are encountered.

This perplexing behavior is largely due to changes in the spring rate of the tires and in the flow gain characteristics of the valve under the different load conditions. System damping and valve gain must then be adjusted until acceptable performance over the entire operating range is obtained. System dynamic analysis, using mathematical modeling techniques with a computer, can provide valuable guidance in the initial design stages and important insights during development to aid in finding solutions to these problems.

The left-hand curve in figure 12-26 shows the generally accepted, typical effective coefficients of friction for rubber tires on dry concrete as a function of the kingpin offset-to-tire width ratio. The kingpin torque T required to turn the wheel under a vertical load W can be calculated by the following equation:

$$T = Wf \sqrt{\frac{I_o}{A} + e^2} \qquad (16)$$

where f = effective friction coefficient
I_o = polar moment of inertia of tire print
e = kingpin offset
A = tire print area

This equation assumes a uniform tire print pressure. If a better description of the tire print pressure distribution is available, a more accurate value can be sub-

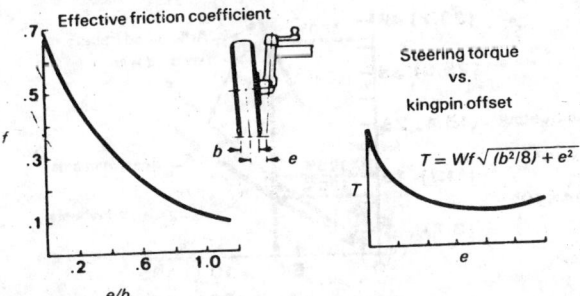

FIGURE 12-26 Typical curves based on rubber-tired vehicles on dry concrete. (From Wittren 1975.)

stituted for I_o/A. If the exact shape of the tire print is unknown, an approximation can be made by assuming it to be circular with a diameter equal to the nominal tire width, b. Since I_o/A then equals $b^2/8$, equation 18 becomes

$$T = Wf \sqrt{\frac{b^2}{8} + e^2} \qquad (17)$$

The change in steering torque as a function of kingpin offset is shown by the right-hand curve in figure 12-26. No values are assigned, since a different curve exists for each combination of W, b, and e. The curve shows that an optimum kingpin offset, e, exists for a minimum value of torque T. Caution must be exercised here, however, since larger offsets contribute to the transmissibility of shock loads into the linkage. Also, the optimum offset varies with different surface conditions. A small kingpin offset or even centerpoint steering in which the offset e approaches zero is better suited for heavily loaded tires in loose soil where rolling resistance is high. If the steered wheel is also powered or is equipped with brakes, a small or zero offset is also advisable to reduce feedback of forces into the steering system.

It soon becomes apparent that the classical theory becomes severely taxed by the large number of variables affecting the kingpin torque on farm tractors. For this reason, there is no substitute for test measurements on the actual tractor or on a similar prior model.

The required work output of the steering actuator for a specific design can be accurately determined only if complete details of the entire steering geometry as well as the kingpin torque are known. Typical work output values are plotted in figure 12-27 for both Ackerman and articulated-type steering systems on agricultural tractors. These values will normally provide acceptable performance for all but stationary steering under maximum load conditions.

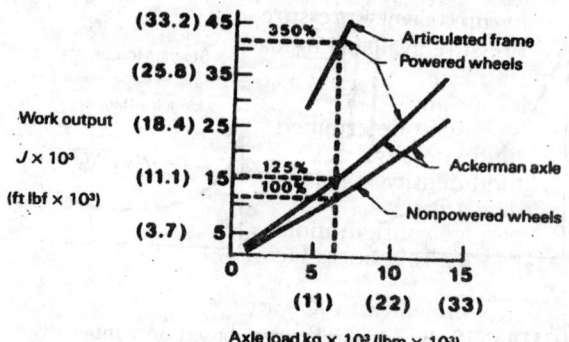

FIGURE 12-27 Actuator work output versus axle load.
(From Wittren 1975.)

Referring again to figure 12-27, it will be noticed that the energy required for steering an articulated-frame tractor is about 3.5 times that for an Ackerman-steer-type tractor with the same axle load.

Surface and Soil Conditions   The operation of farm tractors presents a nearly infinite variety of surface and terrain conditions. These variables have some nominal effect on system dynamic stability but are more influential on power requirements, handling characteristics, and shock loading. Farm tractors are operated on surfaces ranging from ice and snow to dry concrete; from loose sand to heavy, sticky soil; from level fields to bordered and bedded crop lands; and from public roads to stump-studded pastures. Steering system power levels must be tailored to the axle load and surface combination that requires the highest expected turning torque.

Sizing of Components   If the steering system is designed to deliver the required work output at a normal steering rate of one lock-to-lock turn in about 5 s with a 15 percent pressure drop between the pump and the load, then a 15 percent torque reserve will be available at the steering stall point. Assuming that a 5-s turn is one half of the maximum desired rate, the available kingpin torque will be significantly less during a fast 2.5-s turn. How much less depends on the system flow restrictions between pump and load. The overall pressure drop in a typical hydraulic circuit containing lines, passages, control valves, and fittings approximates the general equation

$$P_2 - P_1 = KQ^x$$

(18)

where P_2 = pump delivery pressure
        P_1 = pressure available to load
        Q = flow rate
        x = a constant
        K = a constant determined by:
                fluid viscosity
                fluid density
                flow areas
                passage configurations
                passage roughness factors

Of course, the problem with K is that it does not remain constant very long in a typical hydraulic system, and power steering is no exception. The best a designer can do is to analyze the extreme conditions and the norm and then proceed to make the least painful compromises in the design. Volumes of literature exist, dealing with the calculation of pressure drops in hydraulic systems. Using these references in conjunction with computer simulation techniques allows any degree of design analysis sophistication the designer desires or has the time for.

Measurements on typical hydraulic circuits can be used to predict pressure drops in future similar circuits. Test results for one such circuit indicated that equation 18 can be expressed as

$$P_2 - P_1 = KQ^{1.72} \tag{19}$$

for an oil with a viscosity of 75 SUS (19×10^{-2} stokes) and specific gravity of 0.855 at an operating temperature of 80°C. The value of K will depend largely on the size of flow passages provided, for it is predominantly a function of the effective composite flow area for the entire system.

Equation 19 emphasizes the extreme importance of the proper sizing of hydraulic lines and components. Considering the earlier example of a 15 percent pressure drop at a 50 percent steering rate, equation 19 predicts a 50 percent drop at the full rate. Alternately, a 10 percent pressure drop at half steering rate will reduce the loss to 33 percent at full rate. Typical measured pressure drop values for tractor power-steering systems range between 30 percent and 40 percent for a lock-to-lock turn in 2 to 3 s.

In power-steering design, sizing of the various system elements means providing the correct match between the component and its required function. These functions can be summarized as follows:

1. Generation of fluid power
2. Transmission of fluid power
3. Control of fluid power
4. Application of fluid power

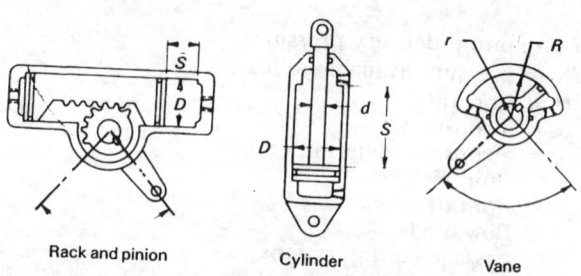

| Rack and pinion | Cylinder | Vane |

FIGURE 12-28 Actuator types. (From Wittren 1975.)

Actuators   Although the physical flow of fluid and energy in a steering system is from the pump to the actuator, the designer must proceed in the opposite direction because all are slaves to the load. After the load has been determined, either by calculation (equation 17) or by measurement, it must next be translated into terms of actuator work output. The location of the actuator in the linkage will determine the mechanical efficiency between the output member and the tire print. Overall efficiencies of manual steering gears and their associated linkage are generally in the 40 to 70 percent range, depending on the types of antifriction means employed.

Typical efficiencies for actuators usually fall between 80 percent and 95 percent, with the cylinder types being somewhat higher, on the average, than the rotary rack and pinion or vane types (fig. 12-28).

Actuator output travel (stroke or angular rotation) is governed to some extent by steering geometry limitations; however, it has been recommended that a stroke-to-bore ratio for cylinders between 5 and 8 be selected, if possible, to maintain adequate column strength and a favorable servo-valve amplification to linkage deflection relationship.

Linkage backlash, especially after significant wear, is equally important to elastic deflection in the selection of stroke-to-bore ratios. These recommended ratios can be extrapolated to rotary-type actuators, although in their case, externally applied column loads do not occur. In addition to actuator displacement, the working pressure available at the piston or vane face must be known or selected. If the steering is to be powered from an existing pressure source, then rates and flow losses must be analyzed at this point. If some latitude in pump pressure is possible, a desired load pressure can be chosen and actuator design continued or selected on the basis of the following equations. A constant torque, T, and effective pressure, p, are assumed:

$$D = \sqrt{\frac{4aT}{\pi SpE_aE_1} + d^2}$$

(20)

where D = piston diameter, mm
a = lock-to-lock steering angle, radians (at kingpin or hinge)
T = required kingpin or hinge torque, Nmm
S = piston stroke, mm
p = effective pressure at piston face, MPa
E_a = overall actuator efficiency
d = piston rod diameter, mm, where applicable
E_1 = linkage efficiency

In terms of displacement, the equation becomes

$$V = \frac{aT}{pE_aE_1} \qquad\qquad (21)$$

where V = actuator displacement, mm³
The appropriate multipliers must be applied to accommodate multiple actuators and/or number of torque loads involved. If the effects of kingpin inclination, caster, and camber are significant, a modified value of aT should be determined and substituted. The effects of caster and camber are usually small; however, large kingpin inclination angles produce a significant lifting action to the axle, which adds to the kingpin torque when going into a turn.

Control Valves   Both spool- and poppet-type control valves are used in agricultural power-steering systems. Poppet valves are used more often in closed-center than in open-center systems because of their superior sealing capabilities. Owing to its relative simplicity, the spool-type valve is by far the most common type used in power-steering systems today.

The size of a directional control valve used for a given flow rate will be influenced by the type of hydraulic system. Open-center, constant-flow systems normally require larger-diameter valves than do closed-center systems to provide low flow losses in the neutral or centered position while maintaining an acceptably small valve travel and deadband. In closed-center systems, only that flow required by the load passes through the valve; therefore, higher intermittent flow losses, occurring only during steering maneuvers, can be tolerated.

Directional, flow, and pressure control valves contain both passage- and orifice-type restrictions. Except at extremely small openings, flows through valve-metering apertures and across seats conform closely to orifice characteristics. Experience shows that the geometry of sharp-edged orifices has no significant influence on pressure-flow relationships under turbulent flow conditions.

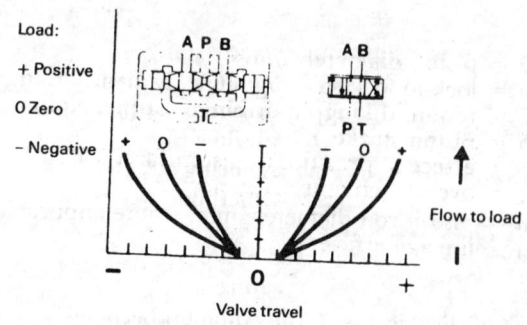

FIGURE 12-29 Flow gain from a closed center valve.
(From Wittren 1975.)

In addition to the sizing of valve dimensions to obtain acceptable pressure losses at maximum flow, the valve design must also provide a flow gain characteristic, the ratio of flow increment to valve travel at a given pressure drop, to maintain good control over the complete operating spectrum of loads, temperatures, and steering rates.

Fortunately, power-steering systems can tolerate considerable input to output error or phase shift in comparison to high-precision servo systems for machine tools, for example. Smoothness of operation and reasonable costs are more important than precision and can be provided with valves manufactured to practical tolerances.

Two types of four-way spool valves are commonly used in position servo systems. These are:

1. Open center, in which sufficient metering land underlap is provided to allow the maximum desired flow to pass through the valve in the null position without excessive pressure drop. These valves are low in cost because tolerances on land locations and, to some extent, radial clearances are less critical as used in power-steering systems.
2. Closed center, where positive overlap is provided to reduce high-pressure leakage in the null position (see fig. 12-29). Land locations and radial clearances are much more critical than on the open-center valve owing to the tight sealing requirement and the sensitivity of steering control characteristics to the port opening timing sequence.

The mechanical or hydraulic ratios between steering wheel rim and control valve travel are usually sufficiently high that valve operating forces do not significantly affect steering effort, assuming that hydraulically balanced or pilot-operated valves are used. When flow rates approach 25 gpm (95 L/min), however, such as on large articulated tractors, valve flow forces must be examined more closely. Since steady-state flow forces always tend to close the valve, they can contribute valuable stabilizing effects to the system. For this reason, some flow metering methods increase the centering flow force with beneficial results; but with high flow rates, flow force compensation may be required to maintain acceptable steering efforts. The rather complex theory of flow forces and compensation methods for control valves is well documented (Merritt 1967).

Whatever type of valve is used in a power-steering system, it is important to match the total valve travel required for the maximum flow, maximum load condition to the desired steering wheel error at this condition. This phase error is more noticeable on open-center systems because of the accompanying increased deadband, but it can also produce undesirable effects on closed-center systems during rapid steering reversals with heavy loads. Excessive steering-wheel-to-valve travel ratios become very noticeable during the manual operating mode without power since the entire valve travel then becomes lost motion. A total steering wheel rim travel of 12 in. (300 mm), equal to approximately 90° rotation on a wheel 16 in. (400 mm) in diameter, is thought to be about the acceptable limit. With total valve travel typically in the 0.020- to 0.50-in. (0.5 to 13 mm) range, ratios of 20 or less will remain within this limit. A valve deadband, while operating under power, from approximately 1 to 3 in. (25 to 76 mm) at the steering wheel rim is suggested. Beyond these limits, the steering either becomes too sensitive to small unintentional inputs or can initiate tractor wander and overcorrection tendencies due to excessive lost motion.

Considerable improvement in the control characteristics of both open- and closed-center power-steering systems should be obtainable by the use of a closed-center valve equipped with a flow compensator spool and bypass or pressure beyond port.

Since the pressure drop across the directional control spool metering land remains essentially constant for any combination of supply and load pressure, steering rate becomes a function of valve opening only.

It is rare that a new power-steering system is designed and assembled without encountering system instability at some operating condition in spite of any and all analytical techniques employed. Since valve damping is easily added or increased, this is usually the first thing to try. It is advisable to design the valve with this in mind by providing end chambers that can be converted to dashpots by restricting flow into and out of the chambers. Preferably, this should be possible without interfering with the normal flow paths through the valve.

If valve dashpots become too restricted, the limited rate of valve movement will manifest itself at the steering wheel as a hesitation or momentary "hang-up" during quick steering-wheel movements.

If stability cannot be achieved without excessive damping, then modifications in other parameters such as flow gain, relative to the input and output, should be explored. This may require changes in linkage ratios, valve spool metering, and actuator piston area. In addition, changes in spring rates and mass distribution may need to be examined. At this stage of development a valid simulation model and a handy computer can be invaluable. The analysis of hydraulic systems by computer-manipulated mathematical techniques is a rapidly maturing and well-documented engineering discipline. A partial listing of publications in this field is included in the suggested readings. It may also be wise to keep in mind that a system or design never malfunctions; it operates exactly the way you designed or built it. The struggle results from trying to make the design, the hardware, the operating conditions, and your objectives compatible with one another.

Lines, Hoses, and Internal Passages   The proper sizing of fluid passages requires a fairly accurate determination of the pressure losses in the circuit for various combinations of flow rates and operating temperatures. Accuracy is necessary because of the exponential relationship between flow and pressure loss, which leaves little room for error when designing for maximum performance while keeping size and cost of circuit elements within practical limits.

Fortunately, a high degree of similarity exists in hydraulic systems, including power steering, used on agricultural tractors. Because of this, the theory can be reinforced by a large body of operational and laboratory data within definite ranges of temperature, fluid viscosity, passage roughness, and flow rates.

Typical flow velocities on farm tractors range from a low of 1.5 m/s for suction lines to a high of 9.1 m/s for pressure lines, with delivery and return lines falling somewhere in between. Significantly higher localized velocities can occur in the associated fittings, valve apertures, and orifices.

Operating temperatures are typically in the 65° to 93°C range; however, pressure losses at the lowest expected temperatures must also be examined to ensure safe cold-weather start-up and warm-up performance. The viscosity range of oils commonly used for tractor hydraulic systems is not extreme. Viscosities of 70 to 100 SUS at 65°C and 45 to 60 SUS at 93°C will bracket most of them. The spread widens considerably, however, at low temperatures, and significant variations in cold-weather start-up characteristics must be recognized and thoroughly evaluated.

The transition from laminar to turbulent flow is experimentally determined in terms of the dimensionless quantity called Reynolds number. Published results vary, as expected for empirical data, but in general, laminar flow exists up to Reynolds numbers of about 2000 and fully turbulent flow develops somewhere between 3000 and 4000. Smoother pipes with low relative roughness ratios extend the transition zone toward the higher Reynolds numbers. Because of cost, space, and performance trade-offs, turbulent flow nearly always exists in tractor steering systems at the maximum performance levels where pressure losses become most significant.

Knowing Q and v, or u, the following formulas can then be used to calculate circuit pressure losses. Constants are dimensionless and are valid for any system of units. Terminology and units are:

R = Reynolds number, dimensionless
Q = flow rate, cm^3/s
d = pipe passage bore, cm
u = absolute viscosity, dyne · s/cm^{2*}
L = passage length, cm
p = pressure drop, dyne/cm^2
v = kinematic viscosity, cm^2/s
ρ = mass density, dyne · s^2/cm^4
w = specific weight, g/cm^3

$$\text{Reynolds number} = \frac{1.27Q}{dv} = \frac{1.27Qp}{du} \tag{22}$$

Laminar flow: $R < 2000$

$$\Delta p = \frac{40.7\rho v QL}{d^4} \quad \text{or} \quad \frac{40.7uQL}{d^4} \tag{23}$$

Turbulent flow: $R > 3000$ to 4000

$$\Delta p = \frac{0.242(\rho v)^{0.25}\rho^{0.75}Q^{1.75}L}{d^{4.75}} \tag{24}$$

$$= \frac{0.242u^{0.25}\rho^{0.75}Q^{1.75}L}{d^{4.75}} \tag{25}$$

The turbulent flow formulas can be used for various cross-sectional shapes if the hydraulic diameter (D) is substituted for diameter (d). The hydraulic diameter is defined as

$$D = \frac{4A}{S} \tag{26}$$

where A = passage flow area
          S = flow passage perimeter

*1 Pa · s = 10 poises (dyne · s/cm^2); 1 N = 10^5 gm · cm/s^2

For laminar flow, published tables should be consulted for noncircular cross sections (Merritt 1967).

References are also available to calculate secondary pressure losses due to entry and exit conditions, passage expansions and contractions, pipe bends, and fittings (Merritt 1967).

Orifices   Other major sources of pressure losses in circuits are orifices. Pressure drop through a sharp-edged orifice for turbulent flow is

$$\Delta p = \frac{Q^2 \rho}{2 C_d^2 A^2} \tag{27}$$

where A = orifice area
         C_d = orifice discharge coefficient, dimensionless

The orifice discharge coefficient, C_d, normally ranges from 0.6 to 0.8, with average values of 0.6 to 0.65 commonly used. Assuming a discharge coefficient of 0.6 and a typical mass density of 8.41×10^{-5} lb-s^2/in.^{4} (9.19×10^{-4} dyne-s^2/cm^4)*, specific gravity = 0.90, and solving for Q:

$$Q = 92.5 A \sqrt{\Delta p} \text{ in.}^3/\text{s} \tag{28}$$

or

$$Q = 28 A \sqrt{\Delta p} \text{ cm}^3/\text{s} \tag{29}$$

and in round numbers Δp becomes

$$\Delta p = \left(\frac{Q}{100A}\right)^2 \text{psi} \quad \text{or} \quad \Delta p = \frac{Q^2}{9 \times 10^4 A^2} \frac{\text{dynes}}{\text{cm}^2} \tag{30}$$

which, in addition to being easy to remember, is close enough for initial estimates. These equations are valid only for turbulent orifice flow. Laminar orifice flow seldom occurs in mobile hydraulics except possibly at very low temperature and/or small orifice areas.

Orifices should be constructed with a passage length equal to the orifice diameter and entry and exit edges left square with burrs only removed. An average discharge coefficient, C_d, is 0.745 for such an orifice, which is in good agreement with throttling losses through valves using the minimum cross-sectional flow area for A. Equation 29 then becomes

*1 poise (dyne · s/cm^2) = 10 N · s/m^2; 1 dyne = 2.248 × 10^{-6} pound = 10^{-5} newtons

$$\Delta p - 0.9 \frac{\rho Q^2}{A^2} \tag{31}$$

Power-Steering Pumps   The power source for a steering system can be either a separate pump, the central hydraulic system pump, or a combination of both.

High-quality, fixed-clearance gear or vane pumps are used for pressures up to 1500 psi (10.3 MPa). The majority of open-center steering pumps, both on farm tractors and on trucks and automobiles, are of this type.

Pressure-balanced and clearance-compensated vane and gear pumps extend the pressure capability into the 2000 psi (13.8 MPa) to 3000 psi (20.7 MPa) range with latest gear types capable of 3500 psi (24.1 MPa) operation. Most commercially available steering control units are rated at about 2000 psi (13.8 MPa) operation, and these may limit system pressure, if used. Closed-center, self-contained steering systems will allow pressures up to 4000 to 5000 psi (27.6 to 34.5 MPa) with piston-type pumps if other system components and circuitry have matching capabilities. Pressures above 3000 psi (20.7 MPa) are difficult to justify on agricultural steering systems.

Most commercial power-steering pumps are designed to be belt or gear driven and have speed capabilities above the engine speeds typical on farm tractors. Unless the pump is to be driven directly off the crankshaft, the higher permissible speeds should be exploited to reduce size and cost. Common maximum pump speeds are from 3000 to 5000 rpm. The required pump flow is given by

$$Q = 0.260 V/t \text{ gal/min} \tag{32}$$

or

$$Q = 0.06 \, V/t \text{ (L/min)}$$

where V = actuator total displacement volume, in.^{3}(cm^3)
          t = desired steering time for lock-to-lock turn, s

A steering wheel rate of less than one revolution per second is seldom satisfactory, and more than two is rarely desired. A steering time t of 2 to 4 s, lock-to-lock, is typical.

With flow rate Q and time t determined, pump displacement can be computed.

The power requirement of the pump is

$$P = \frac{V \, \Delta p n}{E} \text{ in.lb/s (N} \cdot \text{m/s)} \tag{33}$$

$$= \frac{V \, \Delta p n}{6600 \, E} \text{ hp,} \quad \text{or} \quad \frac{V \, \Delta p n}{10^6 E} \text{ kW} \tag{34}$$

where Δp = pressure rise through pump, psi (Pa)
          E = overall pump efficiency
          n = rev/s

Fixed displacement pumps used in closed-center steering systems require pressure unloading means, usually in combination with an accumulator to prevent excessive heat generation and energy waste during low-duty cycle operation. The accumulator is advisable to prevent rapid on and off pressure cycling and to permit use of a smaller pump. Since this represents an uneconomical solution with very little advantage over the open-center system, it is seldom considered.

The null, or centered position, flow capacity of an open-center control valve is incompatible with good system response over a wide flow range. For this reason, a flow control valve is normally used to limit the maximum flow through the valve to that required for the most rapid desired steering rate. The flow control is usually an integral part of the pump and bypasses excess flow, at higher pump speeds, directly from pump outlet to reservoir. The heat generated can usually be dissipated by properly sizing the reservoir or, occasionally, the bypassed flow is routed through a small heat exchanger.

Noise in Hydraulic Systems

Space does not permit the inclusion of any theories of noise generation in a hydraulic system. Some of the suggested readings at the end of this chapter may be of value to the student who wishes to pursue the subject. In general, the noises generated are caused by either turbulence or cavitation in the fluid. There are also mechanical noises. Many, perhaps all , hydraulic systems need a systematic noise reduction program during their experimental development stage.

Hitches

As with most other design problems, one best hitch system does not exist except, perhaps, for a specific application. The number of uses for tractors has increased markedly in recent years to complicate the problem of arriving at a standard hitch system.

Some of the many factors that should be considered in designing a tractor hitch system are:

1. Effect of the implement, through the hitch system, upon the tractor: safety, traction, stability, and steering

2. Effect of the tractor, through the hitch system, upon the implement: control and safety protection for the operator and the implement
3. Ease of attaching implements
4. Standardizing of the hitch
5. Effect upon power takeoff

Tractor Kinetic Energy

When the tractor is being used to propel an implement that may strike an obstacle—for example, a plow striking a rock—some provision must be made for protecting the implement, tractor, and hitch from excessive stresses. This can be done by several means, such as (1) by energy-absorbing hitch, (2) by breakaway hitches, and (3) by having parts of the implement release when subjected to an overload.

The last method is most commonly used but will not be discussed here. The second method can be conveniently used only for implements drawn by a drawbar and cannot easily be used for implements that are attached with a three-point or other type of integral hitch. The first method of protection is of value on the hitches of drawn implements such as plows.

A moving tractor has kinetic energy, which must be absorbed in slowing down or stopping. In addition to the kinetic energy resulting from the mass of the tractor, the engine, wheels, and transmission have additional kinetic energy as a result of their rotation. For rubber-tired tractors the kinetic energy of revolving parts is approximately equal to 10 percent of the total kinetic energy resulting from the linear motion of the tractor (Clyde 1949). For a wheeled tractor the kinetic energy is approximately

$$KE = \frac{1.1}{2} Mv^2 \tag{35}$$

where M is the mass of the tractor and v is the velocity.

If the tractor is equipped with a linear spring type of energy-absorbing hitch, and if the tractor clutch remains engaged after the implement is stopped, then by the *conservation of energy* equation it is apparent that the kinetic energy in the tractor plus the work being done by the drive wheels while the tractor is stopping must be equal to the increase in potential energy stored in the spring hitch (see fig. 12-30).

$$T_1(x_2 - x_1) + \frac{1.1}{2}Mv^2 = (x_2 - x_1)\left(\frac{P_{max} + T_1}{2}\right) \tag{36}$$

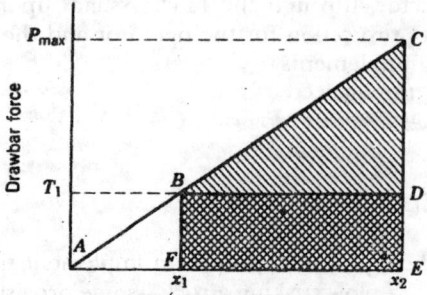

FIGURE 12-30 Graphical representation of energy absorbed by a spring hitch. Area *ACE* is the total potential energy in the spring. Area *FBCE* represents the maximum energy that can be absorbed.

where T_1 = the tractive force before implement stops
$(x_2 - x_1)$ = deformation of spring after the implement strikes an obstacle
P_{max} = maximum force on the hitch
T_2 = the tractive force after implement stops

Obviously the energy to be absorbed by the spring will be less if the clutch can be released at the instant of impact, in which case the tractive force, T_2, will be zero.

In equation 36 we can substitute $P_{max} = kx_2$ and $T_1 = kx_1$, where k is the spring constant.

Other types of energy-absorbing devices have been used but with less success than a spring.

Unfortunately, the value of P_{max} will become prohibitively large for high-speed tractors, or else the distance, $x_2 - x_1$, required to stop the tractor becomes excessively large.

Example   Assume a 2725-kg tractor at a velocity of 1.5 m/s is to be stopped in 15 cm after the implement to which it is attached strikes an immovable object. If the traction, T_1 and T_2, is 18,000 N, the work resulting from traction is 270 J and the kinetic energy is 3065 J. If the tractor speed is doubled (3.0 m/s), then the kinetic energy becomes 12,260 J, while the work resulting from traction remains the same. Obviously, a device to declutch the tractor would not afford much protection to an implement or tractor when the speed is high.

Because the working speed of tractors has been increasing, it would appear that protection will be more logically provided by "break-away" or "break-back" devices on the implements.

Integral Hitch Systems

There have been many tractor hitch systems devised for the purpose of controlling the implement or machine to which the tractor is attached. The two common types are (1) a drawbar with a single-point attachment and (2) a three-point hitch.

It is often desirable to use a hitching system that allows the implement to be controlled more completely than is possible with a single-point drawbar hitch and at the same time provides for rapid hitching and unhitching of the tractor to the implement.

Space considerations preclude a detailed discussion of all the systems that have been and are being used. Because the three-point hitch is used by almost every manufacturer of tractors in the world, it would seem advisable to devote most of the discussion to it.

Three-Point Hitches

Figure 12-31 shows a three-point hitch design of the early 1920s that resulted in limited success. This hitch (Patent No. 1,464,130), developed for the Ferguson-Sherman plow, did not result in any appreciable increase in the rear-wheel reaction because the virtual hitch point, f', was too low. Note that the vector sum of c, the compression force in the upper link, and t, the tension force in the lower link, passes near to O', the instant center of the wheel and soil. It is obvious that the increase in the rear-wheel reaction resulting from Py_f/L_1 (see chapter 10) is approximately zero because y_f is very small.

As the plow in figure 12-31 enters the soil, the virtual hitch point f' continues to rise until the plow reaches equilibrium.

The three-point hitch was developed in 1935 by the late Harry Ferguson (Gray 1954). The dimensions of categories I, II, III, and IV three-point hitches have been standardized by ASAE and SAE. Figure 12-32 illustrates the standard three-point hitch. Details of the standard are shown in ASAE S217.10 (also SAE J715 SEP83). Another standard, ASAE S320.1, describes the three-point hitch standard for lawn and garden tractors having less than 15 kW of power.

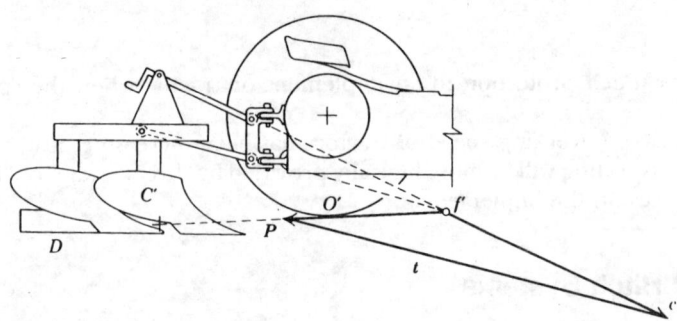

FIGURE 12-31   Early three-point hitch of the free-link type. (From
A. W. Clyde, *Agr. Engr.*, Feb. 1954.)

The design of the three-point hitch has carefully considered the effect on
both the implement and the tractor. Figure 12-33 describes a graphical method
of determining the resultant of all forces acting upon the hitch. Note that a bending force, *b*, exists on the lower link, thus affecting the location of the virtual
center *f'*. With the free-link or floating-type, no bending force can exist except
that resulting from friction in the pin-connected joints. Most tractors employ
some means of holding the lower links in a fixed vertical position (position control) or holding the lower links at a vertical position regulated by the force on
either the upper or lower links (draft control). These systems were discussed in
the section Classification of Hydraulic Controls.

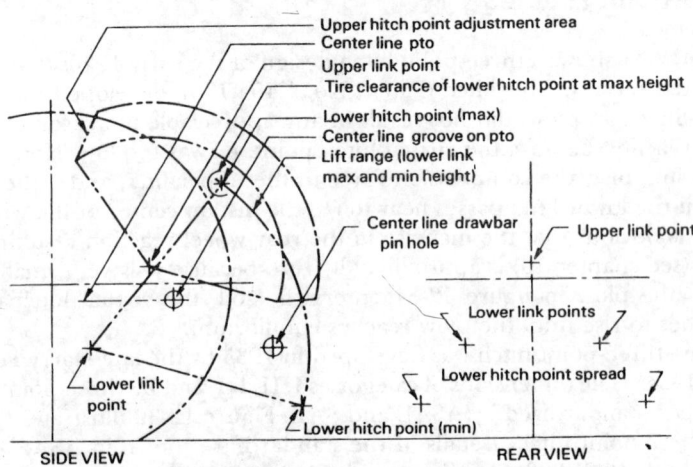

FIGURE 12-32   Standard three-point hitch for agricultural tractors. (From ASAE S 217.10.)

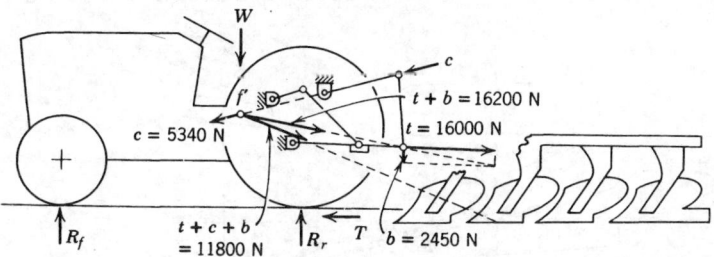

FIGURE 12-33 Method of force analysis on three-point hitch of tractor. (From R. W. Wilson, Paper No. 960, SAE meeting, Sept. 14–17, 1959.)

It is obvious from figure 12-33 that the virtual center, f', of the forces on the hitch system will be higher if a downward bending force exists on the lower links.

If it is desired to determine all the forces in the three links, it is necessary to measure simultaneously the bending and tension forces in the lower links and the compression (or possibly tension) force in the top link, while at the same time recording their position relative to the tractor.

If it is desired to determine the vertical and transverse forces plus the longitudinal forces acting upon a three-point hitch, it will be necessary to equip each of the bottom links with strain gages that will independently measure tension and bending forces. The top link will need strain gages to sense only compression or tension.

The design of three-point hitches provides for lateral motion of the entire hitch system. This allows the tractor to turn easily when pulling a load. If the lower links are designed to converge, the virtual center of the hitch system (as seen from a top view such as fig. 12-34) will be somewhere between the front wheels. It is apparent that the direction of the sum of all forces acting upon the three links of the hitch must be along the line $B' - B''$. The direction of the sum of all forces does not go through the point of intersection, f, of links 1 and 2 but must also consider the force in link 3. Thus the actual line of force in this case passes between the instant center of the lower links, f, and the center line of the tractor.

Quick-Attaching Coupler for Three-Point Hitches

The three-point hitch has many advantages, including extra weight transfer to the tractor for traction, ease of control and movement of the attached implement, and better control of the implement. When the original (Category

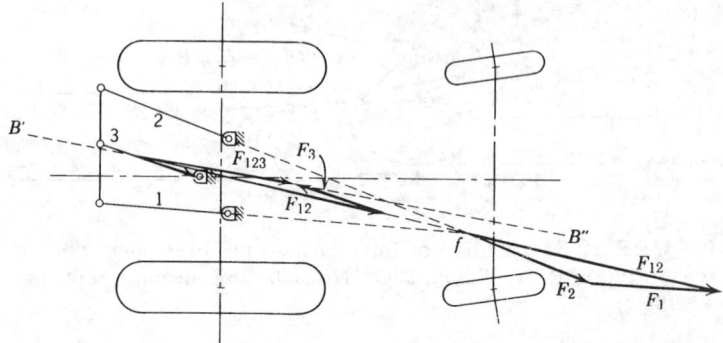

FIGURE 12-34   Method of force analysis on three-point hitch of a turning tractor. Forces shown are those imposed by tractor on hitch links 1, 2, and 3. (From A.W. Clyde, *Agr. Engr.*, Feb. 1954.)

I) hitch was developed, the implements being attached were generally small enough so that the operator could manually move them if the tractor was not in the correct location. With the advent of larger hitches and larger implements, it became much more difficult for the operator to align the implement and the tractor. The quick-attaching coupler is an attachment to the three-point hitch (see fig. 12-35) that allows easier and safer hitching and unhitching, especially with the large tractors. The quick-attaching coupler moves the implement rearwards approximately 10 cm.

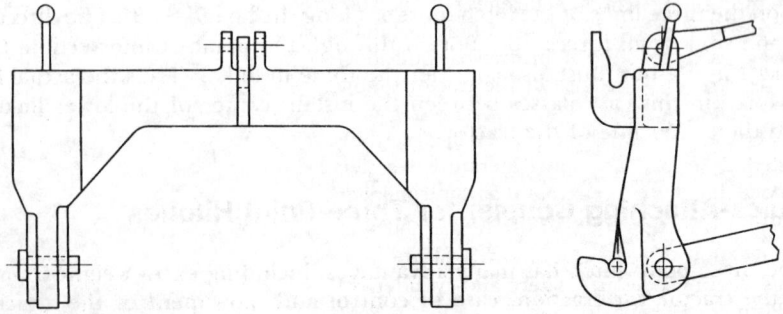

FIGURE 12-35   Standard quick-attaching coupler for three-point hitch on agricultural tractors. (From SAE J909 [Apr. 1980]. Also, ASAE S278.6.)

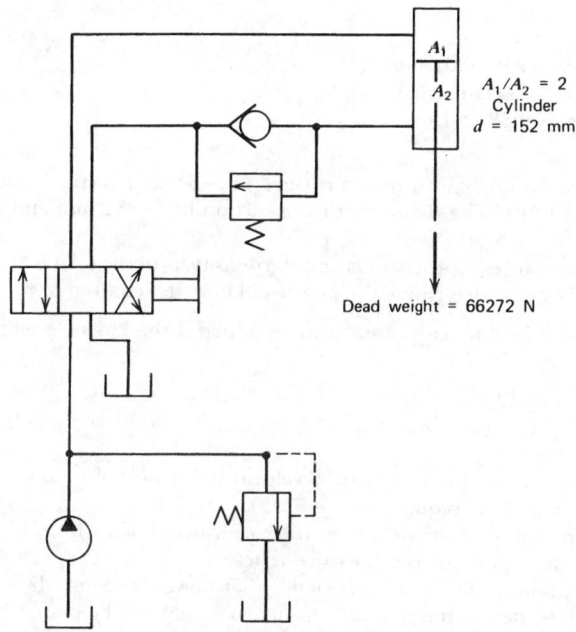

FIGURE 12-36 Simple hydraulic system.

PROBLEMS

1. A rubber-tired tractor is pulling a plow at 6.5 km/h on level ground when the plow hooks a large boulder, causing the unit to stop suddenly. The coefficient of traction is 0.60, the maximum possible drawbar power is 18 kW, the rear wheels weigh 14,700 N, and the front wheels weigh 7350 N.

  (a) Compute the kinetic energy to be absorbed in stopping the tractor.

  (b) If the tractor is stopped (not declutched) in 15 cm by a drawbar spring, what is the increase in the drawbar pull above the maximum tractive ability? Neglect rolling resistance of the tractor. Show all calculations and quote sources of information.

2. For an assigned tractor, obtain descriptive literature of its complete hydraulic system. From the descriptive literature (e.g., technical articles, operator's manuals, and service manuals), prepare a diagram similar to that in figure 12-24 of the complete hydraulic system, using symbols.

3. Compute the force to move a piston in a hydraulic cylinder at a velocity of 25 cm/s. Assume pressure is zero and cylinder is full of oil. Given:

  Piston diameter is 50 mm

  Piston length is 50 mm

Cylinder diameter is 50.01 mm
Oil viscosity, v, is 5 centistokes
Oil density, ρ, is 0.88 kg/L

4. A hydraulic pump has an inlet pressure of -34 kPa vacuum and a discharge pressure of 3440 kPa. The diameter of the suction line is 32 mm and the discharge line is 19 mm.

The pump discharges oil at 1.15 L/s and the density, ρ, of the oil is 0.88 kg/L. Neglect friction and efficiency. Determine the power in kilowatts required to rotate the pump.

5. In problem 4, how much power must be added if the kinetic energy of the fluid is also considered?

6. Refer to figure 12-36.
  (a) What is the minimum pressure setting that would be acceptable for the counterbalance valve?
  (b) If the piston must be capable of a velocity downward of 5 cm/s, what minimum pump capacity is required?
  (c) If the piston must provide a net force downward of 1.11×10^5 N, what is the minimum setting for the pressure relief valve?
  (d) If the pump is 90 percent efficient, what power is required?
  (e) If 3 m/s is the maximum fluid velocity acceptable in the lines, what size tubing should be used?
  (f) What porting configuration would you suggest for the centered position of the valve? Why?

REFERENCES

Clyde, A. W. *Agr. Engr.*, Apr. 1949.

Gray, R. B. *Development of the Agricultural Tractor in the United States, Part II*. American Society of Agricultural Engineers, St. Joseph, MI, 1954.

Merritt, H. *Hydraulic Control Systems*. John Wiley & Sons, New York, 1967.

Wittren, R. A. "Power Steering for Agricultural Tractors." ASAE Distinguished Lecture Series No. 1, presented at winter meeting of ASAE, Dec. 17, 1975.

SUGGESTED READINGS

Blackburn, J., G. Reethof, and J. S. Shearer. *Fluid Power Control*. The M.I.T. Press, Cambridge, MA 1960.

Clyde, A. W. "Pitfalls in Applying the Science of Mechanics to Tractors and Implements." *Agric. Engr.*, vol. 35, Feb. 1954, pp. 79–83.

Cowell, P. A. "Automatic Control of Tractor-Mounted Implements—an Implement Transfer Function Analyser." *J. Agric. Eng'g Res.*, vol. 14, no. 2, 1969, pp. 117–125.

Deere & Co. *Hydraulics, Fundamentals of Service*, 3d ed. John Deere Service Publications, Moline, IL.

Di Stefano, J. J. et al. *Feedback and Control Systems.* Schaum Publishing Co., New York, 1967.

Ernst, W. *Oil Hydraulic Power and Its Industrial Applications*, 2d ed. McGraw-Hill Book Co., New York, 1960.

Fundamental Hydraulics-Components and Circuitry. The Penton Publishing Co., Cleveland, 1971.

Henke, Ross. "Understanding Hydraulic Servosystems." *Machine Design*, Apr. 20, 1972 to May 18, 1972 (in 3 parts).

Herzan, Gene. "Quick Action Couplings: The Tractor to Implement Hydraulic Interface." ASAE Distinguished Lecture Series, no. 8, Dec. 14, 1982. American Society of Agricultural Engineers, St. Joseph, MI.

Julian, A. P. "Design and Performance of a Steering Control System for Agricultural Tractors." *J. Agric. Eng'g Res.*, vol. 16, no. 3, 1971, pp. 324–336.

Martin, H., and T. Keating. "Calculating Hydraulic Servo Frequency Response." *Hydraulics & Pneumatics*, July 1974.

McCloy, D., and H. R. Martin. *The Control of Fluid Power.* John Wiley & Sons, New York, 1973.

Morling, Roy W. "Agricultural Tractor Hitches and Analysis of Design Requirements." ASAE Distinguished Lecture Series, no. 5, Dec. 12, 1979. American Society of Agricultural Engineers, St. Joseph, MI.

"Reducing the Operating Noise of Industrial Hydraulic Systems." Parker Hannifin, Cleveland, Article Preprint PC-1, Oct. 1972.

Stecki, J. S., and Peter Dranfield. "Finding and Fixing Hydraulic Noise Sources." *Machine Design*, no. 13, 1975.

Vennard, John K., and Robert T. Street. *Elementary Fluid Mechanics*, 5th ed., SI Version. John Wiley & Sons, New York, 1975.

Wittren, R. "Techniques for Reducing Hydraulic System Overheating." *Machine Design*, July 1962.

13

TRANSMISSIONS AND DRIVE TRAINS

The tractor drive train has three functions. It transmits power from the engine to the wheels, pto, hydraulic pump, and other auxiliary drives; changes the torque and speed into the torque and speed required by the particular drive; and provides means for operator control with disconnect clutches and speed ratio selection for the wheel and pto drives. The power transmitted by the drive train is approximately constant, changing only by the amount of the small gear and bearing friction and windage losses that occur. For rotating systems, power is a product of torque and speed so that the torque and speed at the engine and wheels are related by

$$T_e N_e (\text{Eff}) = T_w N_w \approx \text{constant} \qquad (1)$$

where T_e = torque at engine
           N_e = rotational speed of engine
           T_w = torque at drive wheels
           N_w = rotational speed of drive wheels

Figure 13-1 shows the constant power curves that are a graphical solution of the equation

$$\text{kW} = PS/3.6 \qquad (2)$$

where kW is the power in kilowatts, P is the drawbar pull in kilonewtons (kN), and S is the speed in kilometers per hour (km/h).

The constant power curves represent 100 percent efficiency of a transmission with infinite gear ratios. Note that lines of increasing power move upward to the right and the pull becomes very large as the speed decreases.

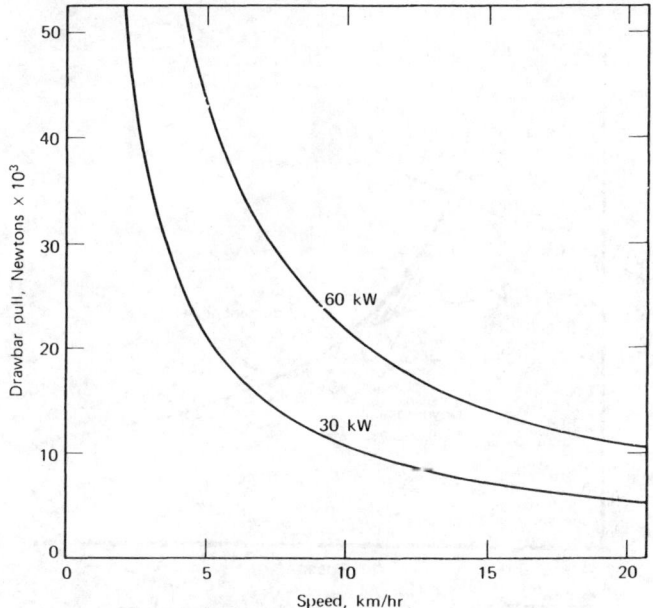

FIGURE 13-1 Speed-pull curves for tractors with no power loss between engine and drawbar.

The tractor tractive capability provides a practical limit on the amount of pull that can be generated and thereby limits drive train torques.

When a limited number of speeds are available, it will not always be possible to use the maximum engine power available, especially if the load is highly variable. This is shown by the performance envelope plotted in figure 13-2, where the lower lines represent the power that can be transmitted from the engine. Moving from right to left, the curves show the pull increasing rapidly with decreasing speed as the engine governor opens to rated engine speed, then slowly increasing as the engine pulls down on its torque curve. Thus the shaded areas under the constant power curve represent areas of operation that are not attainable. Instead, the tractor will operate along the engine limited lines at lower power and lower productivity. This is why the specific speeds provided and the ease with which these speeds can be selected are the most important features determining how well the transmission meets the farmer's needs. To be satisfactory, these features must be provided with good efficiency, durability, and economy. Ideally, any speed from 40 km/h forward through 0 to about 15 km/h in reverse could be selected, under any

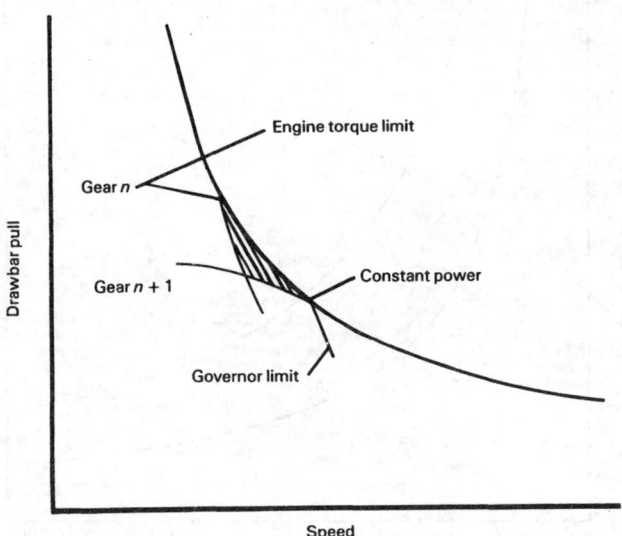

FIGURE 13-2   Tractor performance envelope.

load, with a simple adjustment of a control lever. Engineers and inventors have long sought this type of control with many concepts of mechanical, hydrodynamic, and hydrostatic transmissions, but with limited success owing to cost and functional problems. Consequently, the transmission is a compromise between what is desired and what can be provided with available technology at a reasonable cost. For these reasons, transmissions have evolved from the simple sliding gear transmissions with a few speeds to the more complex synchronized and power shift transmissions with many speeds. In 1941 the average tractor tested at Nebraska had 4 forward speeds. The number of forward speeds had increased to 8 by 1961 and to more than 12 speeds by 1981. Synchronized and power shift speed changing was first offered in the 1950s.

Complete Drive Train

A modern tractor drive train is a complex arrangement of several basic components that are used to control and transmit the power delivered by the engine. Friction or jaw clutches are used to control power by connecting and disconnecting the drive line or to provide braking by locking the drive line to a fixed housing. Friction clutches also provide control for smooth engagement of the drive line and for engagement when the input and output are

FIGURE 13-3 Phantom view of tractor power train from engine to wheels. (Courtesy Deere & Company.)

rotating at different speeds. Shafts are used to transmit power from component to component. Various types of gears are used to modify speed and torque and to redirect or relocate the power output. These basic components provide the building blocks of all tractor drive trains but are often specifically designed for each application to meet minimum size and optimum performance requirements. Browning (1978) provides a comprehensive coverage of the design of these tractor drive train components.

Figure 13-3 shows a typical tractor drive train from the engine to the drive axles. A complete drive train usually consists of the following major components:

1. *Traction and pto friction clutches*—Operator-controlled components used to smoothly connect or disconnect torque-transmitting elements rotating at different speeds.
2. *Transmission*—The operator-controlled speed change portion of the drive train.
3. *Pto drive*—The components that transmit power from the engine to the pto output shaft.

4. *Mechanical front-wheel drive*—The components that transmit power to the front axle drive lines on tractors equipped with front-wheel drive.

5. *Transmission and hydraulic pump drive*—The components that transmit power to pumps for the transmission and hydraulic systems on the tractor.

6. *Spiral bevel gear set*—A special gear set that provides a right-angle drive and a speed reduction.

7. *Differential*—A set of bevel gears that allows power to be transmitted to each axle at different speeds to allow satisfactory steering of the tractor.

8. *Final drive*—A final gear reduction used to provide the proper axle speeds and to reduce torque loads on, and size of, the differential and spiral bevel set.

9. *Individual axle brakes*—A special type of friction clutch, usually located between the differential and final drive, that is controlled by the operator and can be used to restrict or stop the rotation of one or both of the axle shafts.

10. *Rear axles*—The shafts that transmit power to the tractor rear wheels and transmit the wheel loads to the tractor structure.

Transmission Types

All drive trains for today's tractors have the major components listed previously. The features provided by the transmission portion of the drive train, however, are generally quite varied among manufacturers, since the type and configuration of transmission provide unique cost and functional characteristics. The following descriptions are commonly used to distinguish the major types of transmissions in automotive, agricultural, and industrial use:

1. Sliding gear
2. Constant mesh
3. Synchronized or synchromesh
4. Power shift
5. Automatic
6. Hydrostatic
7. Hydromechanical
8. Continuously variable transmission (CVT)

The first three transmission types are often called manual transmissions because the gears are usually engaged with manual effort only. However, power engagement is sometimes used, making the term misleading. The first four transmission types have distinct gear ratios and are sometimes referred to as selective gear transmissions, but this term is obviously not a very explicit description. Each type listed has important distinctive characteristics that need to be considered when selecting a transmission.

A *sliding gear* transmission in which spur gears must be used is shown in figure 13-4. The gear ratio is selected by disengaging the traction clutch and sliding the gear on the shaft until it meshes with a mating gear. The tractor should be stopped when shifting to prevent damage to the gears. This type of transmission is very efficient, but practical applications are limited to small tractors because large gears are hard to move, the large teeth provide too much engagement interference, and the teeth can be damaged during shifting.

In the *constant mesh* transmission the gears are mounted so that they are always in mesh with at least one of the meshing gears free to rotate on the shaft. Splined couplings of various types, usually called shifter collars, are used to engage the gears. Figure 13-5 shows a typical arrangement. Again, it is necessary to disengage the traction clutch and advisable to stop the tractor when shifting. This type of transmission is slightly less efficient than the sliding gear type as a result of the friction between the rotating gears and shafts, but it is used when helical gears are selected to control noise or when it is desired to reduce the shifting effort that would be created by moving large gears. Most tractor transmissions are, at least partially, of the constant mesh type.

A *synchronized* or *synchromesh* transmission has small friction clutches (fig. 13-6), usually cone type, that engage when a shift is initiated. The resulting frictional torque is used to prevent engagement of the shifter collar until the rotational speed of the collar and gear are nearly the same, i.e., synchronized. When synchronization occurs, the frictional torque reduces and the shifter collar can then be engaged with the gear to complete the shift. The advantage of a synchronized transmission is that gear changes can be made easily without damaging the transmission, even when the vehicle is moving. This transmission has long been used in automobiles and is extremely popular on tractors in Europe, where tractors are often used for on-highway transport of heavy loads, making shifting while moving necessary, as pointed out by Ponzio (1973) and Renius (1976).

The *power shift* transmission shown in figures 13-7 and 13-8 uses high-torque friction clutches to engage the gears rather than the splined shifter collars. These transmissions are designed to shift while moving and transmitting power, in contrast to the preceding types of transmissions in which it is necessary to disengage the traction clutch. Power shift transmissions also tend to be more expensive and less efficient than the preceding types of transmissions. Power shift tractor transmissions are described by Erwin and O'Harrow (1958), Harris and Jensen (1963), and Haight (1982). Often a transmission will consist of two or more power shift gears in series with some constant mesh gears to provide a medium-cost transmission with better efficiency and good functional features. Several of these concepts are presented by Elfes (1961), Ronayne and Prunty (1968), Lemke and Rigney (1970), Best (1972), and Hoepfl and Ballendux (1975).

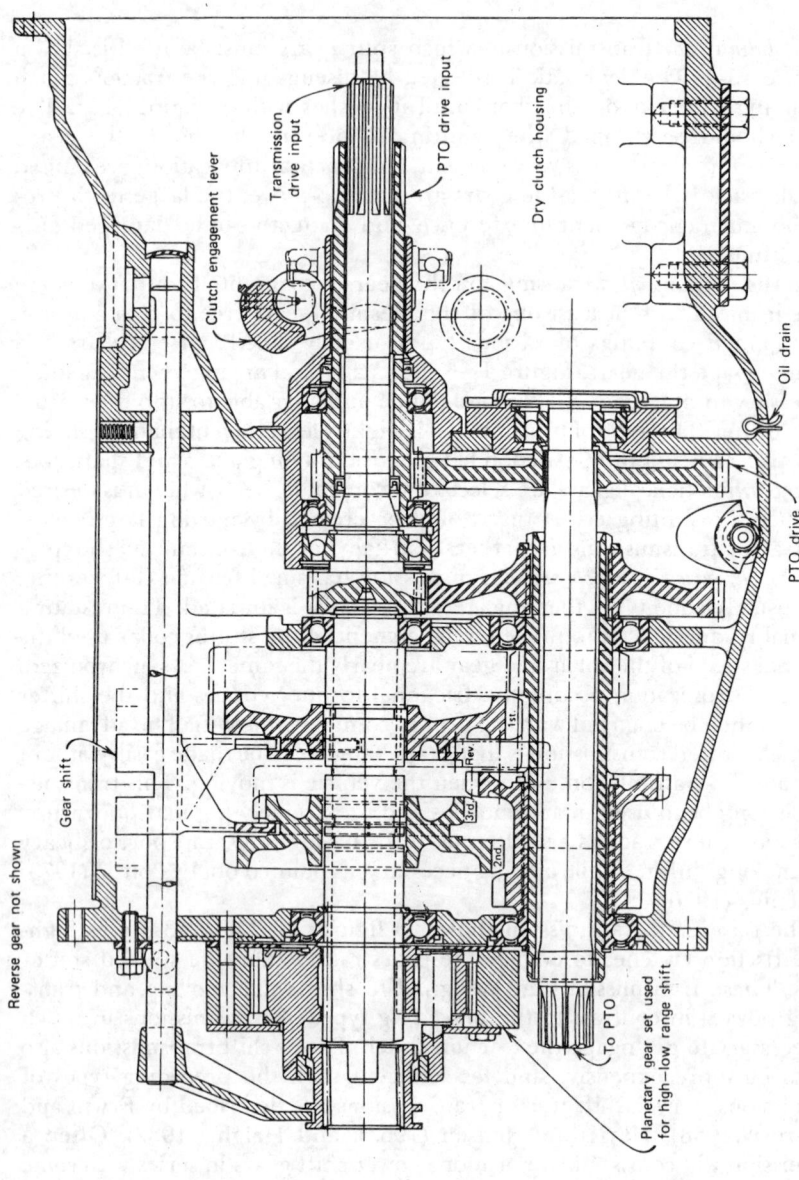

Reverse gear not shown

Gear shift

Clutch engagement lever

Transmission drive input

PTO drive input

Dry clutch housing

Oil drain

PTO drive

3rd

Rev

1st

2nd

To PTO

Planetary gear set used for high–low range shift

FIGURE 13-4 Six-speed transmission. Three sliding gears times planetary high–low range. (Courtesy Massey-Ferguson Ltd.)

366

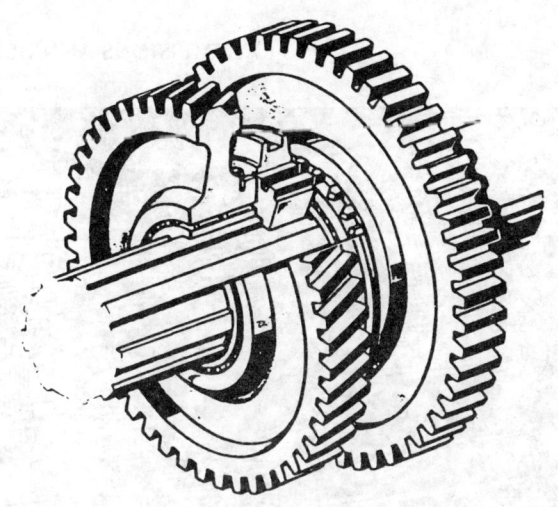

FIGURE 13-5　Constant mesh gears with shifter collar. (Courtesy Zahnradfabrik Passau GmbH.)

FIGURE 13-6　Synchronizer assembly. (Courtesy Zahnradfabrik Passau GmbH.)

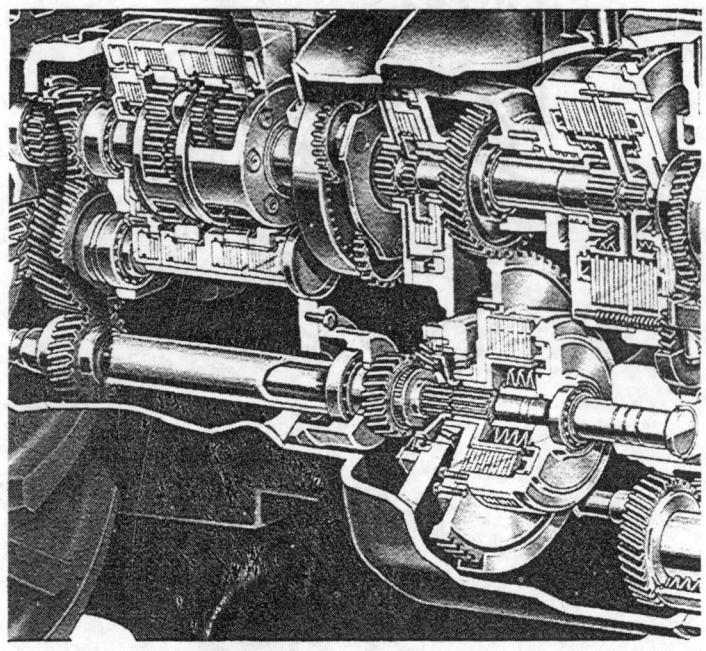

FIGURE 13-7   Sectional view of a power shift transmission with three compound planetary gear sets, four clutches, and five brakes to provide 15 forward and 4 reverse speeds. (Courtesy Deere & Company.)

The term *automatic* has long been used to refer to the two-, three-, or four-speed power shift transmission that is popular in cars and sometimes used in trucks. This type of transmission is coupled to the engine with a torque converter and shifts automatically according to some predetermined logic that is built into the control system. The torque converter is a simple, but inefficient hydrokinetic variable-speed device that provides a smooth transition between speeds and also helps to cushion the power shift clutch engagements.

The *hydrostatic* transmission such as shown in figure 13-9 provides infinitely variable speed selection from reverse to forward as described by Asmus and Borghoff (1968). However, these transmissions are quite costly, very inefficient, and often noisy. Typical performance characteristics of a hydrostatic pump and motor are shown in figure 13-10. Primarily because of the low efficiency, these transmissions have not been found satisfactory for tractors, except for small lawn and garden units in which efficiency is less important. Hydrostatic transmissions have also been successful on crawler trac-

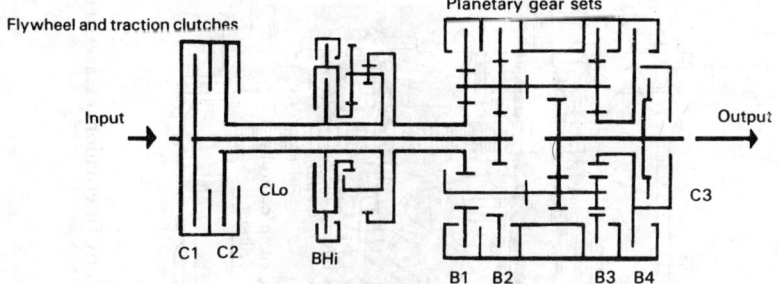

FIGURE 13-8 Schematic diagram of the planetary transmission shown in figure 13-7.

Clutch and Brake Engagement Chart

Gear	Engaged Elements
4R	C2 - BHi - B2 - B3
3R	C2 - CLo - B1 - B3
2R	C1 - CLo - B2 - B3
1R	C1 - CLo - B1 - B3
1	C1 - CLo - B1 - C3
2	C1 - CLo - B2 - C3
3	C1 - CLo - B1 - B4
4	C2 - CLo - B1 - C3
5	C2 - BHi - B1 - C3
6	C2 - CLo - B2 - C3
7	C2 - BHi - B2 - C3
8	C2 - CLo - B1 - B4
9	C2 - BHi - B1 - B4
10	C2 - CLo - B2 - B4
11	C2 - BHi - B2 - B4
12	C1 - C2 - CLo - C3
13	C1 - C2 - BHi - C3
14	C1 - C2 - CLo - B4
15	C1 - C2 - BHi - B4

tors in which the hydrostatic transmission provides excellent productivity. They are also popular on combines in which variable-speed control is very important, and they are used in many other applications in which flexibility and control are more important than efficiency.

The *hydromechanical* transmission can take on a variety of configurations but is characterized by parallel hydrostatic and mechanical power paths as discussed by Kress (1968). This type of transmission has not been widely used because it tends to be complex and expensive with disappointing efficiency. Hunck (1968) describes the performance of a small tractor with a hydro-

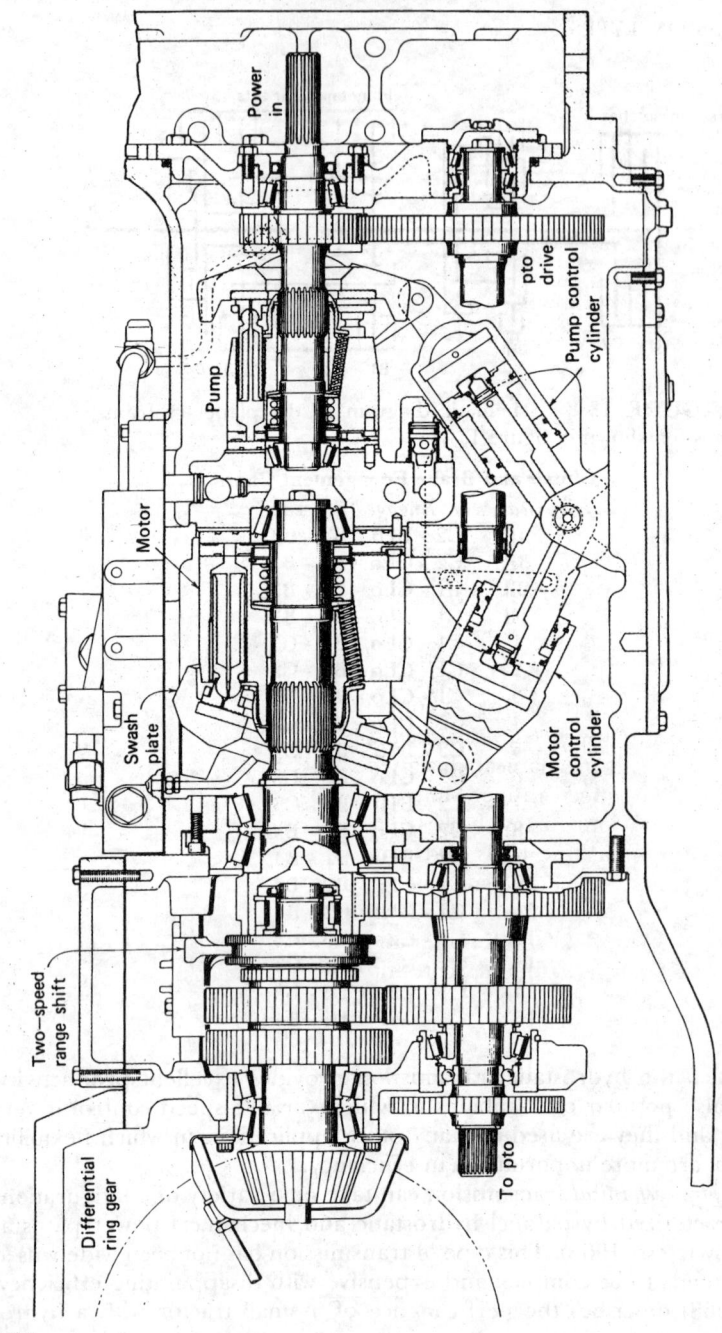

FIGURE 13-9 Cross-sectional view of a heavy-duty hydrostatic transmission. (Courtesy International Harvester.)

Power in

Pump

Motor

Swash plate

Two-speed range shift

Differential ring gear

pto drive

Pump control cylinder

Motor control cylinder

To pto

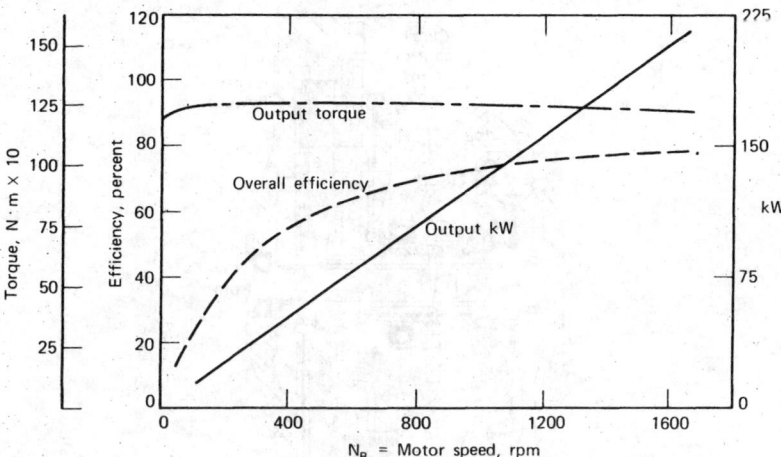

FIGURE 13-10 Typical performance characteristics of a hydrostatic transmission consisting of a variable displacement pump and a fixed displacement motor.

mechanical transmission. MacDonald (1969) and Ross (1972) describe some other hydromechanical transmission concepts for road machinery and trucks.

The *continuously variable transmission (CVT)* type covers a wide range of transmissions, usually mechanical, designed to provide a continuously variable ratio over a limited speed range. Many of these types of transmissions have been proposed, but only a few have been successful in some specific applications, mostly in machine tools and some low-power drives. These transmissions have not yet been used successfully in any kind of high-power mobile vehicles. A considerable amount of development is currently in progress, however. Dickinson (1977) and Kemper and Elfes (1981) describe some tractor CVT concepts. Kraus (1976) and Scott and Yamaguchi (1983) describe some automotive CVT concepts.

Friction Brakes and Clutches

Friction brakes and clutches are required to connect and disconnect drive train components so that the operator can control the tractor power. Tractor drive trains require these elements for the following functions:

1. Traction and pto clutches
2. Pto brake

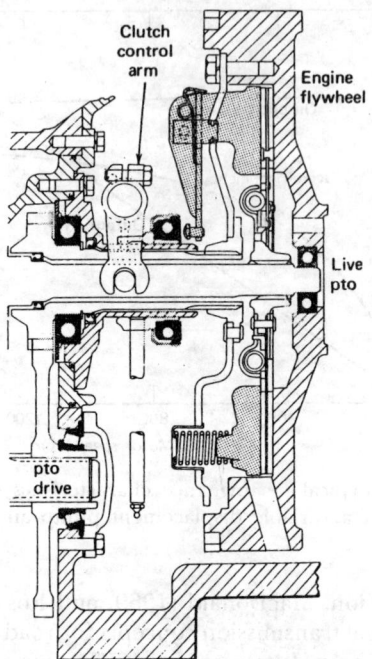

FIGURE 13-11 Cross section of a dry clutch. (Courtesy International Harvester.)

3. Synchronizers
4. Power shift brakes and clutches
5. Differential lock
6. Wheel brakes
7. Mechanical front-wheel-drive clutch

Brakes and clutches can be designed to operate wet (oil cooled) or dry (air cooled) and can be of single- or multiple-disk, caliper, cone, band, or drum configuration. The particular choice depends on the torque capacity and durability required, on the space available, and on whether the element will be manually or power operated.

The disk type of clutch provides generally high capacity and durability and is often used in tractor drive trains. For tractors below about 80 kW, a dry, single-disk traction and pto clutch of the type shown in figure 13-11 is often used because it is low cost, can be manually operated, has good torque

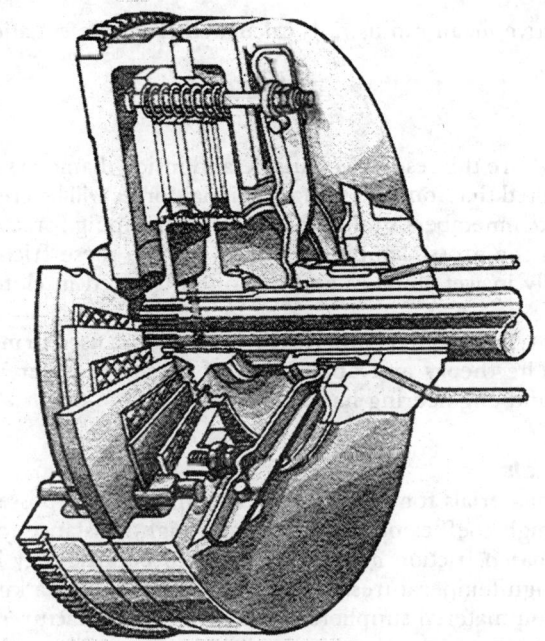

FIGURE 13-12 Sectional view of a wet (oil-cooled) clutch. (Courtesy Deere & Company.)

capacity, and facilitates shifting as a result of having relatively low inertia. Higher power tractors require oil-cooled traction clutches for good durability (fig. 13-12). These are usually operated with hydraulic power. Wet disk clutches and brakes have also been used for all of the other drive train functions listed previously. Multiple disks often are used in wet clutches and brakes to obtain the required torque capacity.

When the facing pressure is assumed to be equalized over the total area, the torque that a disk clutch or brake will develop can be estimated with good accuracy using the equation

$$T = Ffr_mN \tag{3}$$

where
F = effective clamping force
f = coefficient of friction
r_m = mean radius of the clutch facings
N = number of torque transmitting surfaces

The effective mean radius r_m is calculated from the equation

$$r_m = \frac{(D_o^3 - d_i^3)}{3(D_o^2 - d_i^2)} \tag{4}$$

where D_o and d_i are the respective outside and inside diameters of the facing. It should be noted that some resisting frictional forces will be created between the sliding clutch members so that the effective clamping force may be somewhat less than the gross clamping force. However, these friction losses are small, especially in wet clutches and are often ignored in clutch and brake design.

The other types of brakes and clutches are seldom used in modern tractor drive trains. The theory and application of these are covered in machine design books and engineering handbooks.

Friction Materials

Good friction materials for clutch and brake linings must possess good wear properties, a high coefficient of friction, and high resistance to heat. In the past, asbestos-based friction materials were used because they had excellent resistance to high temperatures. However, asbestos is now a known carcinogen, and friction material suppliers and machine manufacturers are striving to use substitute materials. Nonasbestos friction materials are being used in all new applications.

Present-day friction facings are formulated from organic, ceramic, metallic, and other materials using weaving, molding, sintering, and paper-making processes. The "paper" facings made from cellulose fibers with paper-making processes are now widely used in wet clutch and brake applications and have proved to be very durable and relatively inexpensive. Another important advantage of the paper materials is a relatively small difference between the static and dynamic coefficient of friction. The paper materials engage smoothly and are relatively chatter free in wet applications in contrast to the sintered and molded materials, which have wider differences in the static and dynamic coefficient of friction. Although the friction material is important, the material and surface finish of the mating surfaces are almost equally important. Thus, durability testing is required before a final selection of materials can be made. Table 13-1 summarizes typical operating parameters for various friction materials. Many applications exceed these values at maximum torque capacity, but must operate somewhere close to these limits for the thousands of cycles of life necessary for tractor brakes and clutches. The range of maximum values shown in the table is intended to indicate the variation that will occur with the large number of friction material formulations available within each group along with variations in surface finish, energy

TABLE 13-1 Design Data for Brakes and Clutches

Material	Friction Coefficient f	Maximum Facing Pressure (N/mm^2)	Cooling Oil Flow $(L/s/m^2)$	Maximum Energy Rate (W/mm^2)	Maximum Bulk Temperature $(°C)$
Dry operation					
Organic	0.2–0.5	0.1–0.3	Dry	0.5–0.8	150
Cerametallic	0.3–0.4	0.7–0.9	Dry	0.8–2.3	200–250
Wet operation					
Paper	0.09–0.13	0.8–2.5	6–30	0.8–1.2	230–280
Molded	0.08–0.10	1.5–2.5	2.5–30	1.0–1.2	230–280
Filled fluorocarbon	0.08–0.10	1.5–2.5	2.5–12	1.0–3.0	250–300
Sintered metallic	0.04–0.09	1.2–3.5	2.5–12	1.0–2.0	300

SOURCES: Browning (1978), Drislane (1965), Graham (1970), Haviland (1963), Spokas (1968), Wise and Mitchell (1972).

rate, heat absorption capacity, and cooling. The friction coefficient also varies widely with sliding velocity.

Design Techniques

Adequate torque capacity must be provided to quickly cause lockup when the clutch or brake is engaged and to allow a minimum of slippage after lockup due to torque peaks in the drive train (including the torque reserve of the engine). This requires an excess of capacity, usually called reserve. Generally, a reserve of two to three is provided. When engagement is operator controlled, such as with traction clutches or tractor brakes, the clutch or brake can easily be caused to absorb excessive amounts of energy, but most operators are aware of this and can attain satisfactory durability from tractor clutches and brakes.

In addition to furnishing adequate torque capacity, it is important to provide for an adequate heat sink to absorb the energy generated during engagement and then to provide for adequate cooling. Most of the heat generated during the engagement is conducted into the mating surfaces, even in oil-cooled clutches as shown by Tataiah (1972). Thus, the mating plates must be thick enough to absorb this energy and remain at a temperature below the maximum allowable facing temperature. Tataiah (1972) also shows that the separator plates normally used are thin enough that the heat will be conducted into the plates at a nearly uniform temperature when the engagement period is in excess of a few tenths of a second. Therefore, most of the cooling will take place after the engagement. It has been found that even air-cooled clutches and brakes will have better durability if cooling is provided for through air circulation. However, adequate clutch and brake durability will be obtained only if provision is made to remove the total heat in the same cycle it was generated.

A slipping clutch will generally generate heat at a nonuniform rate, making a precise calculation of the total heat generated difficult. A rigorous analysis of clutch temperatures can be made using mathematical methods such as those presented by Newcomb (1960, 1961) and Tataiah (1972) or finite element methods such as presented by Dundore (1968). However, these methods are complex and time-consuming, unless a computer program is available, and not really necessary, since a good estimate of the total heat generated can readily be calculated using estimates of the average torque and average relative slip velocity and the formula

$$Q = TNt/9.5493 \tag{5}$$

where Q = the heat generated in a slipping clutch, J (joules)
         T = the average slip torque, N · m
         N = the average slip, rpm
         t = the total slip time, seconds

When the heat generated in the clutch is known, the bulk or average temperature reached by the heat-absorbing elements of the clutch can be calculated from

$$\theta = Q/(m \, c_p) + \theta_0 \tag{6}$$

where θ = the bulk temperature, °C
         Q = the heat generated as defined previously, J
         m = the total mass of heat-absorbing elements, kg
         c_p = the specific heat of heat-absorbing elements, J/(kg °C)
         θ_0 = the bulk temperature before engagement, °C

The heat-absorbing elements of the clutch are directly adjacent to the slipping surfaces where the heat energy will quickly penetrate. Most nonmetallic friction materials are poor conductors of heat and, therefore, will not absorb a significant amount of the heat energy.

It is also important to limit the maximum rate at which energy is generated (W/mm²), as indicated in table 13-1. This limitation will hold for very short engagements at high facing pressures and high sliding velocity where the rate of heat generation can be extremely high. Although the maximum rate will occur at initial engagement at full pressure when the slip velocity is the highest, the average rate for the engagement generally should be used. The energy rate can be calculated from

$$E_r = T_a N_a / (9.5493 \; A) \tag{7}$$

where E_r = the energy rate, W/mm^2
         T_a = the average torque, N · m
         N_a = the average slip, rpm
         A = the total facing area, mm^2

Spur and Helical Gears

Most transmission and drive train gears transmit power between parallel shafts and are spur or helical (see fig. 13-5). Spur gears have teeth that are parallel to the axis of rotation of the gear, whereas the teeth of helical gears are at an angle to the axis of rotation. Spur gears are somewhat less expensive to manufacture than helical gears and often can be supported in the transmission much more simply and inexpensively because there is no axial thrust force as there is with helical gears. However, helical gears gradually transfer the tooth loads from one tooth to the next and generally generate much less noise.

An involute tooth profile is used. This profile is the form that is generated by a point on a string as the string unwraps about a cylinder (fig. 13-13). It is the preferred profile because constant rotational velocity is transmitted. Even though a perfect involute profile will transmit constant rotational velocity, the machining and heat-treating processes will leave some profile errors that will result in impact stresses and noise. Figures 13-14 and 13-15 show some of the terminology used to describe involute gears. The kinematic details of gears are presented more completely by Mabie and Ocvirk (1975).

Bevel Gears

Bevel gears are used to connect shafts with intersecting axes (fig. 13-16). Usually the shafts intersect at right angles, but there are many bevel gear applications that require shaft angles greater or less than a right angle. The pitch surface of a bevel gear forms a cone. Several types of bevel gears are available. The *straight bevel* gear is the bevel gear equivalent of the spur gear, with load transferring abruptly from tooth to tooth. Therefore, straight bevel gears are noisy at high power and high speed. *Spiral bevel* gears have curved teeth that are angled like helical gear teeth so that the load is gradually transferred from tooth to tooth. Therefore, spiral bevel gears will transmit high loads at high speed with much less noise. The *Zerol bevel* gear is a special type of spiral bevel with zero spiral angle. A *hypoid* gear is similar to the spiral bevel gear, but the shafts do not intersect. Hypoid pinions are stronger and

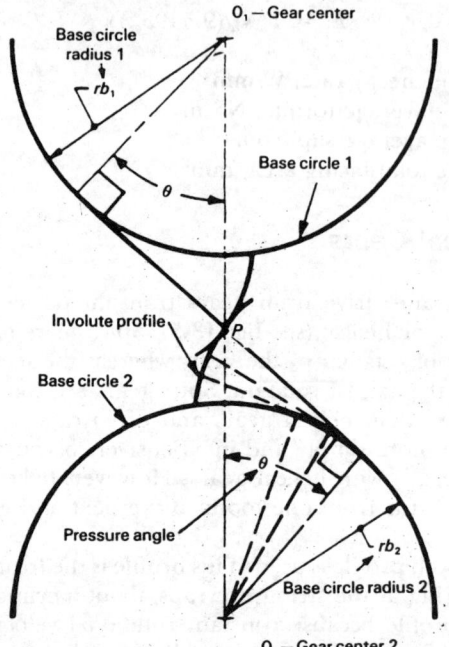

FIGURE 13-13   Involute tooth profile. (From Browning 1978.)

can be used for higher gear reductions than the spiral bevel gear, but they have more sliding between the teeth and are therefore somewhat less efficient. Hypoid gears are often used in automotive-style axles because the hypoid pinion offset and small size allow more interior clearance.

Gear Design

Gear design is an iterative process involving adjustment of the shaft center distance, the number of teeth in both gears, gear width, pressure angle, helix or spiral angle, and other parameters to optimize the gear set. A gear set is normally considered optimized when both the tooth bending and surface compressive stresses are near the maximum allowable, although special consideration may be given to noise control. Generally, the desired gear ratio is known so that when a nominal tooth size or module (pitch in English units) and center distance are selected, the basic gear size as represented by the

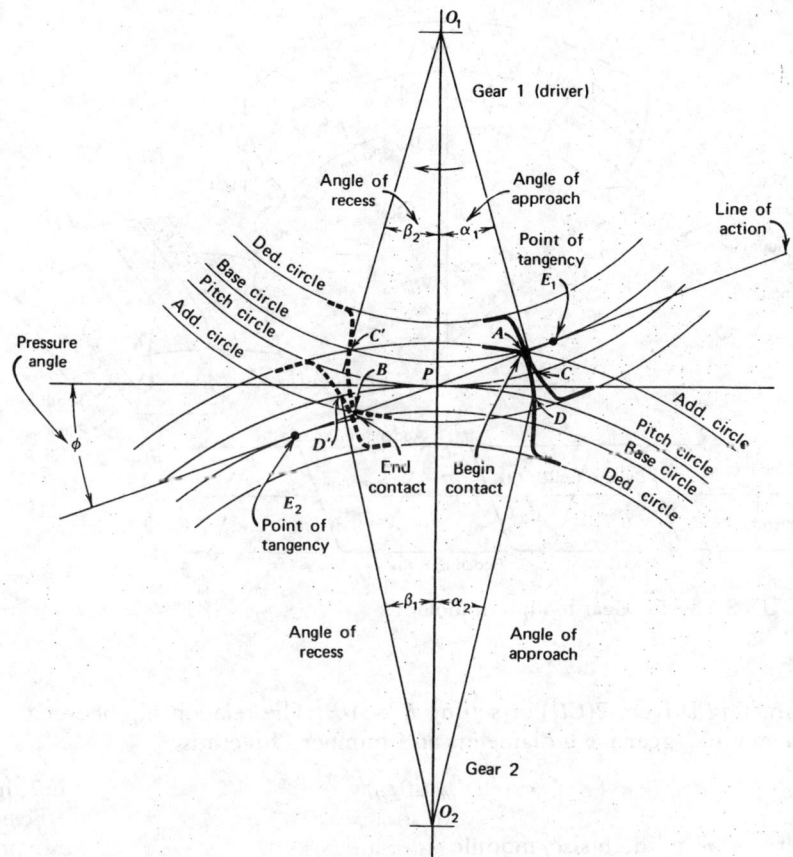

FIGURE 13-14 Gear-tooth pressure angle.

tangential rolling diameters called the pitch diameters and the number of teeth can be determined. The pitch diameter of the pinion, or smaller gear, for spur and helical gears is

$$D = 2(CD)/(1 + R) \qquad (8)$$

where D = the pitch diameter of the smaller gear
      CD = the shaft center distance
      R = the gear ratio greater than 1

It follows that the pitch diameter of the larger gear can be calculated by

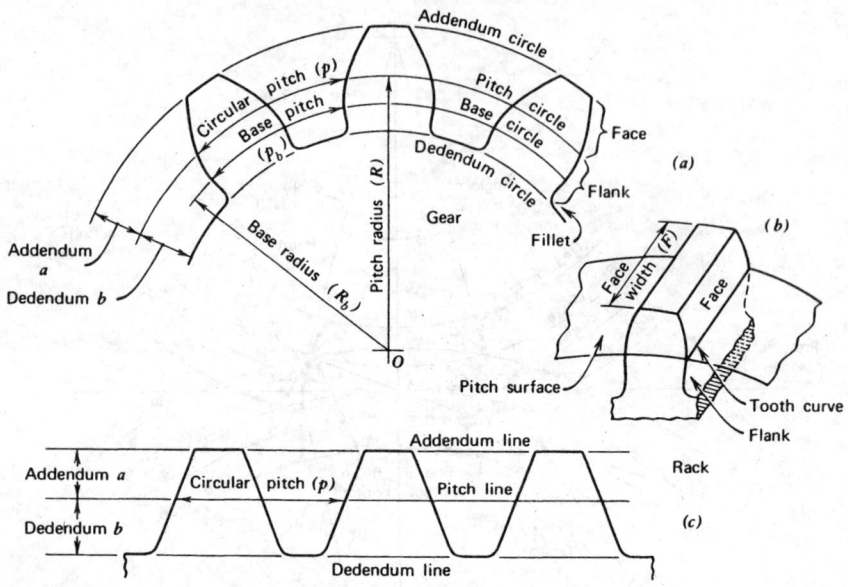

FIGURE 13-15   Gear-tooth terminology.

subtracting D from $2(CD)$ or setting $R = 1/R$. The relationship between the gear module, gear pitch diameter, and number of teeth is

$$m = D/N \tag{9}$$

where m = tooth size, module
D = nominal pitch diameter
N = number of teeth

For helical gears, m is the module in the transverse plane and is related to the normal or cutter module m_n by

$$m = m_n/\cos \psi \tag{10}$$

where ψ = the helix angle

The general range of tooth modules for various gears in the drive train is as follows:

1.   Transmission gears—4 to 5
2.   Power shift planetary gears—2.5 to 3.5

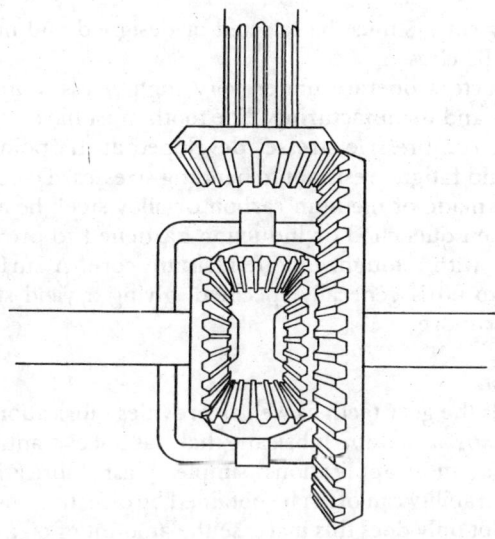

FIGURE 13-16 Conventional bevel gear differential.

3. Spiral bevel gears—8 to 12
4. Final drive gears—5 to 7

Usually cutters of nominal size and pressure angle are used, but often center distances and tooth numbers are chosen so that the *nominal* pitch diameters will not touch, as shown in figure 13-13. This only means that the operating or *working pitch diameter* is different from the *nominal pitch diameter* shown in equation 9. There is a corresponding *working pressure angle*, which is given by

$$\phi_w = \cos^{-1} \frac{m \ (N1 \ + \ N2) \cos \phi_t}{2 \ CD} \tag{11}$$

where ϕ_w = the working pressure angle
        m = the gear module in the transverse plane from equation 10
$N1, N2$ = the number of teeth in gear 1 and gear 2
    CD = shaft center distance
    ϕ_t = the transverse pressure angle = $\tan^{-1}$ (tan ϕ_n/cos ψ)
    ϕ_n = the nominal or cutter pressure angle

Gear geometry and stress calculations are too complex to be done manually now that computers are readily available. Even with computers doing the mathematical work, gear design is complex and involved and is often done by a specialist. For, in addition to gear geometry and stresses being calculated, hob and shaving cutters must be selected or designed and material and heat treatment must be chosen.

Gears in tractors operate under very high stresses and require high-quality materials and manufacturing. The tooth must have very hard surfaces to withstand the compressive stresses developed at the point of contact, yet must be tough and fatigue-resistant to bending stresses. To obtain these properties, gears are made of medium carbon or alloy steel, heat treated by carburizing, and then quenched or induction hardened to provide a very high strength surface with a tough, fatigue-resistant core. A surface hardness of Rockwell C 56 to 60 is generally specified, giving a yield strength of 2050 MPa (N/mm^2) or more.

Gear Lubrication

Lubrication cools the gear teeth as well as provides lubrication. Generally, the oil must be an *extreme pressure* lubricant that has special antifriction and antiweld agents. For most applications, simple splash lubrication is sufficient. But increased durability can often be obtained by directing cooling lubrication onto the gear. Not only does this increase the amount of oil available, but also the cooling action on the gear may increase the thickness of the oil film adhering to the gear tooth, which will decrease wear and surface fatigue. Because Hertzian stresses as high as 1200 MPa exist at the tooth contact, normal hydrodynamic lubrication does not exist. However, oil under extreme pressure between elastic surfaces has been found to increase spectacularly in viscosity so that it becomes like cold asphalt. This is known as elastohydrodynamic (EHD) lubrication and explains why gears and bearings under high contact stress can operate with very little sign of mating surface contact (Dudley 1965).

Gear Durability

Gears fail from bending fatigue and surface wear and fatigue (pitting). Bending fatigue is characterized by an endurance limit below which the gear will have infinite life. Schilke (1967) points out that the bending endurance limit is about one million cycles, which is less than 20 hours for gears running at a modest 1000 rpm. Thus, most tractor gears must be designed to bending stress levels less than the endurance limit.

In most cases, surface wear occurs when lubrication is inadequate or marginal. Slow-speed-tractor final drive gears running in low-viscosity oils will not develop an adequate oil film to completely prevent wear. However, the rate of wear can be controlled to yield adequate life by minimizing the surface loads and amount of surface sliding.

Surface pitting is a complex phenomenon that appears to be affected by the oil film thickness, amount and direction of sliding, surface finish, steel composition, heat treatment, and contact stress. These effects are discussed in some detail by Bowen (1977) and MacPherson (1977). In practice, an empirical method based on allowable stress levels or a known S/N curve is used to select the allowable surface compressive stress level.

Gear Efficiency
Gear mesh power losses are a major source of loss in transmissions. Radzimovsky (1956) shows how gear losses can be expected to vary from about 1 to 2 percent for external gears and 0.5 to 1.5 percent for internal gears, depending on the number of teeth in the pinion and the gear ratio. An average efficiency of 98.5 percent is often used for external gears and 99 percent for internal gears.

Gear Noise
Gear noise occurs when errors in the gear tooth profile and changes in tooth stiffness at the gear mesh create cyclic forces that are transmitted through the shafts and bearings to the housings. The cyclic forces cause the housing walls and surrounding air to vibrate to produce noise. Often, the frequency of these forces will correspond to a natural frequency of the housing, resulting in a particularly loud and objectionable noise level. Gear noise can be minimized by designing stiff housings with curved surfaces to minimize noise amplification and by providing uniform and low-stiffness gear teeth to minimize cyclic force generation. Spur and helical gears should be designed with a total contact ratio (the measure of amount of tooth contact) equal to an integer and with adequate tip relief. Finally, the gear errors (deviations from the ideal geometry) need to be minimized (Welbourn 1977).

Planetary Gear Systems

Figure 13-17 shows several types of planetary gear systems in which a cluster of gears, called planet gears, orbit about a center or sun gear. Planetary gear systems are often used in tractor drive trains to change gear ratios (figs. 13-17 and 13-18) and carry heavy loads. Because the gear loads are shared about equally at three or more meshes, the gear systems are much more compact than simple gear reductions. The simple planetary arrangement (figs. 13.17[a] and 13.19) is used as a final drive on most tractors over 75 kW owing to its high reduction, compactness, and high load capability. When used to change gear ratios in a power shift transmission, a planetary gear system also helps to provide a continuous flow of power during the shift. This is because the speed of the input and output planetary members cannot change without accelerating a third member, which provides a reaction torque and transmittal of power through the planetary system.

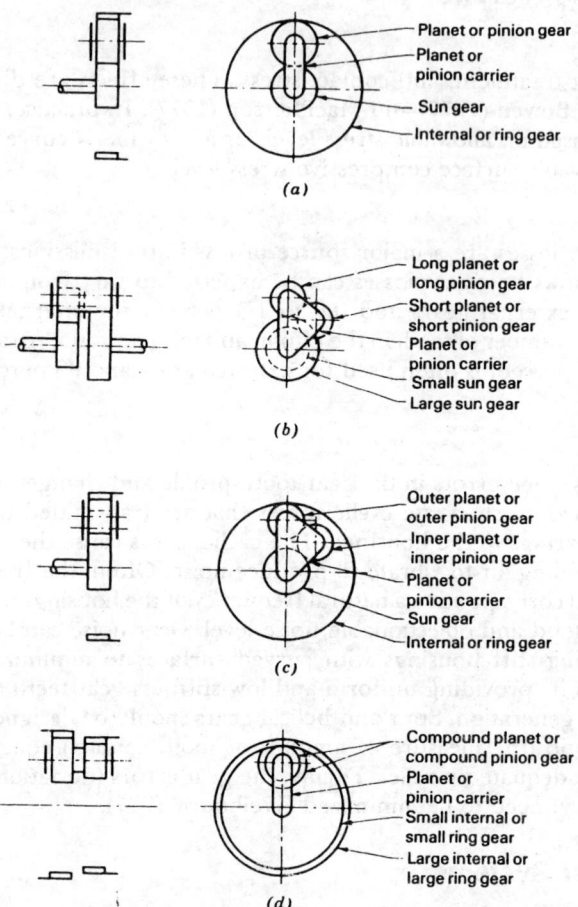

FIGURE 13-17 Planetary gear terminology and arrangments. (*a*) Simple planetary arrangement; (*b*) external planetary arrangement; (*c*) dual planetary arrangement; and (*d*) compound planetary arrangement. (From SAE J646 AUG83 "Planetary Gear Terminology.")

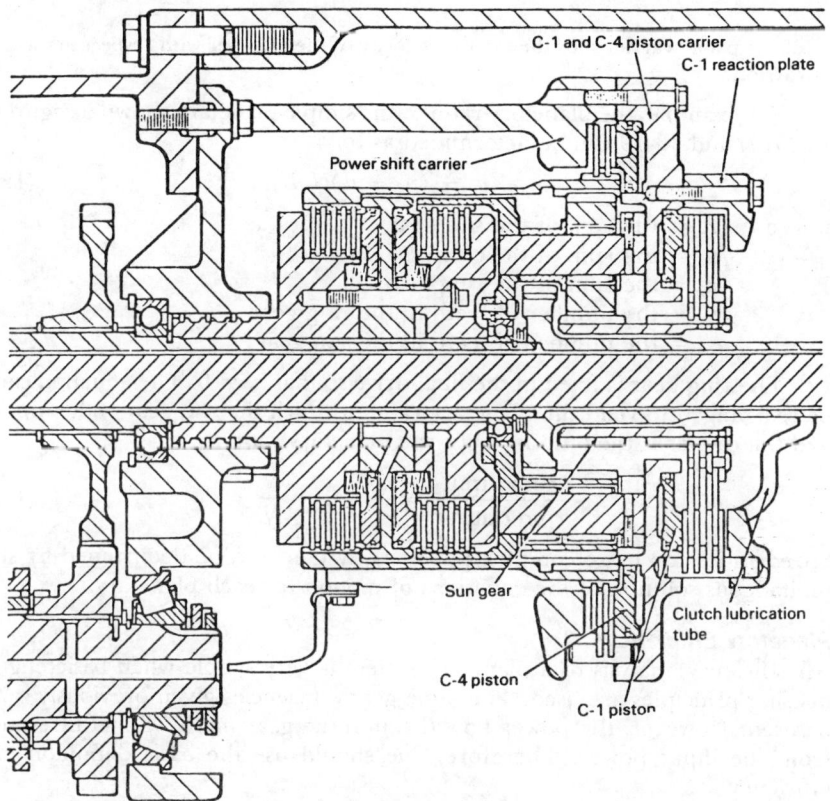

FIGURE 13-18 Planetary gear set with two clutches and two brakes to provide three forward speeds and one reverse speed. (Courtesy J. I. Case.)

Planetary design requires some special techniques to calculate gear ratios and efficiency and to ensure the gears will mesh properly. The large amount of literature available on planetary systems indicates that planetaries are considered challenging to design. However, only a few basic engineering principles of statics and kinematics are needed to design planetary systems.

Planetary Ratio

Engineers use many techniques (Mihal 1978; Hanson 1981) to determine the planetary ratio. A simple and effective method is to use the kinematic principle:

The absolute velocity of any planetary member is equal to the absolute velocity of the planet carrier plus the relative velocity of the member with respect to the carrier.

For example, the planetary ratio of the simple planetary shown in figures 13-17(a) and 13-19 can be determined as follows:

$$n_s = n_c - (n_r - n_c)N_r/N_s \qquad (12)$$

where
n_s = the rpm of the sun gear
n_c = the rpm of the carrier
n_r = the rpm of the ring gear
N_r = the number of teeth in the ring gear
N_s = the number of teeth in the sun gear

The ring gear is fixed in the final drive so that $n_r = 0$. The equation can be rearranged to yield the planetary ratio in which the sun gear is the input and the carrier is the output. Thus the planetary ratio is

$$\frac{\text{rpm input}}{\text{rpm output}} = \frac{n_s}{n_c} = 1 + \frac{N_r}{N_s} \qquad (13)$$

Speed ratios for more complex planetary systems can be determined by simultaneous solution of the equations of motion for each planetary.

Planetary Efficiency

An efficiency analysis of a planetary system is very simple when basic engineering principles are used. The same gear efficiencies given previously can be used. However, the power flow through the gear mesh will be different from the input power. Therefore, one should use the basic definition of efficiency

$$e = \frac{\text{power out}}{\text{power in}} \qquad (14)$$

where
power out = output torque × output rpm
power in = input torque × input rpm

Thus, for a planetary final drive with sun input, carrier output, and fixed ring gear

$$\text{Power out} = T_c n_c$$
$$\text{Power in} = T_s n_s$$

where T represents torque, n represents rpm, the subscript s represents the input sun gear, and c represents the output carrier as defined previously for equation 12. Thus, from equation 14, the efficiency equation is

$$e = \frac{\text{power out}}{\text{power in}} = \frac{T_c n_c}{T_s n_s} \qquad (15)$$

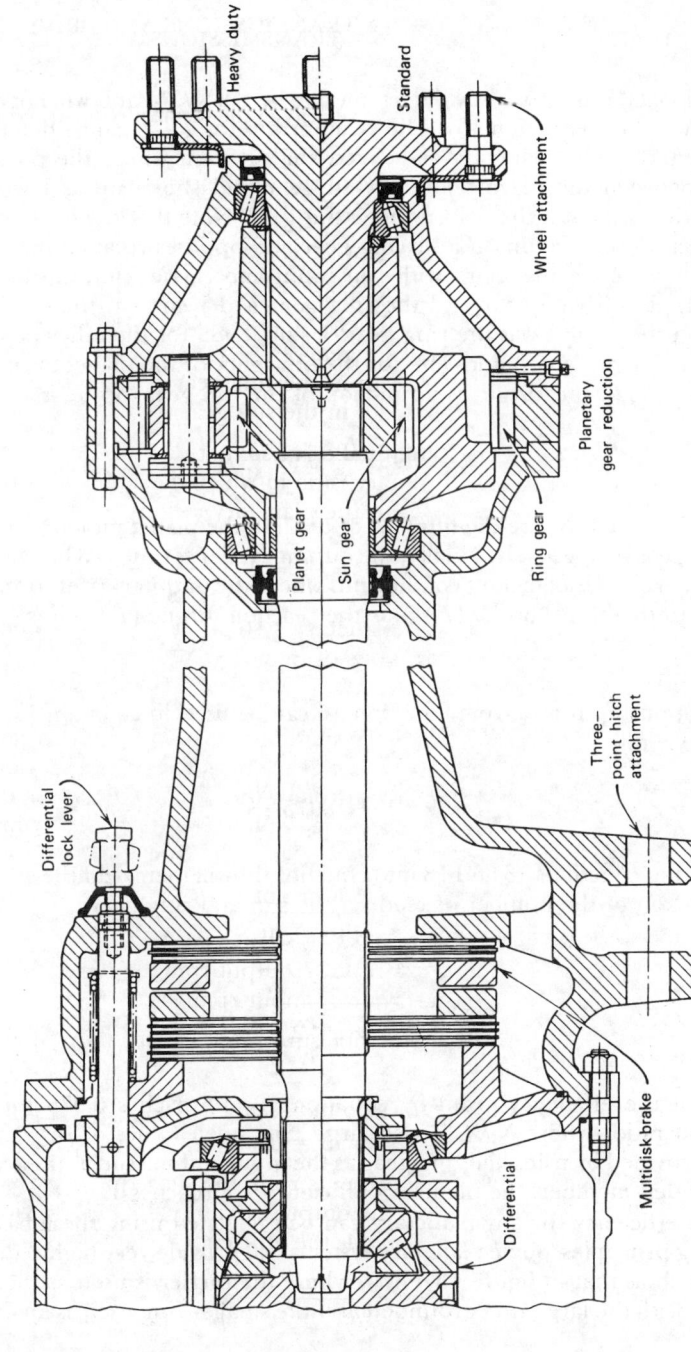

FIGURE 13-19 Cross-sectional view of the rear axle for a 60-kW tractor showing outboard planetary final drive and inboard multidisk brake. (Courtesy Massey-Ferguson Ltd.)

Heavy duty

Standard

Wheel attachment

Planetary gear reduction

Planet gear

Sun gear

Ring gear

Differential lock lever

Three—point hitch attachment

Differential

Multidisk brake

The ratio n_s/n_c was previously determined to be $1 + N_r/N_s$ and will not change with power loss (equation 13). The relationship for T_c/T_s can be determined by considering the planetary as a fixed shaft system where the power loss occurs between the driving and driven gear and shows up as lower than theoretical torque on the driven or output gear. Note that the torque acts in the direction of rotation on a driving gear and opposes rotation of a driven gear. For a simple planetary with sun input and carrier output, the sun is obviously the driving gear and the ring gear is driven as if the carrier was fixed and the output was the ring gear. Using 0.985 for the efficiency of the sun to planet gear mesh and 0.99 for the efficiency of the planet to ring gear mesh, the ring gear torque as a function of the sun gear torque is

$$T_r = 0.985 \, \frac{N_p}{N_s} \, 0.99 \, \frac{N_r}{N_p} T_s \tag{16}$$

where N_p, N_r, and N_s are the number of teeth in the planet pinion, ring gear, and sun gear, respectively. Finally, the output carrier torque can be calculated by considering the planetary equilibrium where the output carrier torque (T_c) is equal to the input torque (T_s) plus the reaction torque (T_r). Thus

$$T_c = T_s + T_r \tag{17}$$

The relationship for T_r from equation 16 can be used in equation 17, which will reduce to

$$\frac{T_c}{T_s} = 1 + 0.975 \, \frac{N_r}{N_s} \tag{18}$$

Substituting equations 13 and 18 into equation 15 will then yield the efficiency as a function of the number of teeth in the sun and ring gears

$$e = \frac{1 + 0.975 \, \dfrac{N_r}{N_s}}{1 + \dfrac{N_r}{N_s}} \tag{19}$$

Note that the efficiency given by equation 19 approaches 0.975 for large reduction ratios where N_r/N_s is also large. For small values of N_r/N_s, which can be provided by using the ring gear as the input and the sun as the reaction or grounded member, the planetary efficiency approaches 1.

The efficiency for any planetary can be calculated using these basic engineering principles. Some planetary systems will provide very high reduction ratios, such as that in figure 13-17(d), where the carrier is used as the input member with the large ring grounded and the smaller ring used as an output.

However, the gears must also carry high torque at high speeds so that the mesh power is high and the corresponding power losses are high.

Planetary Assembly

Only certain gear tooth combinations can be used in planetary gearing for proper assembly and function. A detailed analysis of planetary assembly is presented by Myers (1965). The tooth relationship for the simple planetary system shown in figure 13-17(a) is

$$\frac{N_s + N_r}{K} = \text{Integer} \tag{20}$$

where K is the number of equally spaced planets. The number of teeth in the planet pinion is not important for assembly, as long as it will mesh properly with both the sun and the ring gear. Compound planetary gear systems are much more complex, except when the number of teeth in all the sun and ring gears is divisible by the number of planet pinions. Simple inspection will show that proper assembly will be obtained when this is the case. Other combinations may be available that will assemble properly, but the correct indexing of the planet gears must be found by trial methods, as described by Myers.

Differentials

A differential such as shown in figure 13-16 is necessary to allow tractors to be steered. The differential assembly normally includes a large spiral bevel gear that drives the differential housing and the small bevel gears. The torque driving the larger output bevel gears must be the same on each side to maintain the smaller bevel gears in static equilibrium. Thus, the output gears can rotate at any speed as required for turning but will deliver equal torque to both wheels. Note that when turning occurs, the power delivered to the wheels will not be equal. When one wheel is locked by braking, all of the power is delivered to the opposite wheel.

Since the torque on both output gears must be the same, the torque is limited by the wheel having the least tractive capability. Thus most tractor differentials can be locked to deliver as much torque as possible to each wheel (fig. 13-20). The need for the differential is readily apparent, however, when steering is attempted with the differential lock engaged.

Differentials are really a special type of planetary gear system where bevel gears are used instead of spur or helical gears. Differentials can (and are) constructed from systems such as those in figure 13-17(b) and 13-17(c). Thus

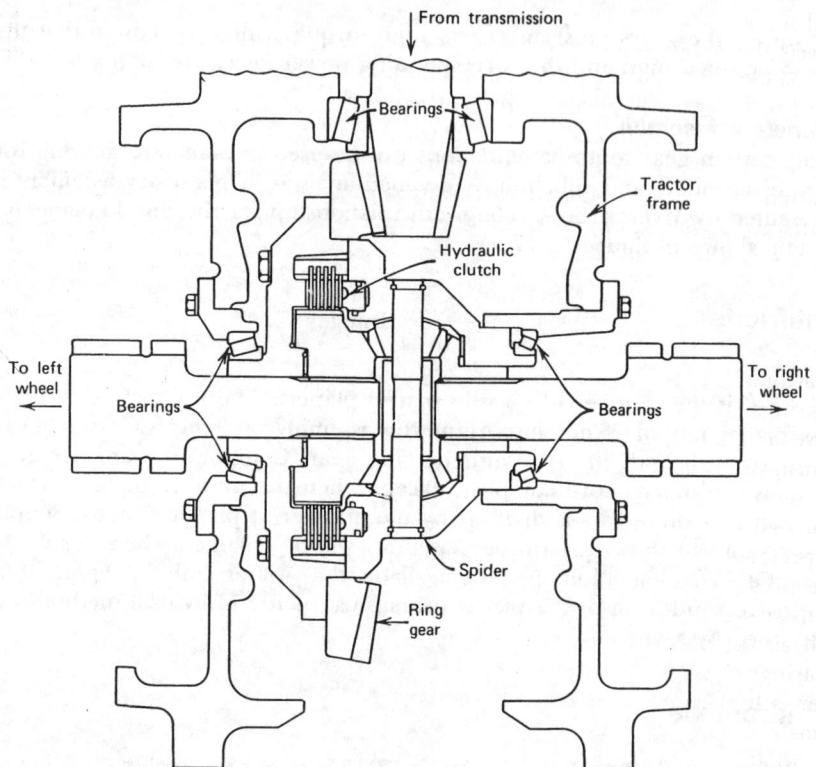

FIGURE 13-20   Differential with hydraulic lock. (Courtesy White Farm Equipment.)

the kinematic relationships given for planetary systems can also be applied to differentials.

Transmission Drive Shafts

Transmission drive shafts are used to transmit gear loads to the bearings and to transmit power from one component to another. These shafts are subjected to peak torques due to engine torque fluctuations, clutch engagements, power shifting, and shock loads, which can significantly exceed the nominal torques. In addition, the shafts often have sharp steps, holes, splines, and keyways that provide stress concentrations. Also, these shafts often carry combined bending and torsional loads. The designer must consider these factors as he or she selects the shaft size.

Since the highest stress for bending or torsion occurs at the surface and decreases uniformly to the center of the shaft, a high carbon steel is often used that is induction hardened to provide increased strength toward the surface. However, many shafts require an extremely hard surface for wear resistance, which is obtained by carburizing a lower carbon steel, but these shafts are not as strong as the higher carbon induction-hardened shafts.

Antifriction Bearings

Tractor transmissions use ball, straight roller, taper roller, and needle roller bearings. Ball bearings are slightly more efficient than roller bearings but have a much lower load capacity and are quite sensitive to debris from transmission gears and other sources. The shielded type of bearing should be used next to gears. Ball bearings are commonly used in the smaller tractors of European and Japanese manufacture. The taper roller bearing is preferred for most high load applications, especially where thrust loads are present from helical and bevel gears or other sources, but a more expensive mounting arrangement is usually required. Taper roller bearings will also tolerate the normal debris found in transmissions. Needle roller bearings have small-diameter rollers and are often used where the diameter is limited, such as between concentric shafts or in planetary gear systems.

Ball, straight roller, and needle bearings are manufactured in cage or full complement types. The full complement bearing has the annular space between the inner and outer race completely filled with balls or rollers with a minimum amount of running clearance between the rollers. The rollers can rub against adjacent rollers so that some frictional losses are generated. The full complement of rollers gives a high rated capacity for the given bearing size. The caged type of bearing has a cage to separate the rollers to minimize the roller friction. Although the rated capacity of the caged bearing is lower than for the full complement bearing, a caged bearing is more tolerant of poor operating conditions and will often have a longer actual life than the full complement bearing.

Extensive bearing testing under controlled conditions has established an empirical relationship for the life of bearings of the form

$$\text{Life} = \frac{C \, (R/P)^n}{N} \tag{21}$$

where   Life = expected life of 90 percent of the bearings, generally
                    called the $L10$ or $B10$ life
               C = a constant for the particular type of bearing used
               R = the capacity rating of the bearing
               P = the load applied to the bearing
               N = rotational speed, rpm
               n = 3.33 for most bearings

This relationship provides a convenient method for selecting bearings. However, lubrication and mounting conditions are very important factors and should be considered before the final bearing selection is made. Some bearing suppliers have analytical methods for considering these factors.

Generally, bearings are loaded in several gears with different loads and for different fractions of the total operating time. The *weighted bearing life* can be calculated by first calculating the life of the bearing for each operating condition. The weighted life can then be calculated from

$$\frac{1}{L10} = \frac{t_1}{L_1} + \frac{t_2}{L_2} \cdots \frac{t_n}{L_n} \tag{22}$$

where $L10$ = the weighted life for 90 percent of the bearings
        $t_1 \cdots t_n$ = the decimal fraction of operating time for life $L_1 \cdots L_n$

Drive Train Speeds and Loads

The speeds and loads used to design a tractor drive train are based on the manufacturer's historical experience with older tractors currently operating in the field. For adequate field durability, the drive train must be designed for most of the heavy operating conditions that will be encountered around the world. Most tractors, however, can be expected to be used many hours and at high loads in the 5- to 12-km/h speed range, with few hours and generally lighter loads in the lower and higher speed ranges. For this reason, an estimate of the fraction of time and load or loads for each gear is required so that the gears and bearings can be designed for adequate life.

Relative to the design of power transmission elements for agricultural tractors, the following ASAE standards and engineering data are useful: ASAE S203.10, S205.2, S207.10, S333.1, and D230.4.

Electronic Transmission Controls

Electronic transmission controls can easily be designed to provide very complex algorithms for shifting control, and these controls can also be combined with electronic engine and hydraulic system controls to provide improved performance or operating efficiency. Electronic controls can also be readily tailored to a variety of vehicle installation and operating needs. For these reasons, an increasing number of transmission manufacturers are providing electronic controls. As the cost of electronic controls decreases and reliability increases, more electronic controls will be used. A number of electronic control systems have been introduced on tractors in recent years, as discussed by Oliver and Bullard (1981) and Bischel (1984). Applications in other heavy-duty off-highway transmissions are presented by Jones and Harris (1979), Farber (1982), Chamberlain (1984), Boyer (1985), and Morris and Sorrells (1985).

Torsional Vibration

Torsional vibration problems of varying magnitude are frequently encountered in drive trains, since drive trains often consist of one or more concentrated masses, which are separated by flexible shafts and subjected to many sources of fluctuating torques. Where the concentrated masses are large, as in power shifting transmissions and remote-mounted transmissions, special design attention must be given to the sources of the fluctuating torques and to the vibration characteristics of the drive train. Hopkins (1972) describes some conditions in which vibration problems have occurred in agricultural machinery.

Most torsional vibration problems occur when there is a harmonic fluctuating torque with a frequency near a natural frequency of the drive train. Vibration analysis of a simple spring-mass torsional pendulum without damping shows that the vibration amplitude will be given by the equation

$$\theta = \frac{T}{(K - J\omega^2)} \tag{23}$$

where
θ = the vibration amplitude, radians
T = the amplitude of the harmonic fluctuating torque, N · m
K = the torsional spring rate, N · m/radian
J = the moment of inertia, kg · m^2
ω = the angular frequency of the fluctuating torque, radians/s

It can readily be seen that when ω approaches the *natural angular frequency*, $\sqrt{K/J}$, the vibration amplitude θ will increase dramatically. Thus, it is important to design the drive train so that all of the natural frequencies of the drive train are separated from the exciting frequencies.

A vibration analysis starts with the simple calculation of the system moments of inertia and shaft stiffnesses. The moment of inertia of a mass element with respect to an axis of rotation is defined as the product of the mass of the element and the square of the distance from the element to the axis of rotation. Thus for a mass with elements, dm, the moment of inertia is

$$J = \int r^2 \, dm \qquad (24)$$

where r is the distance from the element to the axis of rotation. Most drive train components are composed of annular sections concentric with the shaft, for which the solution to equation 24 gives the following equation for the moment of inertia

$$J = \rho L \, \frac{(D_o^4 - d_i^4)}{10.186} \qquad (25)$$

where ρ = density, kg/m³
          = 7830 for steel
          = 7090 for cast iron
          = 2710 for aluminum
          = 8080 for bronze
          = 1500 for rubber
     L = axial length, m
     D_o = the outside diameter, m
     d_i = the inside diameter, m

When the section is not concentric with the shaft, the moment of inertia about the axis of rotation is equal to the sum of the moment of inertia about the mass center, as given by equations 24 or 25, and the product of the mass times the square of the distance between the mass center and axis of rotation

$$J = J_o + md^2 \qquad (26)$$

where J_o = the moment of inertia about the mass center
     m = the mass of the section, kg
     d = the distance between the mass center and shaft axis, m

If the section is not circular, it is often easier to calculate the moment of inertia with respect to a plane than an axis. It can be shown that the moment of inertia of a body with respect to any axis is equal to the sum of its moments of inertia with respect to two perpendicular planes that intersect along the axis. Thus, the moment of inertia about the z axis is

$$J_z = \int (x^2 + y^2) \, dm = J_{yz} + J_{xz} \qquad (27)$$

The shaft stiffness can be determined from the equation

$$K = \pi G \, \frac{(D_o^4 - d_i^4)}{32 \, L} \qquad (28)$$

where D_o, d_i, and L are defined previously and G is the modulus of rigidity or shear modulus of elasticity, about 7.6×10^{10} N/m² for steel. For shafts with varying diameters, it can be shown that the reciprocals of the section stiffnesses can be added as with springs in series. Thus

$$\frac{1}{K} = \frac{1}{K_1} + \frac{1}{K_2} + \cdots + \frac{1}{K_n} \tag{29}$$

The axles and final drives form parallel systems of inertias and stiffnesses that can be added directly to obtain the total effective inertia and stiffness. Inertias and shaft stiffnesses must be transferred to a common shaft for analysis. Gear reductions reduce the effective inertia and shaft stiffness by the square of the gear ratio. Thus, in general, the largest effective inertias and stiffest shafts are at the transmission input. Vehicle mass can be converted into an equivalent torsional inertia at the axle shaft with the equation

$$J = mR^2 \tag{30}$$

where m = the vehicle mass, kg
          R = the tire rolling radius, m

Most of the time, drive trains can be reduced to only a few lumped inertias connected by flexible shafts. Small inertias connected by very stiff shafts to a large inertia can be added to the large inertia with negligible loss of accuracy, except where high-frequency vibrations must be considered. Hopkins (1972) shows an example in which there is only a 4 percent difference in the natural frequency calculation between a six mass model of a tractor power train and a simple torsional pendulum model fixed at the flywheel! The extent to which inertias can be lumped depends on experience, knowledge, and judgment of the frequencies that need to be considered and a deep understanding of the vibration characteristics of a multimass system.

The natural angular frequency of drive trains with three or less discrete masses can be calculated from the following equations

Single mass system $\omega_n^2 = K/J$ (31)
(torsional pendulum)

Two mass system $\omega_n^2 = \dfrac{(J_1 + J_2)K}{J_1 J_2}$ (32)

Three mass system $\omega_n^4 + b\,\omega_n^2 + c = 0$ (33)

where $b = -\left[\dfrac{K_1}{J_1 J_2} (J_1 + J_2) + \dfrac{K_2}{J_2 J_3} (J_2 + J_3) \right]$

          $c = K_1 K_2 \left(\dfrac{J_1 + J_2 + J_3}{J_1 J_2 J_3} \right)$

For more complex systems, Holzer table methods can be used to determine the natural angular frequencies as described by Myklestad (1956). The *natural frequency* (f_n) can be determined from the following equation

$$f_n = \omega_n/2\pi \tag{34}$$

The next step in the vibration analysis is to identify the sources of excitation. Williamson (1968) lists the following potential sources:

1. The inertia and gas torque harmonics from the engine
2. Clutch engagements
3. Second order from universal joints in drive shafts
4. Gear tooth meshing harmonics
5. Self-excited nonlinearities
6. Friction (or stick-slip)
7. Hydraulic pressure variations
8. Pulsations due to load variations
9. Surface irregularities through tires or wheels

The first source, owing to the internal combustion tractor engine, is a major source of harmonic torque excitation. The sum of the gas and inertia torque delivered to the engine flywheel results in a fluctuating torque corresponding to the firing frequency with an amplitude about twice the average torque and with many higher-order harmonics. The firing frequency (f) of a four-cycle engine is given by

$$f = Nn/120 \tag{35}$$

where N = the engine speed, rpm
         n = the number of cylinders

For a six-cylinder engine, this frequency is three times the engine rotational frequency and is often referred to as the third-order frequency. There will also be excitational torque fluctuations at multiples of the engine rotational frequency that may also excite the drive train. The other sources of excitation listed previously generally cause much lower torque amplitudes and usually do not cause vibration problems, but the potential for a vibration problem is also much harder to predict.

When the drive train has many natural frequencies, it is impossible to design the drive train so that all are outside of the operating range. It then is important to simulate the actual response of the drive train to some of the significant exciting torques and to detune the drive train so that the small amount of damping in the drive train will be sufficient to control vibration. Although small, there is still a considerable amount of damping in drive trains owing to the presence of lubrication, friction, nonlinearities, and backlash, which can be determined by correlation of the simulated vibration response with measurements. Williamson (1968) states that damping values vary from about 3 percent to 20 percent of critical, with low-frequency modes having the most damping.

One common problem is a "death rattle" that occurs when the engine is turned off. As the engine speed slows down below slow idle, the drive train may pass through one or more torsional natural frequencies that can result in large torsional oscillations and a very disturbing rattling noise. Normally, the drive train is detuned with a spring coupling (erroneously called a "damper") that is easily incorporated into the traction clutch disk. This device also may be used to control vibrations in the operating range.

Computer Simulation

Computer simulation of drive trains and components is used as often as possible to optimize the design before hardware is ordered so as to minimize development time and cost. When the simulation agrees with measured test results, the engineer can be confident that he or she understands the functional relationships involved. Some of the simpler simulation techniques were discussed in earlier sections on clutch, gear, and bearing design. More complex studies of stress, heat transfer, and dynamics require the use of integration techniques to find a solution. Several system modeling programs are available using various mathematical integration techniques for solution of such problems.

Various dynamic modeling techniques are also presented by Koch (1972) on a power train-vehicle model to simulate shifting transients using an analog computer, and by Kar (1982) on simulation of the torque characteristics in drive lines using a simple trapezoidal integration technique. The computer can also be used for other types of simulation, such as presented by Dundore (1968) to predict clutch operating temperatures and by Martin (1969) to predict the performance of hydrostatic and hydromechanical transmissions.

PROBLEMS

1. (a) Determine the power train efficiency of two tractors recently tested at Nebraska. The pto efficiency of tractor drive trains is usually in the range of 92 percent to 95 percent, depending on parasitic transmission losses and the number of gear meshes in the pto drive train. The engine flywheel power delivered to the transmission will therefore be equal to the maximum pto power (rated engine speed—two hours) divided by the efficiency. The efficiency of the drive train, including traction, is equal to the drawbar power (maximum power in selected gears) divided by the engine

flywheel power. (b) Estimate the drive train efficiency by accounting for the losses due
to tire slip and rolling resistance. Assume rolling resistance on concrete is equal to 2
percent of the tractor weight.

2. (a) On a sheet of graph paper, plot a pull-speed curve from a selected Nebraska
tractor test using the *maximum power with ballast*. (b) On the same graph paper, plot
the pull-speed curve from the *varying drawbar pull and travel speed with ballast*. (c) What
is the relationship between the engine torque curve and the Nebraska test *varying
drawbar pull and travel speed with ballast?*

3. A tractor with a 165-cm rear wheel tread width has a center of gravity 97 cm
above the roadway and a mass of 2600 kg. While the tractor is pulling a plow, 70
percent of the tractor mass is supported on the rear wheels. (a) If the right rear wheel
operates in a furrow 15 cm deep, estimate the percent increase in pull obtainable from
a differential lock (assume the coefficient of traction is the same for both wheels). (b)
Repeat part (a), but with the coefficient of traction equal to 0.36 for the left rear wheel
and 0.55 for the right rear wheel. In both (a) and (b), you may neglect any change
in weight transfer from the front wheels.
Hint: With the differential unlocked, both rear wheels transmit equal torque; with the
differential locked, both rear axles turn at the same speed.

4. A tractor with a wheelbase of 195 cm has a tread width (both front and rear
wheels) of 165 cm. If the tractor is negotiating a 5-m (measured from center of turning
along rear axle center line to point midway between rear wheels) turn at 5 km/h and
the rolling radius of the rear wheels is 70 cm, (a) calculate the speed (in rpm) of the
inside rear wheel and the outside rear wheel. (b) Assuming no side skidding, calculate
the steer angle (i.e., angle through which axle centerline rotates when going into the
turn) of the inside front wheel and the outside front wheel. (c) During the turn, what
percentage of the engine power is transmitted through each of the two rear wheels?

5. The Nebraska test data for a tractor show the maximum pto power was 140 kW
at a fuel consumption of 40 L/h. Calculate the overall engine to pto efficiency. The
fuel has a heat value of 44,000 kJ/kg and weighs 0.836 kg/L.

6. Tests show that a tractor drive train with 1.2 square meters of surface area has
an efficiency of 92 percent. The heat dissipated from the surface is 0.05 kW/(m^2 °C),
and the transmission operating temperature is to be limited to 60°C above ambient.
Determine the maximum power that can be transmitted by the drive train without
external cooling.

7. A tractor drive train must provide a minimum ground speed of 2.5 km/h at an
engine speed of 2200 rpm and with a tire rolling radius of 820 mm. If the maximum
final drive ratio available is 6 to 1 and all of the other gear reductions are limited to
a maximum of 5 to 1, how many gear reductions, including the final drive, are necessary
to attain the reduction necessary for first gear?

8. Calculate the drive train efficiency, considering gear losses only, for a tractor
drive train with a 5 to 1 simple planetary final drive, spiral bevel differential drive,
and two helical gear reductions. Assume the gear efficiency is 98.5 percent for external
gear meshes and 99 percent for internal-external gear meshes.

9. A brake is being designed to stop an 8000-kg tractor from a speed of 30 km/h. Determine the effective mass of steel required to absorb the heat generated by braking with a maximum temperature rise of 250°C (the specific heat of steel is 468 J/kg °C).

10. (a) A 25-degree helical gear set with a 17-tooth pinion and 35-tooth gear is designed to operate at a center distance of 120 mm. Determine the operating pitch diameters of the two gears. (b) A five-module cutter with a 20-degree nominal pressure angle is used to manufacture the gears. Determine the working transverse pressure angle.

11. A powershift planetary uses the planetary arrangement shown in figure 13-17(b). The input sun has 39 teeth, the ring gear has 81 teeth, and the carrier is the output member. Determine the input-to-output ratio and efficiency.

12. A straight roller bearing with a rated capacity of 20,000 newtons is used to carry a load of 25,000 newtons at 1000 rpm for 20 percent of the time, a load of 10,000 newtons at 2500 rpm for 40 percent of the time, and no load the remaining time. The bearing constant is 1.5×10^6. Determine the weighted $L10$ life for the bearing.

13. (a) Assume the vibration characteristics of a tractor drive can be modeled with a three mass–two shaft system in which the engine flywheel torsional inertia is 2.5 kg · m², the effective torsional inertia of the transmission is 0.07 kg · m², and the tractor mass of 6500 kg is effectively coupled to the drive train through the rear axles and tires. A hollow drive shaft with an outside diameter of 45 mm, an inside diameter of 30 mm, and a length of 300 mm connects the transmission to the engine. The tractor is equipped with dual rear tires with a torsional stiffness of 210,000 N · m/rad for each tire and has 100-mm-diameter axle shafts that have an effective length of 600 mm. The drive train reduction ratio is 25 to 1, and the tire rolling radius is 800 mm. Determine the two natural frequencies.
(b) Calculate the natural frequency of a two mass system consisting of the engine and transmission inertias only.
(c) Determine the corresponding engine speed at which the firing frequency of a six-cylinder, four-cycle engine would equal the natural frequencies determined in (a).

REFERENCES

Asmus, R. W., and W. R. Borghoff. "Hydrostatic Transmissions in Farm and Light Industrial Tractors." SAE Paper 680570, 1968.

Best, W. A. "John Deere Power Trains—Quad-Range and Perma-Clutch." SAE Paper 720795, 1972.

Bischel, B. J. "The 4994 Tractor Steering and Transmission Control System." SAE Paper 841153, 1984.

Bowen, C. W. "The Practical Significance of Designing to Gear Pitting Fatigue Life Criteria." ASME Paper 77-DET-122, 1977.

Boyer, R. C. "Digital Electronic Controls for Detroit Diesel Allison Off-Highway Transmissions." SAE Paper 850782, 1985.

Browning, E. P. "Design of Agricultural Tractor Transmission Elements." ASAE Distinguished Lecture Series, Tractor Design, no. 4, winter meeting of ASAE, 1978.

Chamberlain, G. "Hi-Tech Controls Give Earthmovers Competitive Edge." *Design News*, May 1984, pp. 58-62.

Deere & Co., *Power Trains, Fundamentals of Service*, 5th ed. John Deere Service Publications, Moline, IL, 1986.

Dickinson, T. W. "Development of a Variable Speed Transmission for Light Tractors." SAE Paper 770749, 1977.

Drislane, E. W. "Cermetalix Clutch Material." SAE Paper 650683, 1965.

Dudley, D. W. "Elastohydrodynamic Behavior Observed in Gear Tooth Action." *Elastohydrodynamic Lubrication*. Proceedings of the Institution of Mechanical Engineers, vol. 180, Pt. 3B, 1965, pp. 206–214.

Dundore, M. W. "Clutch Energy—A Criteria of Thermal Failure." SAE Paper 680582, 1968.

Elfes, L. E. "Tractor Transmission with On-the-Go Shifts." SAE Paper 391A, 1961.

Erwin, R. L., and C. T. O'Harrow. "A New Tractor Transmission." ASAE Paper 58-607, 1958.

Farber, A. S. "Electronic Transmission Controls for Off-Highway Applications." SAE Paper 820920, 1982.

Graham, G. R. "Friction Materials for Heavy Duty Oil Cooled Clutches." SAE Paper 700741, 1970.

Haight, R. E. "The John Deere 15-Speed Power Shift Transmission." SAE Paper 821063, 1982.

Hanson, R. A. "Calculation of Planetary Gear Speeds Using a Building Block Approach." SAE Paper 810992, 1981.

Harris, K. J., and J. K. Jensen. "John Deere Power Shift Transmission." SAE Paper 739A, 1963.

Haviland, M. L., et al. "Surface Temperatures and Friction in Lubricated Clutches." SAE Paper 642B, 1963.

Hoepfl, J. R., and G. M. Ballendux. "Allis-Chalmers Power Shift Transmission—A New Option for the Models 7040 and 7060 Agricultural Tractors." SAE Paper 750858, 1975.

Hopkins, R. B. "Torsional Vibrations in Agricultural Machinery." ASAE Paper 72-555, 1972.

Hunck, G. "Hydrostatic Power-Splitting Transmissions for Wheeled Tractors—Application and Observed Performance." SAE Paper 680380, 1968.

Jones, W. R., and R. W. Harris. "Expanding the Versatility of the Powershift Transmission by Incorporating Electrohydraulic Control Systems." SAE Paper 790884, 1979.

Kar, M. K. "Simulation of Torque Characteristics in Drivelines with Universal Joints." SAE Paper 821027, 1982.

Kemper, I., and L. Elfes. "A Continuously Variable Traction Drive for Heavy-Duty Agricultural and Industrial Applications." SAE Paper 810948, 1981.

Koch, L. G. "Power Train-Vehicle Modeling to Simulate Shifting Transients of Off Highway Vehicles." SAE Paper 720044, 1972.

Kraus, J. H. "An Automotive CVT." *Mechanical Eng'g,* Oct. 1976, p. 38.

Kress, J. H. "Hydrostatic Power-Splitting Transmissions for Wheeled Vehicles, Classification and Theory of Operation." SAE Paper 680549, 1968.

Lemke, J., and J. C. Rigney. "The Case RPS 34 Power Shift Transmission and Its Controls." SAE Paper 700740, 1970.

Mabie, H. H., and F. W. Ocvirk. *Mechanisms and Dynamics of Machinery,* 3d ed. John Wiley & Sons, New York, 1975.

MacDonald, J. G. "Hydromechanical Transmissions as Applied/to Mobile Equipment." SAE Paper 690358, 1969.

MacPherson, P. B. "The Pitting Performance of Hardened Steels." ASME Paper 77-DET-39, 1977.

Martin, L. S. "The Development of a Digital Computer Program for Analyzing the Performance of Hydrostatic and Hydro-Mechanical Transmissions." SAE Paper 690566, 1969.

Mihal, D. "Planetary Speeds Made Easy—A Practical Method." SAE Paper 780785, 1978.

Morris, H. C., and G. K. Sorrells. "Electronic/Hydraulic Transmission Control System for Off-Highway Hauling Vehicles." SAE Paper 850783, 1985.

Myers, W. I. "Compound Planetaries." *Machine Design,* vol. 37, no. 20, Sept. 1965, p. 134.

Myklestad, N. O. *Fundamentals of Vibration Analysis.* McGraw-Hill Book Company, New York, 1956.

Newcomb, T. P. "Temperatures Reached in Friction Clutch Transmissions." *J. Mech. Eng'g Sci.,* vol. 2, no. 4, 1960, pp. 273–287.

Newcomb, T. P. "Calculation of Surface Temperatures Reached in Clutches When the Torque Varies with Time." *J. Mech. Eng'g Sci.,* vol. 3, no. 4, 1961, pp. 340–347.

Oliver, R. J., and R. N. Bullard. "Productive and Reliable Tractors for the 80's from International Harvester." SAE Paper 810912, 1981.

Ponzio, L. "The New Fiat Model 1300 Tractor—Performance, Safety, Comfort." SAE Paper 730821, 1973.

Radzimovsky, E. I. "A Simplified Approach for Determining Power Losses and Efficiency of Planetary Gear Drives." *Machine Design,* vol. 28, no. 3, Feb. 1956, p. 101.

Renius, K. T. "European Tractor Transmission Design Concepts." ASAE Paper 76-1526, 1976.

Ronayne, R. J., and R. E. Prunty. "Oliver Over/Under Hydraul Drive." SAE Paper 680569, 1968.

Ross, W. A. "Designing a Hydromechanical Transmission for Heavy Duty Trucks." SAE Paper 720725, 1972.

Schilke, W. E. "The Reliability Evaluation of Transmission Gears." SAE Paper 670725, 1967.

Scott, D., and J. Yamaguchi. "CVTs Loom on European Horizon." *Automotive Eng'g,* vol. 91, no. 12, Dec. 1983, p. 57.

Spokas, R. B. "A Wet Clutch for Farm Tractors." SAE Paper 680568, 1968.

Tataiah, K. "An Analysis of Automatic Transmission Clutch-Plate Temperatures." SAE Paper 720287, 1972.

Welbourn, D. B. "Gear Noise Spectra—A Rational Explanation." ASME Paper 77-DET-38, 1977.

Williamson, S. O. "Vehicle Drive-Line Dynamics." SAE Paper 680584, 1968.

Wise, W. R., and J. R. Mitchell. "Gylon Fluorocarbons: New High-Performance Friction Materials for High-Speed, Heavy Duty Wet Clutches." SAE Paper 720365, 1972.

SUGGESTED READINGS

Renius, K. T. "New Developments in Tractor Transmissions." *Grundlagen der Landtechnik,* vol. 34, no. 3, 1984, pp. 132–142.

14

TRACTOR TESTS AND PERFORMANCE

Tractor Performance Criteria

The performance of farm tractors can be expressed in many ways. The criterion that best describes the performance depends largely upon the intended use of the tractor.

Farmers commonly refer to a tractor's size in terms of the number of plows it will pull. Although this system has some obvious weaknesses, it nevertheless is easily understood. The advertising literature on wheeled farm tractors usually states the number of plows each tractor will pull under average to good conditions.

The maximum drawbar pull is often used in comparing or evaluating tractors. Unfortunately, drawbar pull is seriously affected by the soil or test track conditions and also by the gear ratio and the ballast being carried. Since power is a function of both velocity and drawbar pull, it is obvious that the latter only partly describes the ability of a tractor to do work.

Maximum drawbar power is normally the most useful performance criterion for farm tractors.

For the farmer who uses a tractor extensively on machines requiring a pto drive, the criterion of most interest is the maximum pto power that can be developed.

Before 1959, corrected maximum power values were determined by correcting the maximum observed belt (or pto) and drawbar power to standard conditions.* The corrected drawbar power was then multiplied by 0.75 and the belt power by 0.85 to arrive at rated drawbar and rated belt power.

It was assumed that a tractor could continuously maintain its rated power and, therefore, that rated power was a realistic basis for comparison. However, the concept of a "rated power" was more meaningful to engineers than it was to the users and sellers of tractors. Beginning in 1959, the ASAE and the SAE Tractor Test Code and also the Nebraska tractor tests were changed so that corrected and rated hp values are no longer computed, although the atmospheric temperatures and pressures are published.

In 1971 the Nebraska tractor tests began reporting the sound pressure level in dB(A) at the operator's ear level. Beginning with the Nebraska Tractor Test No. 1218 in 1976, all data were reported in metric (SI) units as well as in customary units. In 1977 drawbar performance with radial tires provided supplemental information in addition to bias tire test data. The Nebraska Tractor Test Lab in 1978 started to test four-wheel-drive tractor models that were derived from two-wheel-drive tractors by the addition of a front-wheel-drive assist (FWA). In 1983 performance for tractors claiming "constant" power over a speed range was started. Three-point hydraulic lift tests were first reported in 1984.

The fuel consumption is an important criterion that can be used to indicate directly or indirectly the efficiency of the tractor.

The torque curve, or "lugging ability" in layperson's language, is a way of measuring the stability or pulling ability of an engine as the engine is slowed down because of increased load. For tractors the drawbar pull versus speed, for a single gear and open throttle, is the most useful method of interpretation, since this method also considers the effect of the transmission and traction.

The previously mentioned criteria plus the methods by which they are determined are discussed in the remainder of this chapter.

Power Measurement Methods

To understand power and its measurement, certain terms must be defined and clearly understood.

> *Power* Power is a rate of doing work. A unit of power is a newton meter per second (watt).
> *Kilowatt* A unit of power equal to 1000 N m of work per second.
> *Brake power* The power output of the engine crankshaft. The engine may be stripped of part or all of its accessories.
> *Power-takeoff power* The power delivered by a tractor through its pto shaft.

*Standard atmospheric conditions were temperature, 15.5°C, and pressure, 1.013×10^5 Pa.

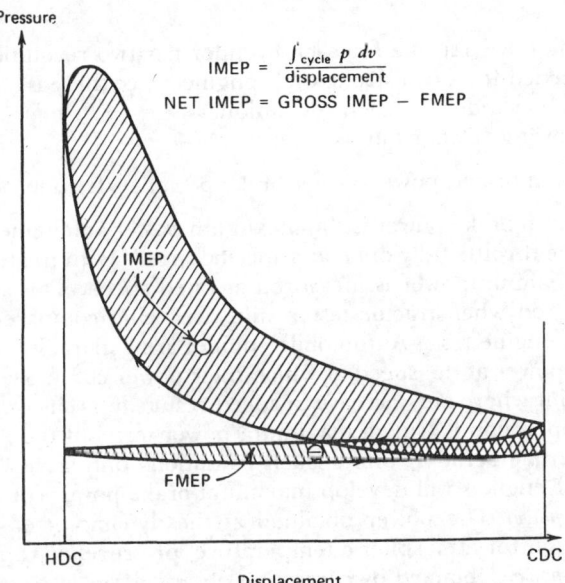

FIGURE 14-1 Definition of the indicated work on a P-V diagram.

Drawbar power The power of a tractor measured at the end of the drawbar.

Friction power The power required to run the engine at any given speed without production of useful work. It is usually measured with a suitable electric dynamometer that runs or "motors" the engine. It represents the friction and pumping losses of an engine.

Indicated power The power calculated from the $\int P\,dV$ (see fig. 14-1) represents the indicated power available at the piston and is calculated as follows:

$$ip = \frac{PLANn}{60 \times 10^{12}} \tag{1}$$

where ip = indicated power, kW
P = mean effective pressure, Pa
L = length of the stroke, mm
A = area of piston, mm^2
N = speed, rpm
n = number of cylinders

For a four-stroke-cycle engine, N must be divided by 2, since there is but one power stroke for each cylinder per two revolutions. The 2 is not needed for a two-stroke-cycle engine, since there is a power stroke for each cylinder for each revolution.

The following relationship is true:

Gross indicated power = net brake power + friction power   (2)

Maximum brake power is the maximum power an engine will develop with the throttle fully open at a specific speed. With the tractor engine the maximum power is measured at rated speed. This should be remembered when tractor test results are being compared with automotive engine tests. Automobile engines are generally rated for maximum power at the speed at which their power curve peaks, that is, at the point where any increase in speed results in no increase in power. Such operation is truly a maximum power test, but the engine would remain in operation under such conditions only a very short time. Tractor engines will develop maximum brake power continuously.

Observed power The power obtained at the dynamometer without any correction for atmospheric temperature, pressure, or vapor pressure.

Corrected power Obtained by correcting observed power to standard conditions* of sea-level pressure (1.013×10^5 Pa), 15.5°C temperature, and zero vapor pressure.

Kilowatt-hour One kilowatt working for one hour. It is 3.6×10^6 joules of work.

Dynamometer An instrument for determining power, usually by the independent measure of force, time, and the distance through which the force is moved. Dynamometers may be classified as brake, drawbar, or torsion, according to the manner in which the work is being applied. Also, they may be classed as absorption or transmission, depending on the disposition of the energy.

Absorption Dynamometers

Prony Brake

An absorption dynamometer measures the power applied and at the same time converts it to some other form of energy, usually heat. A Prony brake, the most elementary form of absorption dynamometer, is sketched in figure 14-2. The brake is essentially an arrangement whereby wooden blocks a can be clamped more or less tightly around the engine pulley b by means of the handwheel c. When the engine turns in the direction shown by the arrow, the lever arm d presses upon the scale e, causing a reading, F, on the scale.

*Standard conditions listed previously are those formerly used by the Nebraska tractor tests. SAE J816 standard conditions are slightly different.

—

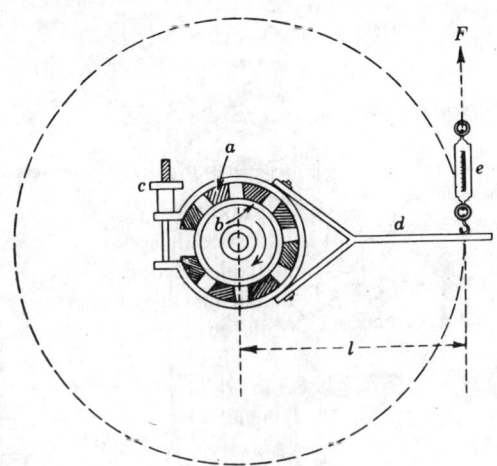

FIGURE 14-2   Prony brake.

A clear conception of the principle of operation of the Prony brake is essential for an understanding of nearly all dynamometers. Assume that the wheel b is locked and that some friction is applied by means of the handwheel c. Then let a force F be applied to the arm d and let the arm be rotated for one revolution along the path of the dotted line. The work done per one revolution is then F times a distance of $2\pi\ell$. Now assume that the arm d is fixed and that the wheel b is rotated one revolution inside the blocks. The work done in either case is necessary to overcome the friction between the blocks and the wheel; hence the work in the second case is again $2\pi\ell F$. Now if the wheel is rotated n times in one minute, the work accomplished will be $2\pi\ell Fn$, or

$$\text{Power} = \frac{2\pi\ell Fn}{60,000}, \text{ kW} \tag{3}$$

where ℓ = meters
        F = newtons
        n = revolutions per minute

Torque is measured by the Prony brake and similar dynamometers. In the preceding equation, torque is equal to Fl.

$$\text{Power} = \frac{2\pi nT}{60,000}, \text{ kW} \tag{4}$$

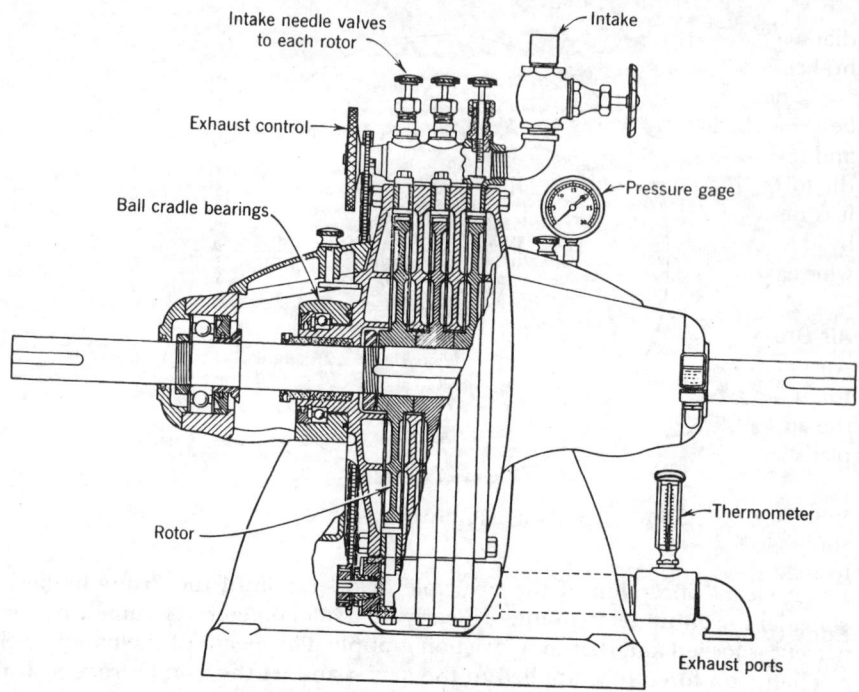

FIGURE 14-3   Hydraulic dynamometer. (Courtesy Taylor Dynamometer and Machine Co.)

where T = torque in newton meters

The Prony brake is not entirely suitable for power versus speed determinations of an internal combustion engine since the torque versus speed curves of the brake and engine are approximately the same. Therefore, speed control may be difficult.

Hydraulic

The hydraulic dynamometer (fig. 14-3) also operates on the principle of converting work into heat. The working medium, usually water, is circulated within the housing and because of friction comes out at a higher temperature than when it entered. The outer case, which is free to rotate about the shaft, is connected to and restrained by the torque arm. With the exception of bearing friction, the torque produced is equal to that supplied to the dynamometer. The power-absorbing capacity for any given type of design varies approximately as the cube of the speed of rotation and the fifth power of the diameter (Culver 1937). The power equation is the same as for the Prony brake.

The accuracy of the hydraulic-brake dynamometer can be expected to be somewhat better than that of the Prony brake and should fall between it and the cradled electric type. An error of about 0.5 percent is probable. Since the torque on a hydraulic dynamometer increases with the cube of the speed, it is desirable for handling open-throttle tests where the speed is controlled by the load. Also, the danger of an engine "running away" is eliminated, whereas this is always a danger with the Prony brake.

Air Brake

Air or fan brakes are useful only in loading engines for run-in purposes and for rough testing at relatively high speeds. The power that is transmitted to the air by the fan depends on the size of the plates, the distance at which the plates are located from the center of rotation, and the cube of the rpm. The probable error of the fan brake may be as much as 20 percent (Marks 1958), since the brake is affected by the temperature and pressure of the air. Brakes should be individually calibrated. To obtain the power, it is then only necessary to read the rpm of the fan.

Eddy-Current

An eddy-current dynamometer consists essentially of a rotor operating in connection with a stator. The rotor is a solid steel casting with tapered teeth projecting in the form of poles. The stator, which is magnetically coupled as to torque with the rotor, is equipped with coils for direct field excitation and is mounted in bearings, thus permitting it to rotate.

Because the eddy-current unit is an absorption dynamometer, a coolant is circulated to carry away the heat generated in the machine. Torque is adjusted by controlling the field excitation, this field excitation being only a fraction as large as the total power input to the machine.

Electric Direct-Current Dynamometers

The direct-current cradle-mounted dynamometer (fig. 14-4) is a shunt-wound generator with separate field excitation. The field frame is free to rotate, and since any effort to turn the armature causes the field to attempt to revolve, the resultant torque will cause a force to be registered on the scales. The accuracy is independent of the electrical efficiency of the machine. Accuracy within 0.25 percent is possible. This type of electric dynamometer can usually be arranged to operate as a motor. In either arrangement, the power put into or developed by the unit is

FIGURE 14-4   DC electric dynamometer. (Courtesy General Electric Co.)

$$\text{Power} = \frac{2\pi \ell F n}{60,000}, \text{kW} \tag{5}$$

which is the same as for the Prony brake (equation 3).

Shop-Type Dynamometers

It is often desirable to measure the pto power of a tractor in the field or in an implement dealer's repair shop. Since dynamometers, such as are shown in figure 14-3 or 14-4, are much too expensive and difficult to use except in laboratories, there have been developed several inexpensive and portable devices generally classified as shop-type or agricultural dynamometers. An example is shown in figure 14-5. This type of dynamometer is used primarily as an indicator of the condition of the engine. It is also used in the process of adjusting or tuning an engine and in indicating to customers the improvement in a tractor engine as a result of an overhaul, maintenance, or adjustment.

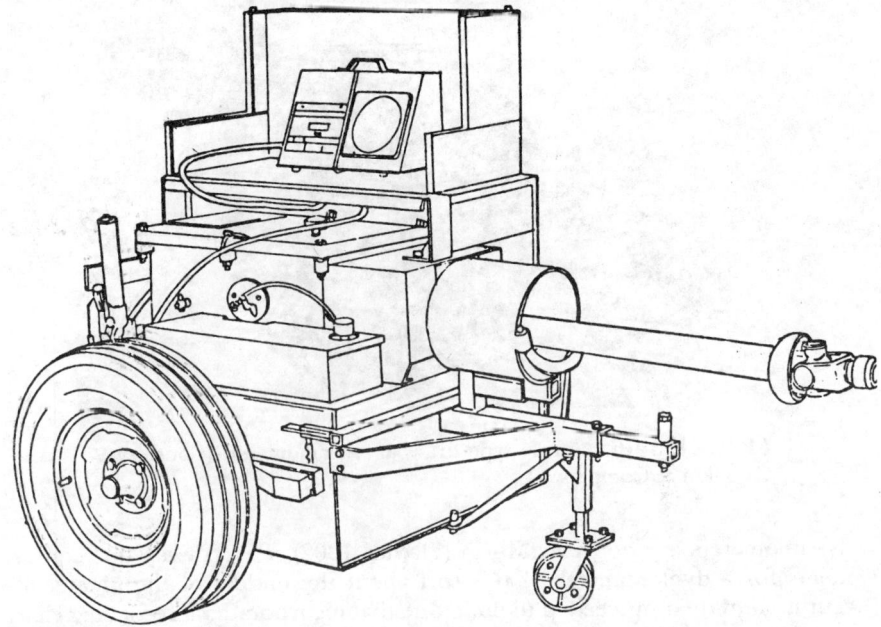

FIGURE 14-5 Shop-type or agricultural dynamometer. (Courtesy A. W. Dynamometer Co.)

The principle of operation of shop-type dynamometers is generally similar to a hydraulic or a Prony brake dynamometer. Hydraulic pumps are also sometimes used. Since shop-type dynamometers are not constructed with the precision nor maintained with the care of laboratory dynamometers, they cannot be expected to have the accuracy attained by dynamometers used in engineering laboratories.

Shop-type dynamometers generally employ a pressure gage to measure the force on the resisting torque arm. The pto speed is usually measured by a direct-reading speed indicator. In some cases power is read directly from a pressure gage, with a correction for pto speed.

Drawbar Dynamometers

Drawbar dynamometers are commonly employed to determine the drawbar pull of power units or to ascertain the draft of field implements. Drawbar

FIGURE 14-6   Spring-type drawbar dynamometer. (Courtesy W. C. Dillon & Company.)

dynamometers are not new. Morin (Flather 1902) set forth certain requirements for a dynamometer of this sort about the middle of the nineteenth century and then proceeded to build one that incorporated a recording chart and was successfully used in draft tests.

Spring Dynamometer

The simplest and most obvious type of drawbar unit consists of a spring that elongates under tension or shortens under compression. A typical example is shown in figure 14-6. Such a dynamometer is suitable for rough measurements of forces; because of rapid variations in loads such as are commonly found in connection with agricultural implements, the actual load at any one instant can only be approximated.

Hydraulic Drawbar Dynamometer

The dynamometer car used for testing tractors at the University of Nebraska (see fig. 14-7) uses a hydraulic cylinder to transmit the drawbar force to the dynamometer car. The pressure is measured by a pressure transducer, the signal from which goes to a recorder and a computer. A hydraulic cylinder for measuring drawbar pull has an advantage over a spring dynamometer in that the fluctuations can be damped by a throttling valve. The drawbar car shown also includes instruments for measuring tractor velocity, tractor wheel slip, engine rpm, fuel consumption, sound level pressure, temperatures of air intake, hydraulic fluid, coolant, fuel intake, and fuel return.

FIGURE 14-7   Tractor test car dynamometer. (Courtesy University of Nebraska, Lincoln.)

Strain Gage Dynamometer
One method of measuring the drawbar pull is by means of a dynamometer that uses electrical resistance strain gages to sense the strain (see fig. 14-8).

Torsion Dynamometers

With the advent of machinery operated by tractor power-take-off shafts a great deal of interest has been displayed in devices for the measurement of power as transmitted by rotating shafts. History records many ingenious arrangements for accomplishing this purpose (Flather 1902).

Torque Meter, Strain-Gage Type
The development of the electrical resistance strain gage has made practical the measurement of torque and force on farm tractors and machines. Many torque meters employing strain gages have been developed. Figure 14-9 shows one such device.

Chassis Dynamometer

The testing of tractors outdoors has some limitations due to weather. One method of avoiding some difficulties of outdoor testing is the use of a chassis dynamometer. The tractor is restrained from forward movement, and the drive wheels are placed on a drum that is part of an absorption dynamometer. Since the tractor is restrained, there is no rolling resistance of the front wheels,

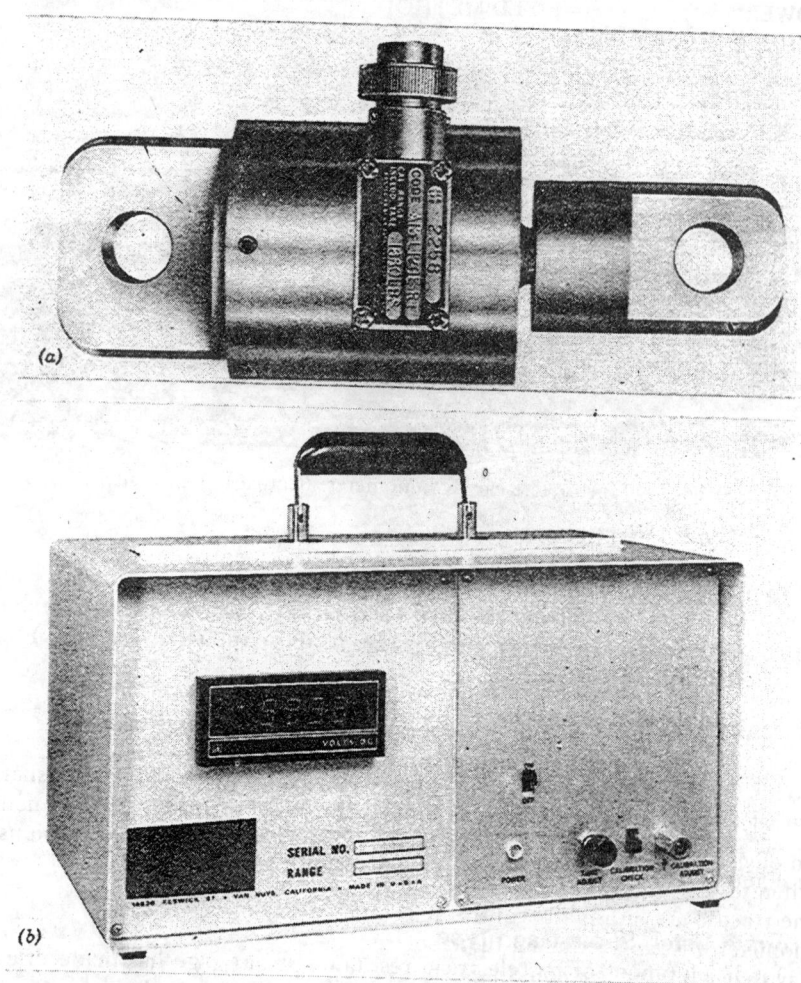

FIGURE 14-8 Drawbar dynamometer using strain gages as sensing elements. (*a*) Strain gage force transducer; (*b*) digital readout. (Courtesy W. C. Dillon & Company.)

FIGURE 14-9 Torque meter using electrical resistance strain gages and slip rings to measure pto torque. (Courtesy Lebow Associates, Inc.)

an obvious disadvantage of the system. Temperature can be better regulated when testing a tractor on a chassis dynamometer. It is also possible to equip the tractor for automatic control and automatic programming of the dynamometer, thereby reducing the labor required and increasing the hours per day that a tractor test could be run.

Power Estimating—Field Method

It is often desired to know the approximate power being developed by a tractor in the field. If the accuracy of a strain-gage type of torque meter is not needed, an estimate of the tractor power output can be obtained by measuring the manifold pressure. A relationship between power and manifold pressure is first obtained by a dynamometer test. The curve is correct only for full-throttle or governor setting. The adjustment and condition of the engine, the air temperature, and the barometric pressure will affect the calibration curve. It is therefore important to

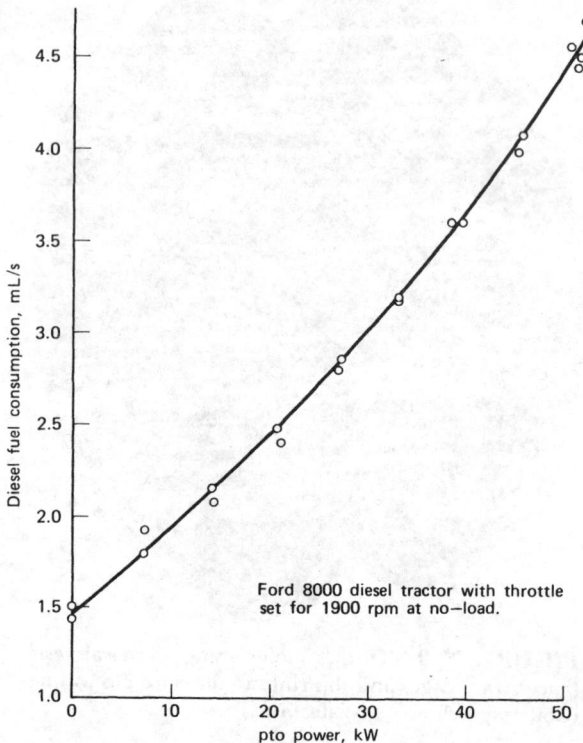

FIGURE 14-10 Calibration curve showing pto power versus fuel consumption for a diesel tractor. (Data from Professor C. B. Richey, Agricultural Engineering Department, Purdue University.)

realize that the manifold pressure method can only be used with confidence if calibration curves are made immediately before and after each field test.

Since the manifold pressure is not controlled on a diesel tractor, a relationship between manifold pressure and power cannot be obtained. For a given no-load engine speed, however, a curve can be plotted of pto power versus fuel consumption. Such a curve is shown in figure 14-10.

Air-Supply Measurement

The measurement of air quantity is important in engine tests. Various schemes have been devised, one of the simplest and most widely used being a bell-

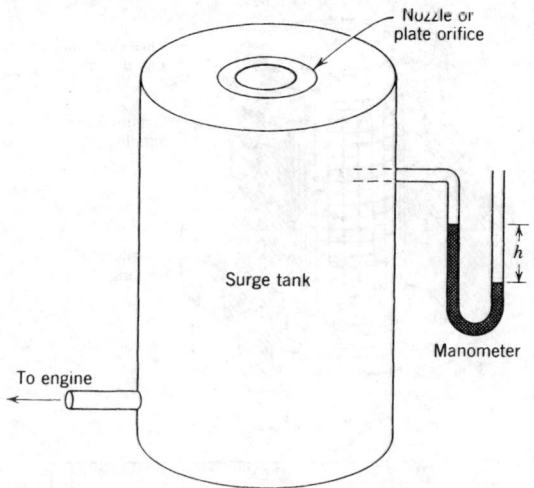

FIGURE 14-11   Method of measuring the airflow into an engine.

mouthed orifice with a surge tank to even out the air pulsations and a sensitive manometer to measure the pressure drop across the orifice (fig. 14-11). Air passes through the orifice into the surge tank and from there into the air cleaner or intake manifold. Orifices of known flow coefficients have been described by Bartholomew (1936).

The mass of dry air flowing through the orifice is computed from

$$M = d_a A C \sqrt{2gh\left(\frac{d_w}{d_a}\right)} \qquad (6)$$

where M = mass of dry air per second, kg
   A = area of orifice, m^2
   C = orifice coefficient
   d_w = density of water, kg/m^3
   d_a = density of air (air and vapor), kg/m^3 at the observed pressure
   h = pressure drop across orifice, mm H$_2$O

If d_w = 997.9 kg/m^3, equation 6 can be simplified to

$$M = 1.63 \, AC \, \sqrt{h \, d_a} \qquad (7)$$

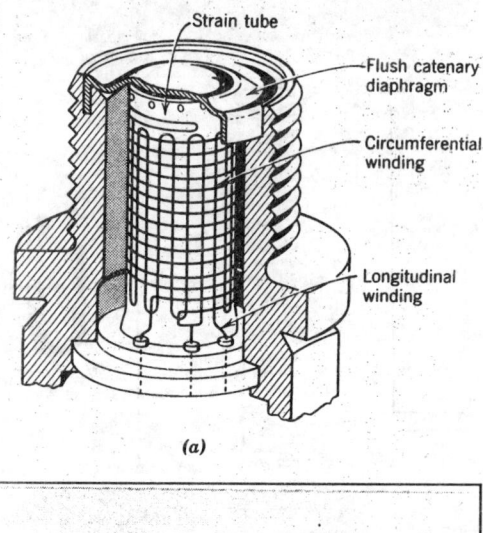

Strain tube

Flush catenary
diaphragm

Circumferential
winding

Longitudinal
winding

(a)

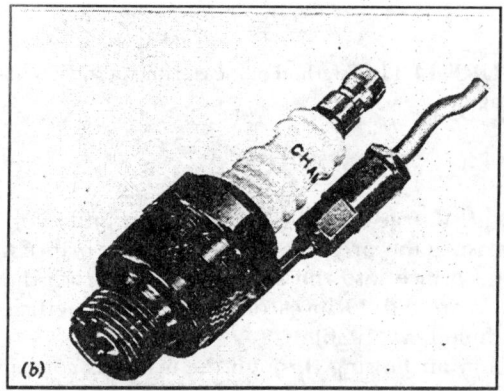

(b)

FIGURE 14-12 Transducers for measuring
pressure inside of a combustion chamber. (Courtesy Beckman Instrument Co.)

Engine Pressure Indicators

The high rotational speeds of internal combustion engines require the use of
electronic measuring devices. One method of measuring the engine cylinder
pressure is by means of a pressure transducer (fig. 14-12) of the unbonded
strain gage or piezoelectric type. A transducer such as that in figure 14-12(a)
is normally installed in a special hole drilled in the cylinder head. Pressure

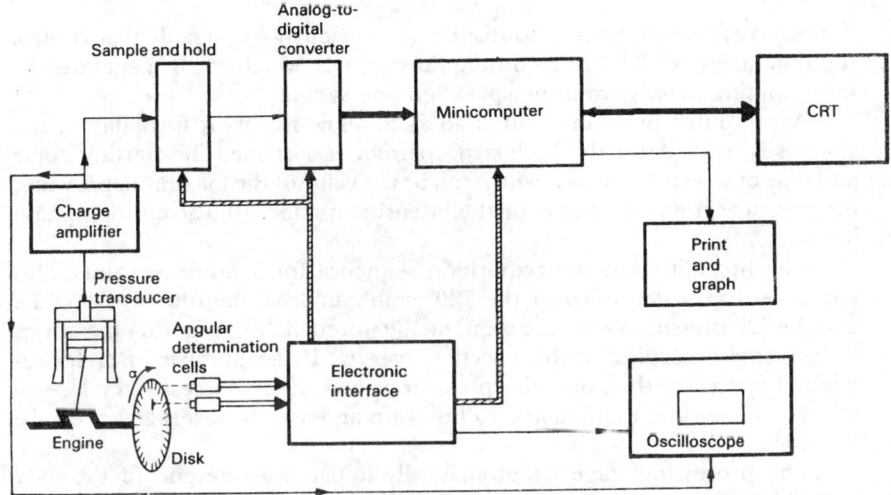

FIGURE 14-13 An on-line data acquisition system for determining a P-V diagram.

transducers using piezoelectric crystals can also be installed in modified spark plugs [fig. 14-12(b)].

The use of electrical resistance strain gages, radio transmission of signals (telemetry), plus the real-time processing of the data (computers) has opened up opportunities for engineers to learn much more when testing tractors and machines. Some techniques are described by Deere & Co. engineers in SAE SP-410 (Society of Automotive Engineers 1976).

When engines were slow, a simple mechanical device called a pressure indicator would make a P-V (pressure-volume) diagram. Figure 14-13 shows the method of obtaining a P-V diagram from a high-speed engine.

Every time the infrared beam (fig. 14-13) passes through one of the 360 slits on the disk, an electroluminescent diode creates an impulse that triggers one pressure measurement. A second diode activated by a single slit allows the determination of the angular position of the crankshaft.

The pressure is measured with a piezoelectric-type pressure transducer that supplies an electrical charge proportional to cylinder pressures, with a charge amplifier transforming it into a voltage.

A sample and hold circuit, which is a transitional analog storage unit, keeps the analog signal constant during the conversion. The converted pressure signals are then fed to a minicomputer. The memory size is sufficient to contain the program that controls the acquisition of the pressure and two tables for the data.

An oscilloscope is used to display the various logic signals that control the system, especially for diagnosing false signals, which can be generated by noise coming mainly from the spark ignition system.

A computer program written so as to allow for great flexibility in the systems is used. After the keyboard operator has defined the starting angle and type of selection criteria, which can be the value of the maximum pressure, the pressure at a given angle, or the intensity of knock, the acquisition phase is begun.

The first slit starts the acquisition sequence for a pressure value. This sequence is repeated to cover the 720° crank angle of the four-stroke cycle, and the 720 pressure values are sequentially stored in the temporary memories to be tested according to the selection criteria. If the number of cycles requested is greater than one, the pressure values of the successive cycles satisfying the test are accumulated to build up an ensemble-averaged pressure cycle.

The processing stage is automatically initiated at the end of the data acquisition. First, some checks on the acquired data are performed. Next, the number of cycles tested to satisfy the selection criteria is given, as are the mean and standard deviation values. Finally, the IMEP (indicated mean effective pressure) is calculated and the results printed out.

Fuel Flow Meter

The variable-area type of flow meter (fig. 14-14) provides rapid and instantaneous readings on the rate of flow of liquids or gases. An annular aperture or orifice exists between the head of the float and the inside of the wall of the tapered tube in which the float travels. The upward and downward forces acting on the float are in equilibrium so that the float assumes a definite elevation at a given flow rate. Since the net weight of the float is the same at all elevations, the pressure drop across the float must also be constant. Therefore, increasing flow rate causes the float to move to a higher position with a correspondingly greater flow area. For a detailed discussion of the theory and operation of variable-area meters, the reader is referred to a discussion of the theory of the rotameter (Fischer and Porter Co.).

The Nebraska Tractor Tests

The 1919 session of the Nebraska legislature passed "A Bill for an Act to provide for official tests for gas, gasoline, kerosene, distillate or other liquid-fuel traction engines in the State of Nebraska and to compel the maintenance of adequate service stations for same (Brackett 1931). The bill was introduced

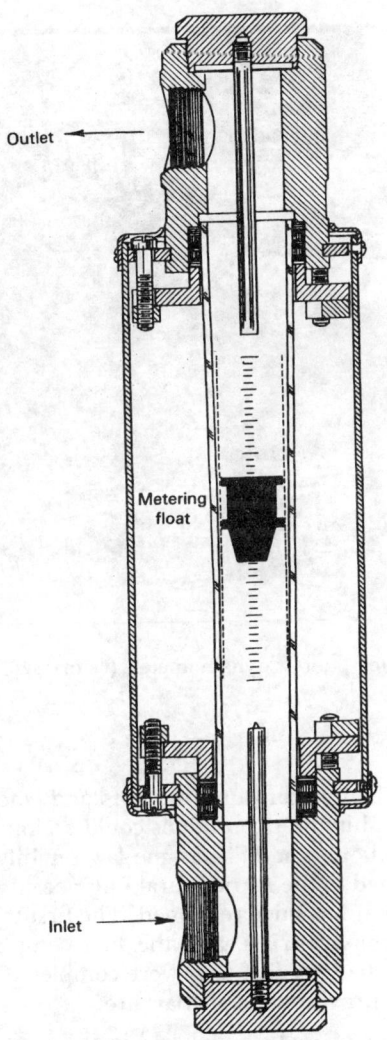

FIGURE 14-14 Fuel flow meter.
(Courtesy Fisher-Porter Co.)

421

FIGURE 14-15 Tractor pto test dynamometer. (Courtesy University of Nebraska, Lincoln.)

by a farmer who had bought and used tractors and whose experiences and observations convinced him that conditions could be improved by the enactment of proper legislation. The bill became law on July 15, 1919, and the testing work was assigned to the Agricultural Engineering Department of the State University, where it has since remained. The first tractor submitted was withdrawn after preliminary trials, and the first complete official test was completed in April 1920. Sixty-five tests were completed in 1920.

Stated briefly, the provisions of the law are:

1. That a stock tractor of each model sold in the state shall be tested and passed by a board of three engineers and the State University management.

2. That each company, dealer, or individual offering a tractor for sale in Nebraska shall have a permit issued by the Nebraska Department of Agriculture, Division of Weights and Measures. The permit for any model of tractor will be issued after a stock tractor of that model has been tested at the University and the performance of the tractor has been compared with the claims made for it by the manufacturer.

3. That a service station with full supply of replacement parts for each model of tractor shall be maintained within the confines of the State and within reasonable shipping distance of customers.

Test codes SAE J708 (July 1983) and ISO 789/1-1981 are technically equivalent to the Nebraska Tractor Test Code.

A manufacturer wishing to have a tractor tested files application with the Agricultural Engineering Department of the University. This application includes detailed specifications of the tractor to be tested and a test fee. A representative of the tractor manufacturer is present during the test to assist with preliminary details and to make decisions involving permissible choices in the test.

The engine oil for the test is selected by the manufacturer's representative and is of the viscosity specified in the application. The fuel is the lowest grade recommended by the manufacturer for the tractor being tested. Premium fuel is used only if the manufacturer has specified it as essential to the successful operation of the engine.

A sample of a tractor test and test explanation is shown in table 14-1. The tests are divided into four main parts: a limber-up period with the representative of the manufacturer operating the tractor; the power-take-off tests; the drawbar tests; and the three-point hitch test.

The three-point hitch hydraulic lift test was incorporated as part of the basic test in 1984. The center of gravity of the load is at a point 610 mm behind the attachment points when the lift frame mast is vertical. The maximum mass lifted through the operating range of the hitch and associated time are determined. Front and rear axles are blocked and restrained.

The drawbar dynamometer used for the Nebraska tractor tests is shown in figure 14-7, and the pto dynamometer setup is shown in figure 14-15.

Text continues on p. 430.

TABLE 14.1 An Official Nebraska Tractor Test Report: Nebraska Tractor Test 1546—Case 1594 Synchromesh Diesel 12-Speed

POWER TAKEOFF PERFORMANCE

Power Hp (kW)	Crank shaft speed rpm	Fuel Consumption			Temperature °F (°C)			Barometer inch Hg (kPa)
		gal/hr (l/h)	lb/hp.hr (kg/kW.h)	Hp.hr/gal (kW.h/l)	Cooling medium	Air wet bulb	Air dry bulb	

MAXIMUM POWER AND FUEL CONSUMPTION

		Rated Engine Speed—Two Hours (PTO Speed—1123 rpm)						
85.90 (64.06)	2300	5.178 (19.602)	0.421 (0.256)	16.59 (3.268)	192 (88.9)	64 (17.8)	75 (24.1)	28.78 (97.20)

(continued)

TABLE 14.1 Continued

POWER TAKEOFF PERFORMANCE

Power Hp (kW)	Crank shaft speed rpm	Fuel Consumption			Temperature °F (°C)			Barometer inch Hg (kPa)
		gal/hr (l/h)	lb/hp.hr (kg/kW.h)	Hp.hr/gal (kW.h/l)	Cooling medium	Air wet bulb	Air dry bulb	

MAXIMUM POWER AND FUEL CONSUMPTION

		\multicolumn{7}{c}{Standard Power Take-off Speed (1000 rpm)—One Hour}	
80.83 (60.28)	2048	4.744 (17.957)	0.410 (0.249)	17.04 (3.357)	192 (89.0)	64 (17.8)	75 (24.1)	28.74 (97.03)

VARYING POWER AND FUEL CONSUMPTION—Two Hours

75.08 (55.99)	2364	4.555 (17.241)	0.424 (0.258)	16.48 (3.247)	191 (88.3)	64 (17.8)	76 (2.42)	....
0.00 (0.00)	2425	1.371 (5.189)	....	....	186 (85.3)	63 (16.9)	75 (23.6)	....
38.24 (28.51)	2409	2.832 (10.719)	0.517 (0.315)	13.50 (2.660)	187 (86.1)	63 (17.2)	74 (23.3)	....
86.38 (64.41)	2300	5.186 (19.632)	0.419 (0.255)	16.66 (3.281)	192 (88.9)	63 (17.2)	74 (23.3)	....
19.18 (14.30)	2414	2.084 (7.889)	0.759 (0.461)	9.20 (1.813)	187 (85.8)	63 (17.2)	74 (23.3)	....
56.86 (42.40)	2387	3.614 (13.679)	0.444 (0.270)	15.73 (3.100)	188 (86.7)	63 (17.2)	74 (23.1)	....
Av 45.96 Av (34.27)	2383	3.273 (12.391)	0.497 (0.303)	14.04 (2.766)	188 (86.9)	63 (17.3)	74 (23.5)	28.70 (96.92)

The following performance figures apply to tractors after chassis S/N 11221501.

DRAWBAR PERFORMANCE

Power Hp (kW)	Drawbar pull lbs (kN)	Speed mph (km/h)	Crank-shaft speed rpm	Slip %	Fuel consumption			Temp. °F (°C)			Barom. inch Hg (kPa)
					gal/hr (l/h)	lb/hp.hr (kg/kW.h)	Hp.hr/gal (kW.h/l)	Cooling med	Air wet bulb	Air dry bulb	

		\multicolumn{9}{c}{Maximum Available Power—Two Hours 7th (L2) Gear}	
72.17 (53.82)	5113 (22.74)	5.29 (8.52)	2300	6.98	5.063 (19.166)	0.490 (0.298)	14.26 (2.808)	191 (88.1)	39 (3.6)	47 (8.3)	29.19 (98.55)

		\multicolumn{9}{c}{75% of Pull at Maximum Power—Ten Hours 7th (L2) Gear}	
58.81 (43.86)	3911 (17.40)	5.64 (9.08)	2401	5.09	4.338 (16.422)	0.515 (0.313)	13.56 (2.671)	191 (88.1)	29 (−1.5)	33 (0.4)	29.28 (99.87)

TABLE 14.1 Continued

DRAWBAR PERFORMANCE

Power Hp (kW)	Crank shaft speed rpm	Fuel Consumption gal/hr (l/h)	lb/hp.hr (kg/kW.h)	Hp.hr/gal (kW.h/l)	Cooling medium	Temperature °F (°C) Air wet bulb	Air dry bulb	Barometer inch Hg (kPa)
		50% of Pull at Maximum Power—Two Hours 7th (L2) Gear						
40.26	2608	5.79 2424	3.54	3.359	0.583	11.99 191	37	44	29.18
(30.02)	(11.60)	(9.32)		(12.714)	(0.354)	(2.361) (88.1)	(2.8)	(6.7)	(98.52)
		50% of Pull at Reduced Engine Speed—Two Hours 10th (L3) Gear						
40.23	2610	5.78 1403	3.65	2.700	0.469	14.90 187	43	44	28.90
(30.00)	(11.61)	(9.30)		(10.220)	(0.285)	(2.935) (86.1)	(6.1)	(6.7)	(97.57)

MAXIMUM POWER IN SELECTED GEARS

Power Hp (kW)	Crank shaft speed rpm	gal/hr (l/h)	lb/hp.hr	Gear		Air wet bulb	Air dry bulb	Barometer inch Hg (kPa)
67.33	8331	3.03 2354	14.64	4th (L1) Gear	191	35	39	29.23
(50.21)	(37.06)	(4.88)			(88.1)	(1.7)	(3.9)	(98.71)
69.32	7581	3.43 2299	11.67	5th (LS3) Gear	190	36	40	29.24
(51.69)	(33.72)	(5.52)			(87.5)	(2.2)	(4.4)	(98.74)
70.59	6494	4.08 2301	9.27	6th (HS2) Gear	190	36	41	29.23
(52.64)	(28.88)	(6.56)			(87.8)	(2.2)	(5.0)	(98.71)
73.54	5215	5.29 2300	7.08	7th (L2) Gear	191	39	46	29.20
(54.84)	(23.20)	(8.51)			(88.1)	(3.9)	(7.8)	(98.60)
72.84	4189	6.52 2299	5.63	8th (H1) Gear	190	37	42	29.23
(54.32)	(18.63)	(10.50)			(87.8)	(2.8)	(5.6)	(98.71)
70.77	3596	7.38 2300	4.78	9th (HS3) Gear	190	37	43	29.22
(52.77)	(16.00)	(11.88)			(87.5)	(2.8)	(6.1)	(98.67)
70.01	2778	9.45 2299	3.76	10th (L3) Gear	190	37	44	29.22
(52.21)	(12.35)	(15.21)			(87.5)	(2.8)	(6.7)	(98.67)

LUGGING ABILITY IN 7th (L2) GEAR

Crankshaft Speed rpm	2300	2070	1837	1610	1373	1148	914
Pull—lbs (kN)	5215 (23.20)	5572 (24.79)	5908 (26.28)	6086 (27.07)	6083 (27.06)	6090 (27.09)	5823 (25.90)
Increase in Pull %	0	7	13	17	17	17	12

(continued)

TABLE 14.1 Continued

LUGGING ABILITY IN 7th (L2) GEAR

Power—Hp (kW)	73.54 (54.84)	70.31 (52.43)	65.84 (49.10)	59.21 (44.15)	50.49 (37.65)	42.21 (31.48)	32.29 (24.08)
Speed—Mph (km/h)	5.29 (8.51)	4.73 (7.62)	4.18 (6.73)	3.65 (5.87)	3.11 (5.01)	2.60 (4.18)	2.08 (3.35)
Slip %	7.08	7.55	8.09	8.35	8.35	8.49	8.09

TRACTOR SOUND LEVEL WITH CAB	dB(A)
Maximum Available Power—Two Hours	83.0
75% of Pull at Maximum Power—Ten Hours	82.0
50% of Pull at Maximum Power—Two Hours	83.5
50% of Pull at Reduced Engine Speed—Two Hours	78.5
Bystander in 11th (H2) gear	86.0

TIRES, BALLAST AND WEIGHT		With Ballast	Without Ballast
Rear Tires	—No., size, ply & psi (kPa)	Two 18.4–34; 6; 16 (110)	Two 18.4–34; 6; 16 (110)
Ballast	—Liquid (each)	582 lb (264 kg)	None
	—Cast Iron (each)	205 lb (93 kg)	None
Front Tires	—No., size, ply & psi (kPa)	Two 10.00–16; 8; 44 (305)	Two 10.00–16; 8; 44(305)
Ballast	—Liquid (each)	None	None
	—Cast Iron (each)	90 lb (41 kg)	None
Height of Drawbar		20 in. (510 mm)	20 in. (510 mm)
Static Weight with Operator—Rear		7880 lb (3574 kg)	6305 lb (2860 kg)
—Front		3220 lb (1461 kg)	3040 lb (1379 kg)
—Total		11100 lb (5035 kg)	9345 lb (4239 kg)

THREE-POINT HITCH PERFORMANCE

Observed Maximum Pressure psi (kpa)	2325	16030
Location	lift cylinder	
Hydraulic oil temperature °F (°C)	170	77
Location	drain plug	

	Maximum Lift Capacity	Lift Capacity for Transport
QUICK ATTACH	no	

426

TABLE 14.1 Continued

	Maximum Lift Capacity	Lift Capacity for Transport
CATEGORY	11	*not measured
LOAD lbs (*kg*)	6002	*2723*
TIME sec	2.78	
HITCH POINT MOVEMENT in. (*mm*)		
Lowest position	12.0	*305*
Top of timed range	36.0	*914*
Highest position	36.0	*914*
LOAD CG MOVEMENT in. (*mm*)		
Lowest position	11.9	*303*
Top of timed range	36.4	*924*
Highest position	36.2	*919*

*Implement load capacity for transport purposes not specified by manufacturer.

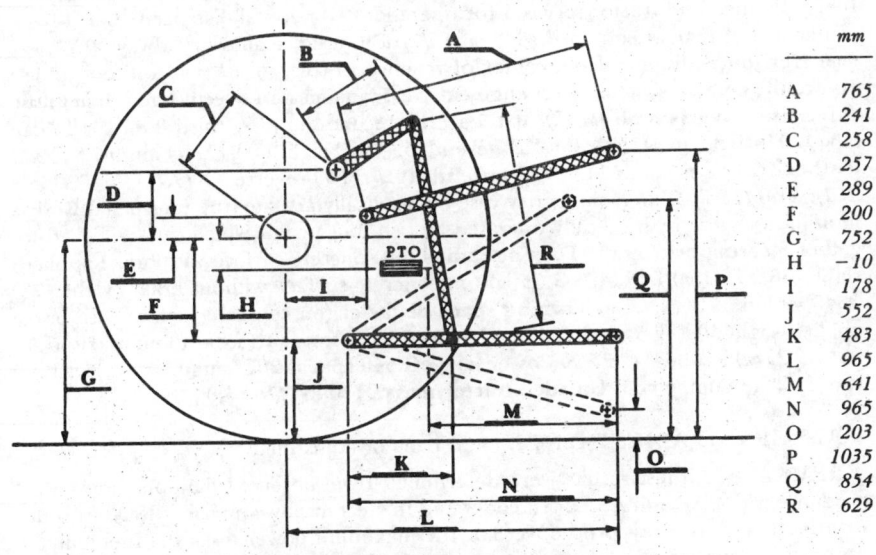

	mm
A	*765*
B	*241*
C	*258*
D	*257*
E	*289*
F	*200*
G	*752*
H	*−10*
I	*178*
J	*552*
K	*483*
L	*965*
M	*641*
N	*965*
O	*203*
P	*1035*
Q	*854*
R	*629*

Hitch Dimensions as Tested — No Load

Department of Agricultural Engineering

Dates of Test: October 24 to November 2, 1984

Manufacturer: J. I. CASE COMPANY, 700 State Street, Racine, Wisconsin 53404

(continued)

TABLE 14.1 Continued

FUEL, OIL AND TIME: Fuel No. 2 Diesel **Cetane No.** 46.8 (rating taken from oil company's inspection data) **Specific gravity converted to** 60/60°F *(15/15°C)* 0.8385 **Fuel weight** 6.982 lbs/gal *(0.837 kg/l)* **Oil SAE 30 API service classification** SF-CD

To motor 3.090 gal *(11.698 l)* **Drained from motor** 2.572 gal *(9.735 l)* **Transmission and hydraulic lubricant** Case TFD fluid **Final drive lubricant** Case ETHB fluid **Total time engine was operated** 36.5 hours.

ENGINE: Make Case Diesel **Type** six cylinder vertical **Serial No.** 330002 11465511 **Crankshaft** lengthwise **Rated rpm** 2300 **Bore and stroke** 3.939″ × 4.500″ *(100 mm × 114.3 mm)* **Compression rate** 16 to 1 **Displacement** 329 cu in *(5392 ml)* **Starting system** 12 volt **Lubrication** pressure **Air cleaner** two paper elements and centrifugal precleaner **Oil filter** one full flow cartridge **Fuel filter** two paper elements with sediment bowl and screen **Muffler** vertical **Cooling medium temperature control** one thermostat.

CHASSIS: Type standard **Serial No.** *154/BEB/11219518* **Tread width** rear 61″ *(1549 mm)* to 85″ *(2159 mm)* front 60″ *(1524 mm)* to 88″ *(2235 mm)* **Wheel base** 100″ *(2540 mm)* **Center of gravity** (without operator or ballast, with minimum tread, with fuel tank filled and tractor serviced for operation) Horizontal distance forward from center-line of rear wheels 33.2″ *(842 mm)* Vertical distance above roadway 36.7″ *(932 mm)* Horizontal distance from center of rear wheel tread 0″ *(0 mm)* to the right/left **Hydraulic control system** direct engine drive **Transmission** selective gear fixed ratio **Advertised speeds** mph *(km/h)* first 1.4 *(2.3)* second 2.3 *(3.7)* third 2.8 *(4.5)* fourth 3.5 *(5.6)* fifth 3.9 *(6.3)* sixth 4.5 *(7.2)* seventh 5.7 *(9.2)* eighth 6.9 *(11.1)* ninth 7.8 *(12.6)* tenth 9.9 *(15.9)* eleventh 11.4 *(18.3)* twelfth 19.7 *(31.7)* reverse 2.3 *(3.7)*, 4.6 *(7.4)*, 5.8 *(9.3)*, 11.6 *(18.7)* **Clutch** single dry disc hydraulically actuated by foot pedal **Brakes** multiple wet disc hydraulically operated by two foot pedals which can be locked together **Steering** hydrostatic **Turning radius** (on concrete surface with brake applied) right 148″ *(3.76 m)* left 148″ *(3.76 m)* (on concrete surface without brake) right 176″ *(4.47 m)* left 176″ *(4.47 m)* **Turning space diameter** (on concrete surface with brake applied) right 309″ *(7.85 m)* left 309″ *(7.85 m)* (on concrete surface without brake) right 364″ *(9.25 m)* left 364″ *(9.25 m)* **Power take-off** 540 rpm at 2077 engine rpm and 1000 rpm at 2048 engine rpm **Unladen tractor mass** 9170 lb *(4160 kg)*.

REPAIRS and ADJUSTMENTS: No repairs or adjustments.

REMARKS: All test results were determined from observed data obtained in accordance with SAE and ASAE test codes and the technically equivalent ISO test codes or official Nebraska test procedure. For the maximum power tests, the fuel temperature at the injection pump return was maintained at 170°F *(76.6°C)*. Seven gears were chosen between 15% slip and 10 mph *(16.1 km/h)*. The drawbar performance figures on this report apply to tractors after chassis S/N 11221501.

We, the undersigned, certify that this is a true and correct report of official Tractor Test No. **1546,** December 3, 1984.

TABLE 14.1 Continued

Report reissued. Supplemental sales permit for Case International 1594 Synciomesh Diesel June 18, 1985.

LOUIS I. LEVITICUS
Engineer-in-Charge
    K. VON BARGEN
    L. L. BASHFORD
    T.L. THOMPSON
        Board of Tractor Test Engineers

Explanation of Test Reports

General

Tractors are tested at the University of Nebraska according to the Agricultural Tractor Test Code approved by the American Society of Agricultural Engineers and the Society of Automotive Engineers or official Nebraska test procedure.

The manufacturer selects the tractor to be tested and certifies that it is a stock model. Each tractor is equipped with the common power consuming accessories such as power steering, power lift pump, generator, etc., if available. Power consuming accessories may be disconnected only when the means for disconnecting can be reached from the operating station. An official representative of the company is present during the test to see that the tractor gives its optimum performance. Additional weight may be added to the tractor as ballast if the manufacturer recommends use of such ballast. The static tire loads and the tire inflation pressures must conform to the specifications of Tire and Rim Association.

Preparation for Test

The engine crankcase is drained and refilled with new oil conforming to specifications in the operator's manual. The operator's manual is also used as the guide for selecting the proper fuel and for routine lubrication and maintenance operations.

The tractor is limbered up for a minimum of three hours on the drawbar. Preliminary adjustment of the tractor is permitted at this time. Any parts added or replaced during the limber-up run, or any subsequent runs, are mentioned in the individual test reports. The tractor is equipped with approximately the amount of added ballast that is to be used during the drawbar runs.

PTO Performance

Power take-off performance runs are made by connecting the power take-off to a dynamometer. During a preliminary power take-off run the manufacturer's representative may make adjustment for the fuel, ignition or injection timing and governor control settings. These settings must be maintained for the remainder of the test. The manually operated governor control mechanism is set to provide the high-idle speed specified by the manufacturer. During the power take-off runs an ambient air temperature of approximately 75°F..(24.0°C) is maintained.

Maximum power is obtained at the rated engine speed specified by the manufacturer with the
(continued)

TABLE 14.1   Continued

governor control lever set for maximum power. This same setting is used for all subsequent PTO runs. Time of the run is two hours. Whenever the power take-off speed during the maximum power run differs from the speeds set forth in the ASAE and SAE standards, an additional run is made at either 540 or 1000 rpm of the power take-off shaft. Time of this run is one hour.

Drawbar Performance

Maximum drawbar power is shown for the normal field speed selected by the manufacturer. All engine adjustments are the same as those used in the power take-off runs. If the manufacturer specifies a different rated engine speed for drawbar operations, then the position of the manually operated governor control is changed to provide the high-idle speed specified.

Maximum drawbar power is determined within the following limits: (1) slip of the drivers must not exceed 15% for pneumatic tires on the concrete test course or 7% for steel cleats on the well packed earthen test course, (2) ground speeds must not exceed 10 miles per hour, (3) safe stability limits of the tractor must not be exceeded, (4) no other operating limit of the tractor must be exceeded. Drawbar load is applied until the manufacturer's rated engine speed is obtained with maximum governor control lever setting. Travel speed, drawbar pull and other data are recorded over two 500-foot straight level areas.

Fuel consumption is determined at the manufacturer's selected travel speed with the drawbar pull set: (1) to give rated engine rpm, (2) 75% of the pull at rated engine rpm, (3) 50% of the pull at rated engine rpm, and (4) maintaining the same load and travel speed as in (3) by shifting to a higher gear and reducing the engine rpm.

This summary shows the drawbar horsepower, corresponding travel speed and fuel consumption at 100%, 75% or 50% loads, and 50% of pull at reduced engine speed.

Also shown is the maximum drawbar horsepower produced for two 500-foot travel sections and corresponding travel speed, the maximum drawbar pull and the corresponding drive-wheel slippage.

In addition, the "lugging ability" obtained by increasing the load so that the full throttle engine speed drops to 80% of rated is given as a percent increase with respect to the pull at rated engine rpm.

The maximum drawbar horsepower and the corresponding speed and drawbar pull for other selected gears or travel speeds are shown in the individual test reports.

Sound Measurement

Sound is recorded during each of the Drawbar Performance runs as the tractor travels on a straight section of the test course. The dB(A) sound level is obtained with the microphone located near the right ear of the operator. Bystander sound readings are taken with the microphone placed 25 feet (7.5 m) from the line of travel of the tractor. An increase of 10dB(A) will approximately double the loudness to the human ear.

Additional Tests

Since 1977 supplemental tests have been added that evaluate drawbar performance with radial ply tires, drawbar performance for rear-wheel-drive tractors equipped with front-wheel-drive assist (FWA), and varying speed test with "constant" power performance.

Drawbar performance is evaluated with bias ply tires in the basic test.

Additional tests with radial ply tires are permitted at extra charge at the option of the manufacturer. The "Maximum Available Power—Two Hours" and the "Maximum Power in Selected Gears" are run for radial ply tires.

Manufacturers of some rear-wheel-drive models provide FWA. If the two-wheel-drive version exceeds 50 percent of sales in Nebraska, the basic test will be conducted on the two-wheel-drive version and supplemental tests on the FWA if claims are made for the FWA or if the manufacturer elects to test the FWA version. These additional tests, with the FWA engaged and without ballast redistribution, are run as follows: A. Maximum available power (two-hour test-fuel run). B. Maximum pull, 15 percent slip limit. C. Maximum power in rated gear. D. Other runs may be required or requested.

A constant power performance claim within an operating speed range for a tractor model will be verified by testing at the pto power outlet whenever possible. Pto power and fuel consumption runs are made at: 1. Maximum (rated) engine speed. 2. Minimum (or lowest) engine speed for which the claim is made. 3. Performance at an intermediate engine speed. If the tractor is not equipped with a pto outlet, suitable drawbar tests are used to determine the constant power performance.

Tractor manufacturers are permitted to select the tractor to be tested. The major reason for allowing the tractor manufacturers to select the tractor to be tested at Nebraska is that the variability is reduced and therefore the performance relative to other tractors will be more exact. An example of the variability of the power output of randomly selected tractors is shown in figure 14-16. The tractors were of the same model and came off the assembly line the same day. The variation in the power output is caused by the normal variation (within tolerance limits) of the size of engine parts, and also the normal variation in the adjustment of the valves, ignition, and fuel systems. A break-in period plus an additional tune-up would reduce the variation as found in figure 14-16. A tractor manufacturer would not allow a tractor from the lower fifth percentile to be tested at Nebraska, but there is a 1-in-20 chance that such a tractor would be selected for the test if it was picked at random from a dealer.

The Nebraska Tractor Test Law was revised on April 14, 1986 (Legislative Bill 768) (Von Bargen et al. 1986). The revision pertains only to agricultural tractors and thus excludes construction, lawn, and garden tractors. The primary provisions are:

1. A stock tractor of each current (new) model agricultural tractor sold or disposed of in Nebraska shall be tested at the University of Nebraska or an OECD (Organization for Economic Cooperation and Development) test station.
2. Official tests shall be conducted according to either the official OECD test

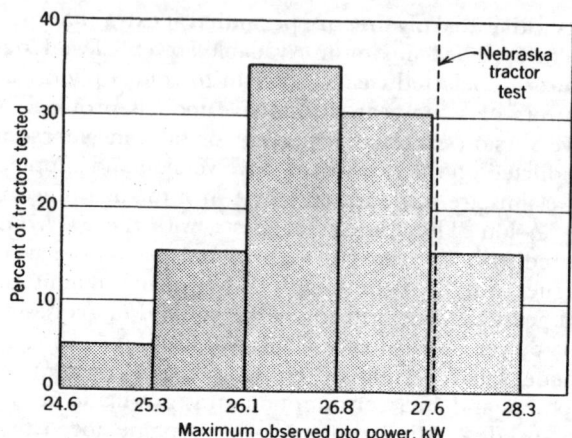

FIGURE 14-16   Distribution of observed pto power
of 20 tractors tested at random off the assembly line.

procedure or the SAE J708/ASAE 209 Agricultural Tractor Test Code at the
election of the manufacturer.
3. Certification that representations for the model have been verified by official
   test results for issuance of a Nebraska sales permit.
4. The availability and accessibility of service and replacement parts shall be
   detailed to the Nebraska Department of Agriculture.
5. Test results shall be published.

Globalization of the Tractor Industry

The engineering design, development, manufacturing, and marketing of agricultural tractors have changed dramatically in recent years and have become
international in scope. Individual nations and organizations develop laws and
standards. Tractor performance test standards have been developed by the
OECD (Organization for Economic Cooperation and Development). Twenty-
six industrialized nations, including the United States, are currently members.
Also, the EEC (European Economic Community), with nine member nations,
develops laws and standards that are primarily designed to protect local mar-
kets. The ISO (International Standards Organization) also develops tractor
performance standards. Several European countries and Japan have OECD
test stations. The National Institute of Agricultural Engineering at Silsoe,
England, carries out OECD, EEC, and ISO tractor tests.

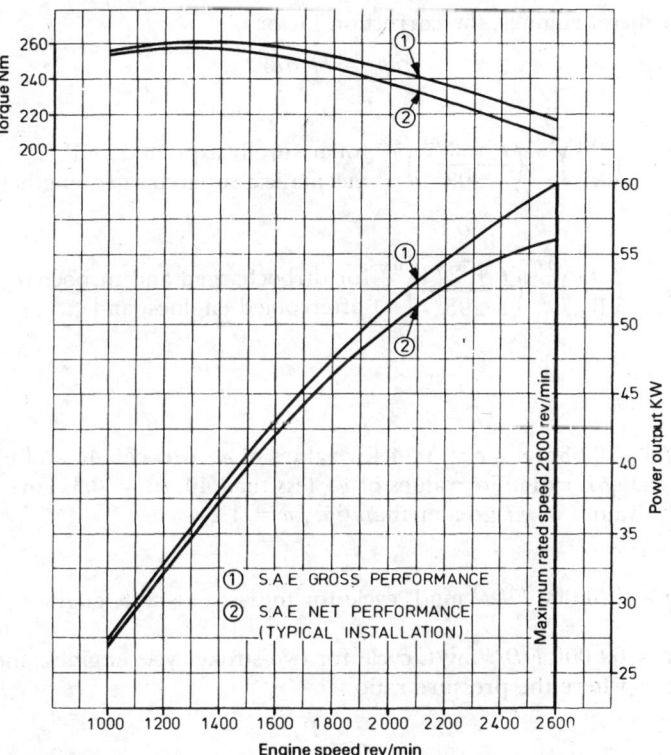

FIGURE 14-17 Performance curves for Perkins 4.236 diesel engine. (Courtesy Perkins Engines Limited.)

Correction for Atmospheric Conditions

Since it is not always possible to test engines under identical conditions of temperature, pressure, and relative humidity, it may be desirable to predict the performance under conditions different from those under which the engine was tested. The usual method of correcting to SAE standard conditions is based on the assumption that at a constant air-to-fuel ratio, indicated thermal efficiency remains unaffected by changes in atmospheric pressure, temperature, and humidity or that the effect is negligible. This assumption is valid only if the range of ambient conditions to be covered is sufficiently small so that engine combustion characteristics are not affected.

For diesel engines, the correction factor is

$$*bp_c = bp_o(fa)^{fm} \tag{8}$$

where

$$fa = \left(\frac{99}{B_{do}}\right)^{1.1} \left(\frac{t + 273}{298}\right)^{1.2} \text{ for naturally aspirated and mechanical supercharged engines,}$$

or

$$fa = \left(\frac{99}{B_{do}}\right)^{0.6} \left(\frac{t + 273}{298}\right)^{0.6} \text{ for turbocharged and turbocharged aftercooled engines, and}$$

where

$fm = 0.036 \times q/r = 1.14$ for values of q/r between 40 and 65 mg/L cycle. For values of q/r less than 40, $fm = 0.3$. For values of q/r greater than 65, $fm = 1.2$,

where

$q = 120\ 000\ F/DN$ mg/L cycle for four-stroke-cycle engines,

or

$q = 60\ 000\ F/DN$ mg/L cycle for two-stroke-cycle engines, and where the pressure ratio

$$r = \frac{P_o}{B_o}$$

In the case of naturally aspirated engines, $r = 1$.

Symbol	Definition	Units
B	Inlet air pressure	kPa
D	Engine displacement	L
F	Fuel rate	g/s
N	Engine speed	r/min
P	Inlet manifold pressure	kPa
fa	Atmospheric factor	
fm	Engine factor	
t	Inlet air temperature	°C
bp	Brake power	kW
Subscripts		
c	Corrected to standard conditions	
d	Dry air condition	
o	Observed at test conditions	

*The correction factor is from SAE J 1349 (June 1983) Engine Power Test Code.

Symbol	Definition		Units
	Standard Inlet Air Conditions		
	Pressure, total	B	100 kPa
	Temperature	t	25°C
	Vapor pressure		1.0 kPa
	Dry barometric pressure	B_d	99 kPa
	Dry air density		1.1572 kg/m^3

Torque Curves

One performance criterion is the "lugging ability" of the engine or, more precisely, the torque curve. It can be expressed as the torque in percentage of the maximum power torque versus engine speed, also in percentages. Or it can be expressed as the engine torque in N · m versus the engine speed in rpm.

A desirable torque curve is one that increases significantly as speed decreases and is therefore stable. Such a torque curve results in a minimum of speed variation in the engine. It is also desirable for the torque curve to peak as far to the left (at lowest speed) as possible. A diesel tractor engine will normally have less speed variation for a given change in torque than a comparable gasoline engine.

Since 1959 the torque curve, as such, has not been reported in the Nebraska tractor tests. Instead, a test called Lugging Ability is used because it combines the performance of the transmission and engine, and to the users of tractors it is more easily understood than a torque curve. Between 1959 and 1976 the test was called Varying Drawbar Pull and Travel Speed.

Engine Performance

Figure 14-17 gives the result of a typical test of a tractor engine whose crankshaft is attached directly to a dynamometer. The same engine when placed in its tractor chassis would have less power through the power takeoff because of losses due to gears, hydraulic pumps, and so forth.

The power ratings of truck and automobile engines usually are the result of dynamometer tests of the engine removed from its chassis. This, combined with the fact that automobile engines are tested at a speed at or near the maximum power speed, accounts for the wide discrepancy in the power output of an automobile engine and the pto power of a tractor having an engine of similar size.

Efficiency of Tractor Engines

An important criterion of engine performance is its thermal efficiency. It can be expressed in percentage, but it is easier to express the efficiency as the ratio of the mass of fuel burned per hour to the pto power. Or it can be expressed as the ratio of the pto power to the volume of fuel burned per hour (kWh/L).

Figure 14-18 gives the fuel efficiency of the tractors tested at Nebraska for gasoline in 1976 and for diesel in 1984. Since 1976 very few gasoline tractors have been tested. The graph represents data from the pto Varying Power and Fuel Consumption Test, which indicates the tractor's ability to convert potential energy (fuel) into useful work. Since the efficiency is expressed as kWh/L, a high number means high efficiency.

From figure 14-18 it is clear that diesel engines have greater efficiency than gasoline engines. Analysis of the data indicates that the diesel engine tractors are approximately 54 percent more efficient than gasoline engine tractors. It is apparent that the fuel efficiencies of both fuel types decrease as the power level percentage of the engine decreases.

The efficiency percentage of a tractor is found from

$$\text{Eff (at pto)} = (100) \frac{\text{output work } [\text{kW} \cdot \text{s}]}{\text{input work } \left[\dfrac{\text{kJ}}{\text{kg}} \cdot \text{kg fuel}\right]} \tag{9}$$

Actual Power Output and Fuel Consumption

It is sometimes desirable, from the standpoint of predicting tractor life, to know something about the actual loads imposed upon a tractor while being used on a farm. Knowledge of the actual tractor power output and speed would then allow design engineers to test tractors under more normal conditions by "programming" similar loads onto a control system for a tractor dynamometer.

Deere and Co. has developed the Dyna-Cart (Van Gerpen and Mayfield 1982), which can be programmed to duplicate the load cycle of most implements. It has an air-cooled eddy-current dynamometer that is controlled by an electronic circuit. The unit is shown in figure 14-19. It is capable of testing tractors from small utility to 150 kW. An important feature is the ability to repeat a varying drawbar load cycle of an implement in the field. This is achieved with a programmer circuit. Load cycles for programming the Dyna-Cart are obtained by operating a tractor and implement in the field. Draft signal and forward speed are radio-telemetered to an instrumentation laboratory, where a digital computer performs the analysis and determines the instructions for the programmer.

Input measurements for a test run are drawbar pull, vehicle speed, test tractor engine speed, and fuel flow. Output is average slip (percent), engine speed (rpm), ground speed (km/h), power (kW), pull (kN), fuel consumption

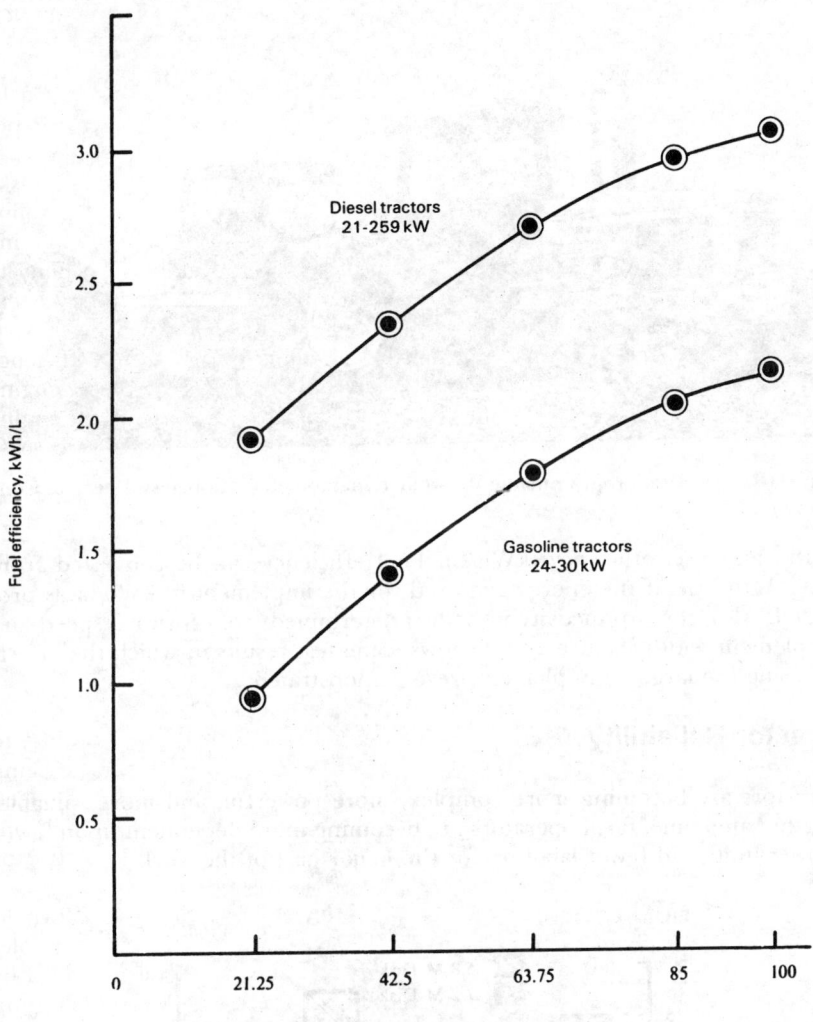

FIGURE 14-18 Average fuel efficiency of gasoline tractors tested in 1976 and diesel tractors tested in 1984 at Nebraska.

FIGURE 14-19   Programmable drawbar dynamometer. (Courtesy Deere & Company.)

(L/h), and fuel efficiency (kWh/L). Fuel efficiency can be converted from kWh/L to ha/L if the energy required for the implement in kWh/ha is provided. Also, the productivity in ha/h is determined from forward speed and implement width. Figure 14-20 shows some test results in which the effects of using too large an implement were demonstrated.

Tractor Reliability

Tractors are becoming more complex, more powerful, and more valuable. At the same time, farm operators are becoming more dependent upon fewer power units and fewer laborers for the major part of the work.

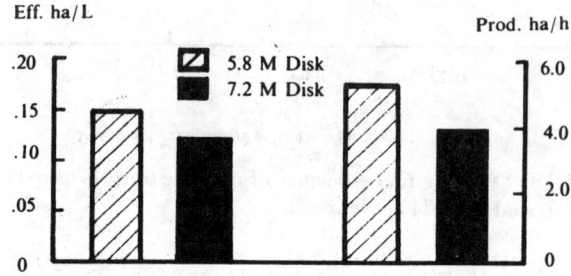

FIGURE 14-20   Effect of oversize implement on efficiency and productivity. (Courtesy Deere & Company.)

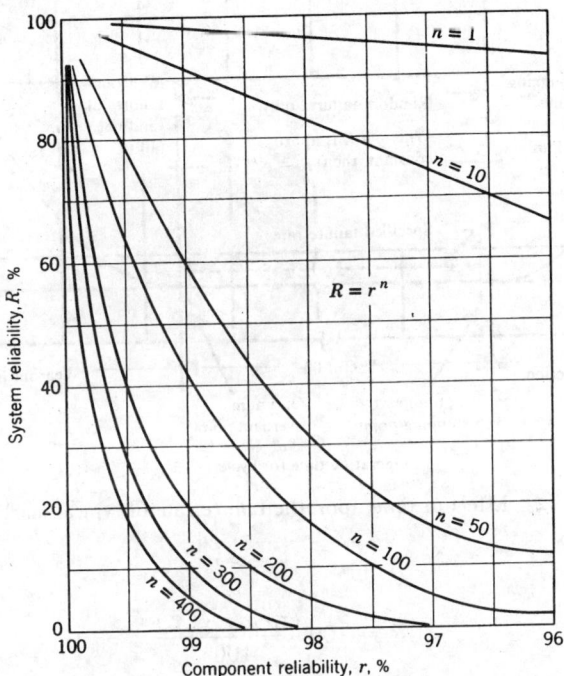

FIGURE 14-21 Effect of series combinations of machine components upon the reliability of the complete system.

As a result, more attention is being given to means of making the complete tractor more reliable. Reliability is the probability that a machine or machine part will perform its specified function for a given period under the specified environment.

If, after a period of use, for example one year, each individual part of a tractor is found to have a reliability r, then the reliability R of the tractor is

$$R = 100(r/100)^n \tag{10}$$

where n is the number of machine parts in series, each of which has a reliability of r percent.

If the reliability of each part is different, then the reliability of the system having the individual parts in series is

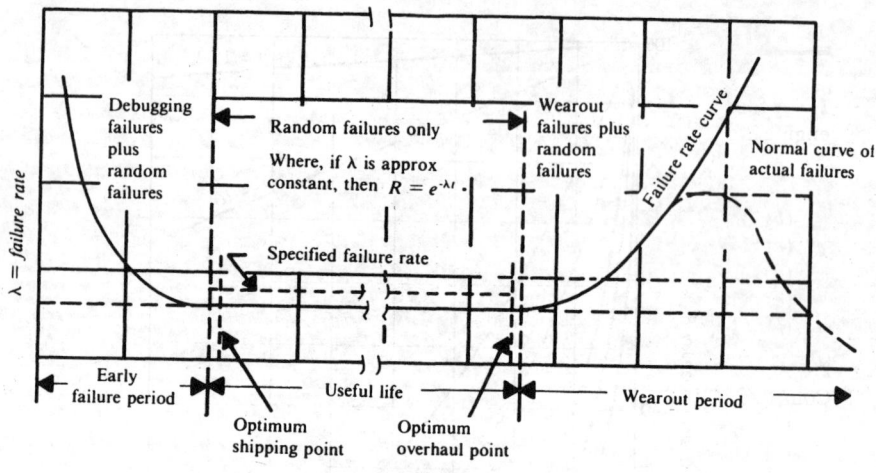

FIGURE 14-22   Effect of time upon the failure rate of typical machine parts.

$$R = 100 \frac{(r_1 r_2 r_3 \cdots r_n)}{100^n} \tag{11}$$

Equation 10 is expressed graphically in figure 14-21. It is obvious that as machines become more complicated—that is, have more parts in series—the reliability of individual parts must increase if the reliability of the complete system is to be kept constant.

The effect of time upon the failure rate of a typical machine part is shown in figure 14-22. The basic reliability equation is shown in equation 12.

$$R = e^{-\lambda t} \tag{12}$$

where R = reliability or probability of success
         t = time in hours

$$\text{MTBF} = \frac{\text{operating hours}}{\text{failures}} \tag{13}$$

where   MTBF = mean time between failures
         λ = failure rate = $\dfrac{1}{\text{MTBF}}$

PROBLEMS

1. A four-cylinder tractor engine has a bore of 98 mm and a stroke of 114 mm. The engine develops 56 kW at 2600 engine rpm when tested on a dynamometer. Compute:

    (a) The length of brake arm in millimeters necessary to give a brake constant, $C = 10^{-4}$, (kW = cnF). See equation 3.

    (b) The brake net scale reading in newtons.

    (c) The torque in newton-meters at the engine crankshaft.

    (d) The engine brake mean effective pressure.

    (e) The brake power per liter of piston displacement.

    (f) The cubic centimeter displacement per minute per brake kilowatt.

2. A tractor operating at a speed of 5 km/h develops a drawbar pull of 9000 N. Compute:

    (a) Speed in meters per second.

    (b) Drawbar power in kilowatts.

    (c) Work at the drawbar per 8-h day.

    (d) Change in speed required to increase the drawbar power to 15 kW.

3. Determine the corrected maximum pto power of a selected tractor tested within the last two years at Nebraska. Correct for ambient air pressure, moisture in the air, and barometric pressure. What percent more (or less) is the corrected pto power? List all pertinent data.

4. On a single sheet of graph paper, plot the following data from one recent Nebraska tractor test. (a) Drawbar power and pull versus speed from the test, Maximum Power with Ballast. (b) Drawbar power and pull versus speed from the test, Varying Drawbar Pull and Travel Speed. Why did the slip essentially remain constant in the second test? What slip resulted in maximum power? Except for the maximum power point the Varying Drawbar Pull Test resulted in less power for a given speed than the Maximum Power Test. Why?

5. A four-cylinder diesel tractor engine has a cylinder bore of 106 mm and a piston stroke of 127 mm. The engine develops 52.5 kW at 2500 rpm when coupled to a dynamometer.

    At the same speed, and with the fuel shut off, the tractor engine required 9.3 kW to motor it (to overcome friction). During the test when 52.5 kW was developed the engine used 18.5 L/h of No. 2 diesel fuel. The fuel contained 39.0 MJ of heat per liter.

Calculate:

    (a) Engine indicated power.

    (b) Indicated mean effective pressure.

    (c) Engine efficiency, net.

    (d) Engine efficiency, indicated.

6. Assume the above engine was actually tested when the temperature was 35°C and the atmospheric pressure was 96 kPa. What is the power corrected to standard conditions?

What would be the power output at 1000, 2000, 3000, 4000, and 5000 m altitude?

Use equation 8 in this chapter to correct to standard conditions. The following equations can be used to correct for altitude.

To correct for altitude

$$p = 101.32 \left(1 - \frac{z}{x} \right) 5.26$$
$$T = 288 (1 - z/x) - 273$$

where x = 44,409 m = equivalent depth of atmosphere
         z = height above sea level, m
         p = pressure at elevation z, kPa
         T = temperature at elevation z, °C

REFERENCES

Bartholomew, Earl. *SAE Trans.*, vol. 31, 1936, pp. 97–98.

Bracket, E. E. *Agric. Engr.*, June 1931.

Culver, E. P. *Mechanical Eng'g*, Oct. 1937.

Flather, J. J. *Dynamometers and Measurement of Power*. John Wiley & Sons, New York, 1902.

Marks. *Mechanical Engineers Handbook*, 6th ed. McGraw-Hill Book Co., New York, 1958.

"On-line Data Acquisition and Reduction in the Field." SP 410, Society of Automotive Engineers, Warrendale, PA, 1976.

"Theory of the Rotameter," Fischer and Porter Co., Catalog 98-Y.

Van Gerpen, H. W., and R. L. Mayfield. "Dyna-Cart: A Programmable Drawbar Dynamometer for Evaluating Tractor Performance." ASAE Paper 82-1056, 1982.

Von Bargen, K., et al. "Nebraska Tractor Testing Beyond 1985." ASAE Paper MCR 86-109, 1986.

SUGGESTED READINGS

Barger, E. L., and W. J. Promersberger. "Dynamometer for Testing Farm Machinery." *Agric. Engr.*, vol. 22, Sept. 1941, pp. 323–324.

Barnes, K. K. "Part Load Fuel Savings." *Implement & Tractor*, Aug. 7, 1969.

Baumeister, T., and L. S. Marks. *Mechanical Engineer's Handbook*, 7th ed. McGraw-Hill Book Co., New York, 1967.

Buckingham, E. "Notes on Small Flow Meters for Air, Especially Orifice Meters." Technologic Paper No. 183, *J. Res. Natl. Bur. Standards*, Dec. 1920.

Candee, R., and F. C. Walters. "History of Society of Automotive Agricultural Tractor Test Code." SAE Paper 800930, 1980.

Carter, A. D. S. *Mechanical Reliability*. John Wiley & Sons, New York, 1972.

Chase, L. W. "Nebraska Tractor Tests, 1917." *Trans. of ASAE*, vol. 11, 1917, pp. 132–158.

Dilworth, J. L. "Characteristics of Exhaust-Gas Analyzers." *SAE Trans.*, vol. 48, 1941, pp. 234–239.

Douaud, A., and D. Eyzat. "DIGITAP—an On-Line Acquisition and Processing System for Instantaneous Engine Data-Applications." SAE Paper No. 770218, presented at the annual meeting, Detroit, Feb. 28–Mar. 4, 1977.

Edwards, S. G. "Dynamic Measurement of Vehicle Wheel Loads Using a Special Purpose Transducer." *General Motors Eng'g J.* Fourth Quarter 1964.

Jervis-Smith, J. F. *Dynamometers*. D. Van Nostrand Co., New York, 1915.

Judge, A. W. *The Testing of High-Speed Internal Combustion Engines*, 4th ed., rev. Chapman Hall and Co., London, 1955.

Kivenson, G. *Durability and Reliability in Engineering Design*. Hayden Book Co., New York, 1971.

Larsen, L. F., and L. I. Leviticus. "Thirty Years of Nebraska Tractor Testing." Paper No. 76-1045 presented at 1976 annual meeting of ASAE, Lincoln, NE.

Lee, G. H. *Experimental Stress Analysis*. John Wiley & Sons, New York, 1950.

"Nebraska Tractor Tests, 1920–1948." *Univ. Nebraska Agric. Expt. Sta. Bull.* 392, Jan. 1949.

Polson, J. A., J. G. Lowther, and B. J. Wilson. "The Flow of Air Through Circular Orifices with Rounded Approach." *Univ. Illinois Eng'g Expt. Sta. Bull.* 207, 1930.

SAE. "Statistics for the Engineer." SP-250. Society of Automotive Engineers, Pittsburgh, 1963.

Splinter, W. E., et al. "Nebraska Tractor Test—Programs and Philosophy." SAE Paper 730763, 1973.

Zoz, Frank M. "Predicting Tractor Field Performance." *Trans. of ASAE*, vol. 15, no. 2, 1972, p. 249.

APPENDIXES

APPENDIX A

STANDARDS FOR AGRICULTURAL TRACTORS

The following standards were selected from the ASAE Standards, 1985. This is a partial list of ASAE standards, many of which have corresponding standards by Society of Automotive Engineers (SAE) and American National Standards Institute (ANSI).

Because of the space required, only the identifying numbers and titles are used. The reader should refer to the most recent set of ASAE standards and to the SAE standards in the latest *SAE Handbook*.

Explanation of ASAE Designation

The letter *S* preceding the numerical designation indicates ASAE standard; *EP* indicates engineering practice; *D* indicates data. A decimal and numeral following the file number indicate the number of times a document has been revised. Thus, ASAE S201.4 indicates standard number 201, four times revised.

ASAE Designation	Title
S201.4	Application of Hydraulic Remote Control Cylinders to Agricultural Tractors and Trailing-type Agricultural Implements
S203.10	Rear Power Take-off for Agricultural Tractors
S205.2	Power Take-off Definitions and Terminology for Agricultural Tractors
S207.10	Operating Requirements for Power Take-off Drives

ASAE Designation	Title
S209.5	Agricultural Tractor Test Code
S210.2	Tractor Belt Speed and Pulley Width
S217.10	Three-point Free-link Attachment for Hitching Implements to Agricultural Wheel Tractors
S219.2	Agricultural Tractor and Equipment Disc Wheels
S276.3	Slow-moving Vehicle Identification Emblem
S277.2	Mounting Brackets and Socket for Warning Lamp and Slow-moving Vehicle (SMV) Identification Emblem
S278.6	Attachment of Implements to Agricultural Wheel Tractors Equipped with Quick-attaching Coupler for Three-point Free-link Hitch
S279.8	Lighting and Marking of Agricultural Equipment on Highways
EP285.6	Use of Customary and SI (Metric) Units
S295.2	Agricultural Tractor Tire Loading and Inflation Pressures
S296.2	Uniform Terminology for Traction of Agricultural Tractors, Self-propelled Implements, and Other Traction and Transport Devices
S304.5	Symbols for Operator Controls on Agricultural Equipment
S310.3	Overhead Protection for Agricultural Tractors—Test Procedures and Performance Requirements
S313.2	Soil Cone Penetrometer
S316.1	Application of Remote Hydraulic Motors to Agricultural Tractors and Trailing-type Agricultural Implements
S318.8	Safety for Agricultural Equipment
S331.3	Implement Power Take-off Drive Line Specifications
S333.1	Agricultural Tractor Auxiliary Power Take-off Drives
S335.3	Operator Controls on Agricultural Equipment
S338	Safety Chain for Towed Equipment
S346.1	Liquid Ballast Table for Drive Tires of Agricultural Machines
S349.1	Test Procedure for Measuring Hydraulic Lift Force Capacity on Agricultural Tractors Equipped with Three-point Hitch
S350	Safety-Alert Symbol for Agricultural Equipment
S351	Hand Signals for Use in Agriculture
S355.1	Safety for Agricultural Loaders
EP363.1	Technical Publications for Agricultural Equipment
S365T	Brake Test Procedures and Brake Performance Criteria for Agricultural Equipment
S366.1	Dimensions for Cylindrical Hydraulic Couplers for Agricultural Tractors
S383.1	Roll-over Protective Structures (ROPS) for Wheeled Agricultural Tractors

The following standards were selected from the 1987 *SAE Handbook*. The SAE standards that follow do not have a corresponding ASAE standard.

SAE Designation		Title

Agricultural, Construction, and Industrial Equipment

J722	NOV84	Power Take-off Definitions and Terminology for Agricultural Tractors
J714	JUN86	Wheel Mounting Elements for Industrial and Farm Equipment Disc Wheels
J1012	MAR80 ·	Agricultural Equipment Enclosure Pressurization System Test Procedure
J1013	JAN80	Measurement of Whole Body Vibration of the Seated Operator of Off-Highway Work Machines
J745	MAR86	Hydraulic Power Pump Test Procedure
J746	MAR86	Hydraulic Motor Test Procedure
J214	MAR86	Hydraulic Cylinder Test Procedure
J874	OCT85	Center of Gravity Test Code
J833	DEC83	USA Human Physical Dimensions
J154a		Operator Enclosures—Human Factor Design Considerations

Power Plant Components and Accessories

J151	MAR85	Pressure Relief for Cooling System
J342	NOV80	Spark Arrester Test Procedure for Large Size Engines
J604	JAN86	Engine Terminology and Nomenclature—General
J631	FEB82	Radiator Nomenclature
J726	MAY81	Air Cleaner Test Code
J922	NOV79	Turbocharger Nomenclature and Terminology
J1004	SEP81	Glossary of Engine Cooling System Terms
J1141		Air Cleaner Elements
J1244	JUN81	Oil Cooler Nomenclature and Glossary
J1260	JUN83	Standard Oil Filter Test Oil
J1349	JUN85	Engine Power Test Code—Spark Ignition and Diesel

Electrical/Electronic Equipment

J56	JUN83	Electrical Generating System (Alternator Type) Performance Curve and Test Procedure
J139		Ignition System Nomenclature and Terminology
J240	JUN82	Life Test for Atuomotive Storage Batteries
J537	JUN86	Storage Batteries
J539a		Voltages for Diesel Electrical Systems
J544	MAY82	Electric Starting Motor Test Procedure

SAE Designation		Title
J548d		Spark Plugs
J771	APR86	Automotive Printed Circuits
J831		Nomenclature—Automotive Electrical Systems
J973a		Ignition System Measurements Procedure
J1253	MAR86	Low Temperature Cranking Load Requirements of an Engine

Fuels and Lubricants

J183	JUN86	Engine Oil Performance and Engine Service Classification
J300	JUN86	Engine Oil Viscosity Classification
J304	JUN86	Engine Oil Tests
J357	JUN86	Physical and Chemical Properties of Engine Oils
J313	JUN86	Diesel Fuels
J1297	JUN85	Alternative Automotive Fuels

Emissions

J35		Diesel Smoke Measurement Procedure
J215	JAN80	Continuous Hydrocarbon Analysis of Diesel Emissions
J244	JUN83	Measurement of Intake Air or Exhaust Gas Flow of Diesel Engines
J254	AUG84	Instrumentation and Techniques for Exhaust Gas Emissions Measurements
J1145a		Emissions Terminology and Nomenclature

Sound Level

J1008	SEP80	Sound Measurement—Self-Propelled Agricultural Equipment—Exterior
J1074	AUG84	Engine Sound Level Measurement Procedure

STANDARD GRAPHICAL SYMBOLS

THE SYMBOLS SHOWN CONFORM TO THE AMERICAN NATIONAL STANDARDS INSTITUTE (ANSI) SPECIFICIATIONS. BASIC SYMBOLS CAN BE COMBINED IN ANY COMBINATION. NO ATTEMPT IS MADE TO SHOW ALL COMBINATIONS.

LINES AND LINE FUNCTIONS		PUMPS	
LINE, WORKING		PUMP, SINGLE FIXED DISPLACEMENT	
LINE, PILOT (L>20W)			
LINE, DRAIN (L<5W)		PUMP, SINGLE VARIABLE DISPLACEMENT	
CONNECTOR			
LINE, FLEXIBLE		MOTORS AND CYLINDERS	
LINE, JOINING		MOTOR, ROTARY, FIXED DISPLACEMENT	
LINE, PASSING		MOTOR, ROTARY VARIABLE DISPLACEMENT	
DIRECTION OF FLOW, HYDRAULIC		MOTOR, OSCILLATING	
LINE TO RESERVOIR ABOVE FLUID LEVEL BELOW FLUID LEVEL		CYLINDER, SINGLE ACTING	
LINE TO VENTED MANIFOLD		CYLINDER, DOUBLE ACTING	
PLUG OR PLUGGED CONNECTION		CYLINDER, DIFFERENTIAL ROD	
RESTRICTION, FIXED		CYLINDER, DOUBLE END ROD	
RESTRICITION, VARIABLE		CYLINDER, CUSHIONS BOTH ENDS	

451

MISCELLANEOUS UNITS		BASIC VALVE SYMBOLS (CONT.)	
DIRECTION OF ROTATION (ARROW IN FRONT OF SHAFT)		VALVE, SINGLE FLOW PATH, NORMALLY OPEN	
COMPONENT ENCLOSURE		VALVE, MAXIMUM PRESSURE (RELIEF)	
RESERVOIR, VENTED		BASIC VALVE SYMBOL, MULTIPLE FLOW PATHS	
RESERVOIR, PRESSURIZED		FLOW PATHS BLOCKED IN CENTER POSITION	
PRESSURE GAGE		MULTIPLE FLOW PATHS (ARROW SHOWS FLOW DIRECTION)	
TEMPERATURE GAGE		VALVE EXAMPLES	
FLOW METER (FLOW RATE)		UNLOADING VALVE, INTERNAL DRAIN, REMOTELY OPERATED	
ELECTRIC MOTOR		DECELERATION VALVE, NORMALLY OPEN	
ACCUMULATOR, SPRING LOADED		SEQUENCE VALVE, DIRECTLY OPERATED, EXTERNALLY DRAINED	
ACCUMULATOR, GAS CHARGED		PRESSURE REDUCING VALVE	
FILTER OR STRAINER			
HEATER		COUNTER BALANCE VALVE WITH INTEGRAL CHECK	
COOLER			
TEMPERATURE CONTROLLER		TEMPERATURE AND PRESSURE COMPENSATED FLOW CONTROL WITH INTEGRAL CHECK	
INTENSIFIER			
PRESSURE SWITCH		DIRECTIONAL VALVE, TWO POSITION, THREE CONNECTION	
BASIC VALVE SYMBOLS			
CHECK VALVE		DIRECTIONAL VALVE, THREE POSITION, FOUR CONNECTION	
MANUAL SHUT OFF VALVE			
BASIC VALVE ENVELOPE		VALVE, INFINITE POSITIONING (INDICATED BY HORIZONTAL BARS)	
VALVE, SINGLE FLOW PATH, NORMALLY CLOSED			

METHODS OF OPERATION		METHODS OF OPERATION	
PRESSURE COMPENSATOR		LEVER	
DETENT		PILOT PRESSURE	
MANUAL		SOLENOID	
MECHANICAL		SOLENOID CONTROLLED, PILOT PRESSURE OPERATED	
PEDAL OR TREADLE		SPRING	
PUSH BUTTON		SERVO	

SOURCE. *Sperry-Vickers Industrial Hydraulics Manual,* 1978.

AGRICULTURAL TRACTOR TIRE LOADINGS, TORQUE FACTORS, AND INFLATION PRESSURES— SAE J709d

1. **Purpose**

   This SAE Standard establishes loadings, tangential pull values, and inflation pressure relationships for tire sizes and ply ratings currently used on agricultural tractors.

2. **General Principles**

   2.1. All agricultural tractor tire loads shown in table C-1 are expressed in pounds. SI units will be added when total agreement on all conversion factors have been finalized.

   2.4. The maximum individual tire loading shall not exceed the tire load versus inflation pressure for its respective ply rating shown in table C-1 for the respective tire size.

3. **Definitions**

   3.1. Maximum Load—Maximum loads on individual tires are determined by considering the maximum axle load on each half of the axle and dividing by the number of tires on that half.

TABLE C-1 Agricultural Drive Wheel Tractor Tires Used in Field Service—Tires Used as Singles Basic Tire Vertical Load and Tangential Pull (Italics) Values for Tire Selection, lbs

Tire Size Designation	Tire Load Limits at Various Cold Inflation Pressures lb/in.^{2}							
	12	14	16	18	20	22	24	26
8.3–24	970 / *910*	1060 / *1000*	1150 / *1050*	1230 / *1160*	1310 / *1230*	1380(4) / *1300*		2390@32(8) / *1980@32*
9.5–16	910 / *780*	1000 / *860*	1080 / *930*	1160 / *1000*	1230(4) / *1060*			
9.5–24	1210 / *1110*	1330 / *1220*	1430 / *1320*	1540 / *1420*	1630(4) / *1500*			
11.2–24	1470 / *1320*	1610 / *1450*	1740 / *1570*	1860(4) / *1670*				
12.4–16	1350 / *1120*	1480 / *1230*	1590 / *1320*	1710 / *1420*	1820 / *1510*	1920 / *1590*	2020(6) / *1680*	2120 / *1760*
12.4–24	1760 / *1570*	1920 / *1710*	2080(4) / *1850*	2230 / *1980*	2370 / *2110*	2510 / *2220*	2640(6) / *2350*	
12.4–28	1880 / *1710*	2050 / *1870*	2220(4) / *2020*					
13.6–24		2270(4)[b] / *1970*						
13.6–28	[a]2210 / *[a]1970*	2420(4) / *2150*						
13.6–38	[a]2470 / *[a]2120*	2810(4) / *2640*	3030 / *2850*	3250 / *3060*	3460 / *3250*	3660(6) / *3440*		
14.9–24		2700 / *2320*	2920 / *2510*	3130 / *2690*	3330(6) / *2860*	3520 / *3030*	3710 / *3190*	3880(8) / *3340*
14.9–26		2790 / *2430*	3020 / *2630*	3240 / *2820*	3440(6) / *2990*			
14.9–28		2890 / *2540*	3120 / *2750*	3340 / *2940*	3560(6) / *3130*	3760 / *3310*	3960 / *3480*	4140(8) / *3640*

Size									
14.9–30		2980	3220	3450	3670(6)	4110	4330	4540(8)	4920@28(10)
		2650	*2870*	*3070*	*3270*	*3860*	*4070*	*4270*	*4130@28*
15.5–38	[a]3000	3160	3410	3660	3890(6)	4270	4500(8)	4710	
	[a]2520	*2970*	*3210*	*3440*	*3660*	*3590*	*3780*	*3960*	
16.9–24	[a]3100	[a]3280	3550	3800(6)	4040	4560			
	[a]2670	*2760*	*2980*	*3190*	*3390*	*3970*			
16.9–26		[a]3390	3660	3920(6)			4800(8)		
		2920	*3150*	*3370*			*4180*		
16.9–28			3780	4050(6)	4310				
			3290	*3520*	*3750*				
16.9–30			3900	4180(6)					
			3430	*3680*					
16.9–34			4150	4440(6)					
			3740	*4000*					
16.9–38			4390	4700(6)	5000	5290	5560(8)		
			3990	*4280*	*4550*	*4810*	*5060*		
18.4–16.1	[a]2370	[a]2600	2810(6)	3010	3200(8)				
	[a]1870	*2050*	*2220*	*2380*	*2530*				
18.4–24			4240	4550	4840(8)				
			3520	*3780*	*4020*				
18.4–26	[a]3710	[a]4060	4380(6)	4780	5000(8)	5280	5560	5830(10)	6580@32(12)
	[a]3150	*[a]3450*	*3720*	*4000*	*4250*	*4490*	*4730*	*4960*	*5590*
18.4–28			4530(6)	4850	5160(8)	5460	5740	6020(10)	6790@32(12)
			3900	*4170*	*4440*	*4700*	*4940*	*5180*	*5840@32*
18.4–30			4670(6)	5010	5330(8)	5630	5920	6210(10)	
			4060	*4360*	*4640*	*4900*	*5150*	*5400*	
18.4–34			4960(6)	5320	5650(8)	5980	6290	6590(10)	
			4410	*4730*	*5030*	*5320*	*5600*	*5870*	
18.4–38			5250(6)	5630	5980(8)	6330	6660	6980(10)	
			4730	*5070*	*5380*	*5700*	*5990*	*6280*	
20.8–34			5560(6)	6440(8)	7250				
			4840	*5600*	*6450*				
20.8–38			6360	6820(8)	7670(10)				
			5660	*6070*	*6830*				

(continued)

Tire Size	Tire Load Limits at Various Cold Inflation Pressures lb/in.^{2}							
Designation	12	14	16	18	20	22	24	26
23.1–26	ᵃ5310	ᵃ5810	6280(8)	6730	7160(10)			
	ᵃ4350	ᵃ4760	5150	5520	5870			
23.1–30			6690(8)					
			5620					
23.1–34			7110(8)					
			6110					
24.5–32	ᵃ6450	ᵃ7060	7640	8180	8700(10)			
	ᵃ5420	ᵃ5930	6420	6870	7310			

ᵃLoads at these inflation pressures are R-3 type when in agricultural service only.

NOTES:

1. Figures in parentheses denote ply rating for which load and pull values are maximum.
2. For transport service and operations that do not require sustained high torque, the following load limits at various speeds apply with no change in inflation pressure.

Max. Speed	% Increase to Loads in Above Table
10 mph	20%
15 mph	10%
20 mph	Same as above table

3. Load ratings for tires used as duals are to be reduced by a factor of 0.88. See complete SAE J709 for more precise ratings.

SOURCE: Reprinted with permission. "Copyright © Society of Automotive Engineers, Inc., 1978, All rights reserved."

APPENDIX D

CONVERSION FACTORS

U.S.-British Units to SI Units

To Convert from	To	Multiply by
(Acceleration)		
foot/second^2 (ft/s^2)	meter/second^2 (m/s^2)	3.048×10^{-1a}
(Area)		
foot^2 (ft^2)	meter^2 (m^2)	9.2903×10^{-2}
inch^2 (in.^{2})	meter^2 (m^2)	6.4516×10^{-4a}
(Density)		
pound mass/inch^3 (lbm/in.^{3})	kilogram/meter^3 (kg/m^3)	2.7680×10^4
pound mass/foot^3 (lbm/ft^3)	kilogram/meter^3 (kg/m^3)	1.6018×10
(Energy, Work)		
British thermal unit (BTU)	joule (J)	1.0551×10^3
foot-pound force (ft · lbf)	joule (J)	1.3558
kilowatt-hour (kw · h)	joule (J)	3.60×10^{6a}
(Force)		
pound force (lbf)	newton (N)	4.4482
(Length)		
foot (ft)	meter (m)	3.048×10^{-1a}
inch (in.)	meter (m)	2.54×10^{-2a}
mile (mi), (U.S. statute)	meter (m)	1.6093×10^3
(Mass)		
pound mass (lbm)	kilogram (kg)	4.5359×10^{-1}
slug (lbf · s^2/ft)	kilogram (kg)	1.4594×10
(Power)		
foot-pound/minute (ft · lbf/min)	watt (W)	2.2597×10^{-2}
horsepower (550 ft · lbf/s)	watt (W) .	7.4570×10^2

To Convert from	To	Multiply by
(Pressure, stress)		
atmosphere (std) (14.7 lbf/in.^{2})	newton/meter^2 (N/m^2 or Pa)	1.0133×10^5
pound/inch^2 (lbf/in.^{2} or psi)	newton/meter^2 (N/m^2 or Pa)	6.8948×10^3
(Velocity)		
foot/second (ft/s)	meter/second (m/s)	3.048×10^{-1a}
knot (nautical mi/h)	meter/second/ (m/s)	5.1444×10^{-1}
mile/hour (mi/h)	meter/second (m/s)	4.4704×10^{-1a}
mile/hour (mi/h)	kilometer/hour (km/h)	1.6093
(Volume)		
foot^3 (ft^3)	meter^3 (m^3)	2.8317×10^{-2}
inch^3 (in.^{3})	meter^3 (m^3)	1.6387×10^{-5}

[a]Exact value

Additional Acceptable Units and Their SI Equivalents

(Frequency)	1 hertz (Hz)	= 1 cycle/second (s^{-1})
(Mass)	1 tonne (metric ton) (t)	= 1000 kilogram (10^3 kg)
(Pressure or stress)	1 bar (b)	= 100 kilopascal (10^2 kPa or 10^5 N/m^2)
	1 pascal (Pa)	= 1 newton/meter^2 (N/m^2)
(Viscosity, absolute)	1 centipoise	= 1 millipascal-second (mPa $\cdot$ s or 10^{-3} N $\cdot$ s/m^2)

SOURCE: From J. L. Meriam. *Statics and Dynamics, SI Version.* John Wiley & Sons, New York, 1978.

INDEX

A

Additives, oil, 194
  gasoline, 67
Adiabatic change, 26
Aftercooling, 129
Air
  required for cooling, 178
  supply measurement, 417
  temperature control, 174
Air cleaners, 163
  dry type, 164
  location of inlet, 163
Air cleaner test procedure, 166
Air supply, measurement of, 417
Alternative fuels, 74
Alternator, 139
Ash content of fuel, 73

B

Balance of engines, 104, 106
Battery, 134
Brake, Prony, 406

C

Calorimeter, bomb, 66
Cams, 93
Carbon residue, 72
Center of gravity, 302
  determination of, 302
Cetane number, 69
Cleaner, air, 163
  tests, 165

Clutches, 371
  friction materials for, 374
Combustion, 54
Combustion chamber design, 94, 112
Compression ignition, 111
Compression ratio, 95
  effect on engine performance, 95
Cone index, 251
Constant-entropy change, 26
Constant-pressure change, 24
Constant-volume change, 24
Control requirements, 226
Controls, hydraulic, 314
Cooling, air, 170
Cooling load, 167, 169
Corrosion test of fuel, 65
Corrosives, anti-, for oil, 69
Cracking fuel, 51
Crank-moment diagrams, 102
Crankshaft, 80
  four-cylinder engine, 80
  six-cylinder engine, 83
Cycle, diesel, 41
  efficiency of, 41
  ideal air-standard, 38
  Otto, 38
  two-stroke, 112
Cycles
  actual, and deviation from ideal, 43
  engine, 37

D

Detergents for oil, 194
Diesel cycle
  efficiency of, 42

Diesel cycle (*Cont.*)
  ideal air-standard, 41
Diesel-cycle work, 43
Diesel engine construction, 111
Diesel fuel additives, 73
Diesel fuel grades, 70
Diesel fuel injectors, 119
Diesel fuel pump, 116
Diesel fuel tests, 67
Differential, 381, 389
Dimensional analysis, 247
Distillation, 61
Draft control, automatic, 329
Draft sensing, 329
Drive wheels, in depression, 284
  soil reaction against, 245
Duel tires, 254
Dynamometer
  absorption, 406
  air-brake, 409
  chassis type, 413
  drawbar, 411
  eddy-current, 409
  electric direct current, 409
  hydraulic, 408, 412
  Prony brake, 406
  shop-type, 410
  spring, 412
  strain gage type, 413
  torque meter, 413

E

Electrical systems, 134
  environmental problems, 152
Energy, 13
  conversion methods, 14
  sources of, 13
Energy changes, 23
Engine balance
  multicylinder engine, 106
  single-cylinder engine, 104
Engine cooling, 167
  heat loss in, 169

Engine design
  general, 78
  stroke-to-bore ratio, 78
Engine flywheels, 102
Engine noise, 129
Engine performance, 435
  correction for atmospheric
    conditions, 433
  effect of compression ratio on, 94
  effect of oil viscosity on, 139
  efficiency, 435, 437
Engine speed control, 156
Engine types, 77
  cylinder arrangements, 80
Entropy, 23
Environmental factors, 203

F

Fan, air delivery versus power, 179
Filters, oil, 199
Flash point, 72
Flywheels, 102
  design of, 103, 104
Forces, static, 278
Four-wheel drive, 254
Free body, 273
Friction materials for clutches, 374
Fuel, 48
Fuel-air ratio requirement, 55
Fuel-combustion chemistry, 54
Fuel flow measurement, 420
Fuel-injection systems, 116
Fuel injectors, 119
Fuels
  atomic weights of, 55
  molecular weights of, 55
Fuel tests and their significance, 56, 70
Future engine, 16

G

Gasoline
  additives, 67

cracked, 54
premium-grade, 63
regular-grade, 63
tests, 56
Gear, 377
tooth design, 379
tooth layout, 380
Generators, 139
Governor
principles of, 157
regulation of, 158
strength of, 158
Gravity, test of, 65–66
Gum content in fuel, 65

H

Heat balance, diesel, 170
Heat transfer in engine, 170
Heat values, 66, 72
Hitch, 350
three-point, 353
Human factors, 203
Human reaction to vibration, 210
Hydraulic controls, 314
accumulators, 320
automatic control, 330
classes of, 328
components, 314
draft control, 329
fluids, 323
motor performance, 315
power steering, 333, 335
symbols for, 314, 448
valves, 321, 343
Hydraulic valve lifter, 88
Hydrometer, 66
Hydrostatic transmissions, 368

I

Ignition quality of fuel, 69
Ignition systems, 140

battery with breaker points, 140
magneto, 145
timing, 145
Indicator diagrams, 38
Inertia force
connecting rod, 98
single-cylinder engine, 100
Inflation of tires, 257
Internal combustion, 37
Isothermal change, 25

K

Kinematic viscosity, 187
Kinetic energy, 351

L

Lubricants, properties of, 183
SAE viscosity ranges for, 190
viscosity, 184
Lubrication systems, 201

M

Mass moment of inertia, 305
Mechanics of tractor, 272
center of gravity, 302
equations of motion, 273
force analysis, two-wheel drive, 278
four-wheel drive, 282
moment of inertia, 305
power effect, 281
stability, 283, 289, 295
three-dimensional analysis, 297
traction, 277
vibratory system, 285
Molecular weight, 55
Moment of inertia, of tractor, 305
Mufflers, 162

N

Nebraska tractor tests, 420
Noise, 206
Nonflow processes, equations for, 33

O

Octane number, 57
  requirements, 58
Oil, 183
  additives, 194, 195
  contamination, 198
  filters, 199
  pour-point depressants, 195
  service classification, 193
Operator space, 226
Operator-machine interface, 214
Otto cycle, 38
Oxidants, inhibitors, for oil, 195

P

Petroleum refining, 51
Piston crank kinematics, 95
Planetary gear transmissions, 369, 383
Polytropic change, 28
Pour-point, of fuel, 73
  oil classification, 188, 193
Power, 404
  estimating, 416
  measurement of, 404
Power correction formula, 433
Power take-off, 363
Power train, 362
  torsional vibration, 393
Pressure indicators, 418
Pressure–time diagrams, 419
Prony brake, 406
Pump, diesel fuel, 116

R

Radiator
  construction of, 174
  design, 171
  performance curves, 173
Refining of petroleum, 52
Reid vapor pressure, 62
Reliability, 438
Response of man's body, 210
Riding comfort, 215
Rolling resistance, 248
ROPS, 231
Rotary engine, 14

S

Safety, 234
Seating, 217
Sensors, 149
SI conversion factors, 454
Soil reactions, 245
Sound control, 224
Spark arrester, 160
Spark plugs, 148
Spatial requirements, 226
Specific heat, 22
Spring-mass system, 218
Stability, 283, 289, 295
Standards, 444
  graphical symbols, 448
Starting motor, 137
Stroke-to-bore ratio, 78
Sulphur test, 64
Swirl chamber, 115

T

Test engine, 58
Theoretical soil thrust, 241
Thermal comfort, 205, 233
Thermodynamic laws, 20

Thermodynamic principles, 20
Three-point hitches, 353
Three-point hitch standard, 354
Tire inflation, 257
Tire load ratings, 257, 450
Tires, 259
    radial ply, 258
Tire tread, 259
Torque curves for engines, 435
Traction, 240
    improvement, 263
    mechanics, 240
    prediction, 241
    testing, 265
Traction aids, 263, 265
Traction performance, 254
    duals, 254
    four-wheel drive, 254
    tandem, 254
Tractive efficiency, 249
Tractor
    development of, 3
    efficiency, 435
    four-wheel drive, 7
    lawn and garden, 5
    performance criteria, 403
    row-crop, 4
    test, 403, 420
    types, 4
    utility, 5
    world variations, 8
Tractor design trend, 16
Tractor tests, 420
Transient and steady state handling,
    289
Transmissibility, 219
Transmission gears, 377
Transmissions, 364
    bearings, 391
    design techniques, 375
    drive shafts, 390
    hydrostatic, 368

Tread design, 259
Trends in tractor performance, 16, 17
Turbochargers, 120
Two-stroke cycle, 112

V

Valve
    arrangement of, 84
    cams, 93
    clearance of, 90
    design, 83
    hydraulic lifter, 88
    opening area of, 91
    rotator, 86
    temperature of, 89
    timing of, 87
Valve materials, 85
Valve-seat insert, 91
Valve seats, 91
Valve timing, optimum, 87
Vibration, 210, 285
    power train torsional, 393
Viscosity
    absolute, 184
    conversion of, 189
    determination of, 185
    index of, 188
    kinematic, 187
    SAE numbers for lubricants, 188
    saybolt, 186
    μ n/p factor, 197
Viscosity-index improvers for oil, 192
Viscosity versus temperature, 190
Visual requirements, 226
Volatility of fuel, 59

W

Weight transfer, 243